Weather Matters

CultureAmerica

Karal Ann Marling and Erika Doss, *Series Editors*

Weather Matters

An American Cultural History since 1900

Bernard Mergen

 University Press of Kansas

A list of permissions can be found on p. 398.

© 2008 by the University Press of Kansas
All rights reserved

Published by the University Press of Kansas (Lawrence,
Kansas 66045), which was organized by the Kansas
Board of Regents and is operated and funded by
Emporia State University, Fort Hays State University,
Kansas State University, Pittsburg State University, the
University of Kansas, and Wichita State University

Library of Congress Cataloging-in-Publication Data

Mergen, Bernard.
Weather matters : an American cultural history since
1900 / Bernard Mergen.
p. cm. — (CultureAmerica)
Includes bibliographical references and index.
ISBN 978-0-7006-1611-4 (cloth : alk. paper)
1. United States—Climate—Social aspects—History.
2. Climate and civilization—History. 3. Climatology—
Social aspects—History. 4. Weather—Social aspects—
History. 5. Human beings—Effect of environment on—
United States—History. I. Title.
QC983.M43 2008
551.60973—dc22

2008022848

British Library Cataloguing-in-Publication Data is
available.

Printed in the United States of America

10 9 8 7 6 5 4 3 2 1

FOR

Alexa and Andrew

Contents

Acknowledgments

I want to thank Jack Borden, founder of For Spacious Skies, whose passion for the sky is inspiring. He put me in touch with several people, none more important than John Day, cloud physicist, author, and photographer. I was fortunate to have had Charles A. Doswell III teach me about the history of severe storm research as we pursued tornadoes across the Midwest. Chuck's perspective on the organization of meteorology in the United States and his sense of humor were greatly appreciated as I sought to understand the cultural meanings of weather.

I am also grateful to James R. Fleming, Karna Small Bodman, Keay Davidson, H. Michael Mogil, Jesse Ferrell, George Hirschmann, Edward Lorenz, and Roger A. Pielke Jr. for sharing their knowledge.

I appreciate the support of the University Press of Kansas, especially Nancy Jackson, who first suggested the book, and her successor, Kalyani Sunethra Fernando, who has seen it to completion.

Claudia Spohnholz—weather-sensitive wife and editor—made innumerable improvements as she read and edited and reread the manuscript. This book would not exist without her contributions. My son, Andrew, and daughter, Alexandra, expressed their interest and support of the project in copious ways. Based on their professional efforts to improve the earth's environment, I forecast clearer skies.

Bernard Mergen
Peregrina, Franklin, West Virginia
Spring Equinox–Summer Solstice 2007

Introduction

The wind is blowing steadily at 70 miles per hour, the temperature is about 10°F below zero, it's 2:30 in the afternoon, January 16, 2003, on top of Mount Washington, New Hampshire. I take a few steps away from the shelter of the observatory, and the wind shoves me forward like an automatic revolving door, only much harder. There's no stepping back. I do a couple of glissades but go where the wind blows. Swinging around, I stumble back under the trellises that protect observers from windblown pieces of ice from the observatory tower. All the antennae are covered with rime flags whose delicate filaments stretch out toward the wind. They look like white iron filings on a magnet. I'm standing at 6,200 feet in the middle of a cloud. It is the beginning of my two-day EduTrip offered by the Mount Washington Observatory.

The meteorological division of the U.S. Signal Service and its successor, the U.S. Weather Bureau, made observations here from 1870 to 1892. The observatory was rebuilt with private funds and has operated as a nonprofit research and educational institution since 1932. The observatory cultivates its mystique as the place with "the worst weather on earth." Its great claim to fame is the 231-mile-per-hour wind recorded at 1:21 P.M. on April 12, 1934, which remains a world's record for a surface station. Wind, cold, and, for a few hours a day, the view of the Presidential Range and the White Mountains are what the observatory is selling. In summer, tourists are allowed to drive up the narrow seven-mile road, but in winter EduTrippers ride up in a big Bombardier snowcat.

Bryan Yeaton, an energetic young man in a bright red parka, greets us in the parking lot at the base of the mountain. Yeaton does education programs for the Mount Washington Observatory and produced *The Weather Notebook,* a nationally syndicated radio show about weather that aired from 1997 to 2005. He introduces the other participants and staff. My fellow worst-weather seekers are a retired couple recently settled in New Hampshire for its outdoor activities and two guys in their forties from Massachusetts. One was a commercial fisherman, the other is a computer programmer; their trip is a Christmas present from their wives. Ken Rancourt is a research meteorologist for the observatory who doubles as the Bombardier driver. His assistant, Wayne, looks like the actor Mark Wahlberg and mentions this resemblance when he is introduced. When he speaks, he reveals several missing teeth. The next day, as I am inching down the icy mountain on rented crampons, he speeds past me on a Flexible Flyer, providing an explanation for his dental condition.

In winter the observatory and adjoining visitors center are usually coated with snow and ice. The buildings resemble a Gothic wedding cake. The instrument rooms and labs are a jumble of records, supplies, and humorous signs that initiate the visitor into the culture of a research institution. "Big Empty Box" is attached to what looks like an air-conditioner, and "Big Metal Box" contains supplies. These are insiders' jokes on stupid questions from visitors. Bryan begins our orientation with similar "ice-breaking" jokes about being the "world's worst weather observer" and "world's highest paid weatherman." He then explains why Mount Washington is important for meteorological observation. Its elevation makes it a kind of permanent weather balloon, transmitting data visually and numerically twenty-four hours a day. Its location exposes it to many of the storm tracks across New England, and New England is the "exhaust pipe" of the nation's weather.[1]

Later in the afternoon, sharing a chair with the observatory's mascot cat, Nin, I look at the logbooks. For almost seventy-five years, visitors have written comments such as "fine day," "high clouds," and "gale winds." Confined by nature to the prosaic, the steady accumulation of such observation ascends to poetry. I glance out the window. The sunset is a soft apricot line on the western horizon. Thus it is with the history of weather. The simplest daily occurrences of sun, wind, clouds compose its raw materials; a hand shielding eyes from the brightness, or a finger pointing to the sky, is the beginning of a weather chronicle.

Weather Matters is about the ways in which Americans of the past century have coped with weather. In that century, meteorology became one of the premier environmental sciences, weather reports emerged as computer-generated works of art, forecasts based on thousands of variables and covering areas as large as the globe and as small as neighborhoods were developed, weather disasters came to be anticipated and the loss of life and property kept to a minimum, and artists and poets struggled to offer alternatives to the reduction of weather to numbers and formulas—but people never stop praying for rain, or sunshine, or snow, or to be spared from tornadoes, hurricanes, droughts, and blizzards. Many of those prayers are mere incantations, others deeply pious appeals to God, while some are spontaneous expressions of faith in the promised order of science. While some believe in miracles, others believe in the miracle of General Circulation Models (the complete statistical description of atmospheric motions over the earth) and Advanced Weather Interactive Processing Systems (AWIPS).

Weather is, as others have noted, well suited to the electronic age—constantly in motion, frequently fast moving (aided by time-lapse photography), ubiquitous, and visually beautiful. In a culture that enshrines free will but expects conformity, weather offers a third alternative, chance, to explain the fate of nature and humanity. Even meteorologists acknowledge that weather is chaotic. Weather reminds us of who we are and what we value. This is why there are so many different kinds of weather, so many different maps of what we think we see.

This book is organized by the perception, marketing, and management of weather in the United States since 1900. The focus is on weather in American life in the past 107 years, with occasional brief excursions into the nineteenth century when it seems necessary to make a point about contemporary issues. By perception I mean, simply, how do we see and understand the largely invisible phenomena we call weather. What tools do we use to describe it? What do we do with what we think it is? How do we depict it?

Marketing, too, is fairly straightforward. I realized when I was reading about why and how the Weather Bureau began to study snow that the bureau, like all government agencies, needed to justify its existence to taxpayers and their congressional representatives. The bureau sold the only thing it had of value—knowledge. Like any business, it sought to create a monopoly on its product and succeeded for many years, partly by controlling the language used to discuss weather. Hence, the emphasis in this book is on words and how they are used to make the ineffable seem normal. Businessmen and businesswomen found ways to market sunshine, wind power, snow, and other "products" of weather.

Management of the weather is here taken to mean all attempts to predict, create, and protect against weather. We manage weather when we dress in the morning, open an umbrella, or write a poem or paint a picture of it.

Obviously, these rough categories overlap in significant ways. As the manuscript grew and I looked for ways to keep it within reasonable limits, seven chapters became five. Sections on drought, floods, climate and climate change, risk and natural hazard mitigation, including weather insurance, were shortened or eliminated. Each deserves its own book, and there are already many books and articles that cover these subjects quite well.[2]

What remains is a cultural history of meteorology and weather, subjects I did not think were covered adequately in the existing literature. For the twentieth century, the only institutional history is Donald R. Whitnah's *History of the United States Weather Bureau,* published in 1961. Kristine C. Harper's doctoral dissertation, "Boundaries of Research: Civilian Leadership, Military Funding, and the International Network Surrounding the Development of Numerical Weather Prediction in the United States" (2003), is indispensable for understanding the origins of the current weather establishment. Frederik Nebeker's *Calculating the Weather: Meteorology in the 20th Century* is a fine survey of the science, but it lacks institutional, political, and cultural contexts. James R. Fleming sets the standard for the history of meteorology and climatology in *Meteorology in America, 1800–1870* and the several volumes he has edited. Mark Monmonier's *Air Apparent: How Meteorologists Learned to Map, Predict, and Dramatize Weather* is a fascinating account of the ways atmospheric scientists visualize their subject. Only two recent books attempt what I would call cultural history of weather, geographer William B. Meyer's *Americans and Their Weather* and writer David Laskin's *Braving the Elements: The Stormy History of American Weather,* neither of which devotes more than seventy pages to the twentieth century. Gary Fine's ethnographic study of the work culture of meteorologists was published just as my book was going to print; it should make important contributions to our understanding of how the weather is perceived, marketed, and managed.[3]

This book is meant to complement the existing scholarship on the history of meteorology and weather in the United States and to strike out in new directions. It also addresses the major questions of environmental history, namely, the ways in which nature in the form of weather has affected humans, and how humans have thought about weather and acted upon it. Chapter 1, "Talking about Weather," introduces the U.S. Weather Bureau in its principal role as forecaster, and the struggle that ensued over definition of terms, public acceptance of meteorological authority, and the meanings of uncertainty. In its complexity and chaotic behavior, weather interacts with three other mysterious human activities—religion, politics, and play—in numerous ways, and weather chatter becomes, by turns, profound, ridiculous, and sublime.

Chapter 2, "Managing Weather," is chiefly concerned with the ways in which weather has become part of the nation's economy. It continues the story of the Weather Bureau's attempt to market its knowledge but adds the history of the American Meteorological Society, founded in 1919, which helped to legitimize the work of the Weather Bureau but also challenged its authority. As the atmospheric sciences grew, new groups with vested interest in the weather emerged: meteorologists in private business, media weathercasters, and promoters of various weather-related businesses. These communities were not fettered by the bureau's quaint notion of public service and aggressively worked to control what the public should think about weather. Some particularly entrepreneurial individuals advocated weather modification through geoengineering projects on various scales and with questionable results. I touch on a few of them to illustrate the most audacious aspect of selling the weather.

Chapter 3, "Seeing Weather," looks at the history of sky awareness, the fascination with clouds, and weather-themed photography, motion pictures, painting, and sculpture. One of the most important developments for understanding our relation with nature in the late twentieth century and early twenty-first century may, I think, be found in the work of artists such as Walter De Maria, James Turrell, and Dozier Bell, who strive to use elements of weather to place an observer in a living landscape created by a fusion of science and art. Philosopher Edward Casey's ideas on the meanings of place offer ways to understand how weather may enable us to achieve a sense of harmony with nature. The chapter concludes with a section on the exhibiting of weather in museums and science centers.

Chapter 4, "Transcribing Weather," examines the work of more than fifty

American poets and novelists who have used weather in various ways to explore the meanings of human existence and our relationships with nature. The pervasiveness of weather in literature and the arts raises in another way the question of obsession. Are we obsessing on the weather or the idea of weather? The answer, of course, is both, and poems by Wallace Stevens, Richard Hugo, Howard Nemerov, Carol Muske, Naomi Shihab Nye, and others lead us into mental and spiritual towers where real and imaginary winds blow. Weather in the work of novelists—George R. Stewart, Gore Vidal, Rick Moody, Jean Thompson, Clint McCown, Paul Quarrington—is the avatar of chance. Necessity, the laws of science and religion, and free will, the cornerstone of Western philosophy, are challenged by the chaos of weather.

Chapter 5, "Suffering Weather," looks at the times and places where weather is most troublesome. Episodes of disaster are just part of the story, however, and I am interested in showing how we accommodate and acclimate to everyday weather on our own and with the help of the National Weather Service and the media. I begin the chapter with an account of my own quest to understand the extreme weather obsession of storm chasing. I look at some of the social and cultural meanings of tornadoes and hurricanes, then explore the impact of everyday weather on our lives. Efforts by the air-conditioning industry and the Weather Service to protect us from any weather-caused discomfort only underscore the highly personal meaning of weather. Each of us experiences weather differently.

This book is full of quirky facts and eccentric personalities. I sought to bring the past alive by quoting from some of the more colorful commentators on weather. Their anecdotes stand alone. Nevertheless, there is coherence to the story I tell, and that is that weather is the part of the physical environment closest to us. We are enveloped by air that is the medium for our life and communication. It is so much a part of us, our thoughts, our languages, our feelings, that we are mostly unaware of it, but we ignore it at our peril. Weather is obviously more than what atmospheric science says it is, but how much more? My argument is that we live simultaneously in at least four weather systems, the one we feel with our senses, another that we learn about in science classes and National Weather Service reports, a third from the media that interprets the first two for our consumption, and a fourth in which our minds work to synthesize our experience and the knowledge acquired from art and literature. We express all these in meaningless banter and eloquent art. The paradox of weather is that it is both quotidian and unique.

As someone once said about sex, weather is what we think about all the time when we're not thinking about something else. Maybe it's also what we talk about when we are thinking of something else. "How's the weather?" often solicits more than meteorological commentary. A word on the wind is sufficient to establish a relationship between speakers, a mood, a tone, an atmosphere within the atmosphere. Weather can be a metaphor and a metalanguage. "Hot enough for you?"

may be an indirect inquiry into sensitive questions of health, well-being, success, or failure. We deem weather a safe topic, unlike religion and politics, but weather is religion and politics. Weather raises the most fundamental questions about the origin and purpose of life, the ability of humans to predict and control nature, and the place of science in public policy. To talk about the weather in the twenty-first century requires us to at least consider the possibility of anthropogenic climate change. Weather is a commodity, its products—water for people and agriculture, solar and wind energy, snow for winter sports and sun for summer—marketed and managed. By some estimates, trillions of dollars are made and lost annually by weather-dependent businesses. Insurance losses from hurricanes, tornadoes, winter storms, and other weather-related disasters are increasing dramatically. Media attention to weather is extensive, compelling, and sometimes misleading. In short, we cannot escape weather even if we want to.

This book began as a challenge from Nancy Scott Jackson, at the time an acquisitions editor at the University Press of Kansas. Why, she asked, are Americans obsessed with weather? I hope this book answers Nancy's question. The preponderance of evidence leads to the conclusion that *some* Americans are obsessed by weather, but fewer than are obsessed by genealogy, NASCAR, or sex. As a nation we are more *possessed* by weather than *obsessed*. We possess weather from within, in our lungs, blood, and mind.

You don't need to have an opinion about it, but if you already love weather, then this book will provide you with some new perspectives on what you already know and give you something to talk about with other weather weenies. If you are indifferent to weather and the people who talk about it, this book offers plenty of evidence for why you are.

1

Talking about Weather

Weather Bureau Weather

"I've been thinkin' it over, Hinnissy, an' I come to th' con-clusion that there's two kinds iv weather, human weather an' weather-bureau weather." This pronounce-ment in *Harper's Weekly,* July 1901, by the shrewd Irish American saloon keeper Mr. Dooley, created by journalist Finley Peter Dunne, is a wonderfully succinct summary of much of the past century's talk about weather. Often dismissed as fu-tile—talking about the weather instead of doing something about it—weather chatter is really as important as intervention, since both human weather and weather bureau weather are the products of past and present conversations about the experience of the atmosphere. Moreover, naming is a kind of control. If we ever stop talking about the weather, we will stop feeling its effects, lose sight of its beauty, and cease trying to understand it and ourselves.

Human weather for Mr. Dooley, as for most of us, is the daily experience of temperature, wind, and sky; weather bureau weather is a mysterious language of "highs" and "lows," "fronts" and "pressure centers," and lines on a map. Dooley trusted the pains in his friend Clancy's leg more than he did weather bureau "pro-fessors" to predict the weather, citing the disastrous McKinley inauguration day forecast as proof. "The weather bureau's weather is on a map, an' our weather is in th' air," Dooley concludes. "That's why th' pro-fissor fails an' Clancy's leg is a gr-reat success. Tis' an out-iv-dure leg."[1] Clancy's leg notwithstanding, Mr. Dooley and his fellow Americans were already deep in a struggle with the institutions of modern science to name and define such natural phenomena as weather, a con-test that continues today.

We talk about the weather with words formed on the tongues of our ancestors, who watched in awe as skies darkened and lightning flashed, but those words are never enough. The process of describing and naming the elements of what we call "weather" is as endless as the weather itself. Moreover, all weather is local, and every language, dialect, and argot has words to praise or curse, to describe and to classify the weather. Naming implies an understanding sufficient to distin-guish one thing from another and a reason for the distinction. How is rain differ-ent from snow, and why is this difference important? Do we experience different kinds of rain, and do we have different names for them? From such simple ques-tions, great taxonomies grow.

Meteorology developed in the United States in response to scientific curiosity, geographic expansion, weather-dependent agriculture and business, national pride, and political and bureaucratic rivalries. Weather data were collected and tabulated on a national scale by scientists at the Smithsonian Institution in the 1850s, then by the U.S. Army Signal Corps. In 1870, Congress created a meteorological service administered by the Signal Corps, then transferred its functions to the Department of Agriculture in 1891, with this mission:

> The Chief of the Weather Bureau, under the direction of the Secretary of Agriculture, shall have charge of forecasting the weather; the issue of storm warnings; the display of weather and flood signals for the benefit of agriculture, commerce and navigation; the gauging and reporting of rivers; the maintenance and operation of seacoast telegraph lines and the collection and transmission of marine intelligence for the benefit of commerce and navigation; the reporting of temperature and rainfall conditions for the cotton interests; the display of frost, cold-wave, and other signals; the distribution of meteorological information in the interest of agriculture and commerce and the taking of such meteorological observations as may be necessary to establish and record the climatic conditions of the United States, or are essential for the proper execution of the foregoing duties.[2]

The Gilded Age legislators could not have made it clearer; agriculture and commerce are mentioned six times (counting "cotton interests"), but meteorology only twice. Forecasts and warnings, not theory and research, are the original purpose of the Weather Bureau, and, although Congress and presidents have tinkered with the bureau's budgets, organization, and the scope of its work for over a century, the bureau preserves most of its original purpose. The 2005 mission statement of the National Weather Service (NWS), successor to the bureau, declares that the National Weather Service "provides weather, hydrologic, and climate forecasts and warnings for the United States, its territories, adjacent waters and ocean areas, for the protection of life and property and the enhancement of the national economy."[3]

Despite the narrow mandate for the Weather Bureau and the paucity of university courses in meteorology available in 1891, most of the bureau's leadership had graduate training in physics or astronomy, often from European universities, and they wished to do research and train a new generation of weathermen. One of their priorities was the establishment of doctoral programs in meteorology in American universities; another was the popular education of the public in the new science. Hence, Mr. Dooley's sobriquet, "professor," for the hapless competitor to Clancy's leg was appropriate. In 1871 the Signal Corps, which trained West Point graduates in the rudiments of meteorology, hired Cleveland Abbe as professor and chief meteorologist. Abbe, who had done graduate work at Russia's

Nicholas Central Observatory near Saint Petersburg before becoming director of the Cincinnati Observatory, established a meteorological laboratory at Fort Myer, Virginia, in 1880. Abbe also founded and edited the *Monthly Weather Review* from 1873 to 1909, providing scientific direction to both the military and the civilian-led Weather Bureau.[4]

Abbe added a second professorship to his title when he began teaching meteorology at George Washington University (then Columbian University) in 1893. H. H. Kimball, a bureau employee, became the school's first Ph.D., in solar physics, in 1910, and LeRoy Meisinger, who carried out upper air pressure research in balloons, became its first Ph.D. in meteorology in 1922. Harvard University and Johns Hopkins University approached meteorology and climatology through geography programs. Abbe was appointed lecturer at Hopkins in 1898, a year before Oliver Fassig received the first Ph.D. in meteorology and climatology from any American university. Fassig wrote books on the climatology of Maryland and Puerto Rico and ultimately became chief of the Weather Bureau's Division of Climatology. Harvard awarded the second and third Ph.D.s in meteorology and climatology to Charles F. Brooks, first secretary of the American Meteorological Society (AMS), and Co-Ching Chu, first director of Academia Sinica's Institute of Meteorology. They received their degrees in 1914 and 1917, respectively. The bureau entered the twentieth century with a very small core of well-trained scientists who found opportunities to do research while making forecasts and issuing warnings for the benefit of agriculture and commerce. Its leadership sought to demonstrate that the scientific study of the atmosphere would lead to better forecasts and replace weather lore with systematically acquired data and testable theories. But selling science to skeptical Dooleys required the bureau's weathermen to become public advocates for the new profession.[5]

Gaining the respect of the scientific community was somewhat easier than convincing the public that it needed a weatherman to tell it which way the wind was blowing. Early advocates of meteorology promoted the accessibility of their activities to invite public participation, often giving the impression that knowledge of advanced physics and mathematics was not necessary. In an 1879 article titled "A Word about Weather," the editor of a privately published journal, the *Meteorologist,* announced confidently: "There is one science which is within the grasp of every mind, and which, to be successfully cultivated, requires no preparation, and furnishes an admirable resource for those who have a taste for the observation of natural phenomena. It is what we may call the science of rain and fine weather, but which now receives the higher title of meteorology."[6]

By 1900 this was the wrong message. What the bureau sought to convey to the public was not so much the accessibility of weather knowledge, though that was important, but the vital role of the bureau in defining and interpreting the weather. Regular notices began to appear in Abbe's *Monthly Weather Review* on the educational activities of the bureau's staff. In April 1900, for example, the section director in Springfield, Illinois, lectured on the thermometer, barometer, anemometer, rain gauge, and weather maps to the Kane County Farmers' Institute and to the Congregational Club. A Weather Bureau observer in Havre, Montana, spoke to grammar school students about weather maps and the work of the bureau. Abbe himself held classes with teachers in Baltimore to encourage them to present meteorology in the lower grades of the public schools. Taking the great Harvard naturalist as his model, Abbe advised:

> Agassiz educated naturalists by giving them a mass of material and asking them to tell him what they saw and what conclusions they could draw. It is not our business to fill the mind with other people's ideas but to make the child's mind develop such ideas as the better judgment of the teacher recognizes as appropriate and true. Meteorology is first to be studied by observation and record; we begin by recording general impressions but gradually develop greater exactness by the help of instruments. . . . There is nothing scientific that man has learned from nature but what may be made comprehensible to the child.[7]

Showing great faith in individual ability, Abbe nonetheless emphasized the work of Weather Bureau employees in universities and colleges. In 1900 it was reported that an observer in Spokane, Washington, lectured at Gonzaga College; Isaac Cline, section director in Galveston, gave a course at the University of Texas; Oliver Fassig was on the faculty at Johns Hopkins; and Alexander McAdie, forecast official in San Francisco, taught at the University of California. Colby College in Maine began offering a meteorology course in 1905. The geographic spread of these early efforts to regularize curriculum and promote a profession of meteorology is important because the bureau was often identified as a federal agency, isolated in the capital, remote from the rest of America.

Other venues for public education were not neglected. With the encouragement of the Weather Bureau, the Brooklyn Institute of Arts and Sciences exhibited barographs, thermographs, psychrometers, and other apparatus and offered an eight-week lecture series on elementary meteorology in the spring of 1900. A year later the bureau mounted a large display at the Pan-American Exposition in Buffalo that included not only instruments but storm-warning flags, maps, and photographs of clouds and lightning. For those who could not visit these exhibits, the bureau prepared a set of lantern slides and made them available to teachers. A particularly whimsical example of the Weather Bureau's effort to capture public attention was its kiosks. From 1909 to 1917, the bureau built and maintained four-sided, seven-foot-tall, cast-iron and concrete structures in the centers of at least twenty-nine major American cities.

As described by D. T. Maring of the Instrument Division, the kiosk had glass cases on three sides for the display of maps, charts, and other two-dimensional in-

Group of men looking at a weather kiosk erected in front of the Federal Building, *Chicago Daily News,* 1909. Chicago History Museum.

formation, while one side held a hygrometer to show the humidity, a Fahrenheit thermometer, maximum and minimum thermometers, a thermograph to show temperatures over a two-week period, and a rain gauge whose tipping bucket was concealed in the roof of the kiosk, masked by four classical pediments. Maring concluded, "While the records obtainable from this apparatus . . . can not, of course, have special meteorological value, and can not well be used for climatological purposes, they may still, at times, serve as a check on the records of the standard instruments of the Service properly exposed. The kiosk records should also prove useful to the local press, and, as giving actual *street* weather conditions, they should be of considerable utility and of more than passing interest and convenience to the general public."[8]

The rise and fall of the Weather Bureau kiosks is itself a kind of barometer of the bureau's efforts to popularize its work, the final words of the article providing the key. Although there is no record of how the public responded to the kiosks, the press was obviously the most influential communication medium of the time.

The bureau's effort to pull weather down to street level, to make it the space in which we dwell and therefore vital to our daily lives, was just beginning, and the kiosks were just one experiment in public relations. As a novelty it was successful in drawing the attention of local newspapers.

Currently, weather occupies half a page in most metropolitan dailies, and *USA Today*'s color weather map has been an icon of weather mania for twenty-five years. In 1900, however, fewer than two dozen newspapers published daily weather forecasts or maps. Moreover, many newspapers treated weather as an opportunity for humor, interviewing obviously crackpot prognosticators and ridiculing the claims of the bureau forecasters. Newspapers, as boosters of local business, were not above publishing fictitious accounts of bad weather in rival regions. There was, quite naturally, an antagonism between the bureau and the press, one that opened a brief career for the otherwise obscure Philadelphia journalist and poet Harvey Maitland Watts. In February 1901, the widely read *Popular Science Monthly* published his essay "The Weather vs. the Newspapers," a 6,000-word screed against newspaper editors who attribute weather to God's will, who underestimate the importance of the weather, who refuse to keep up with the progress of modern science, and who continue to use incorrect terminology for tornadoes, hurricanes, and other weather conditions. While clinging to the belief that American weather was a relatively simple matter of west-to-east cyclonic and anticyclonic winds, Watts found that newspapers and their readers continued to confuse cyclonic with cyclones:

> The spectacle is stultifying, and yet, in the face of this, in the face of the fact that Weather Bureau stations in the great centers of population have been compelled to phrase their forecasts in primer English, because "cyclone" and "anti-cyclone" puzzled the newspapers and frightened the people, whose idea has been formed on newspaper interpretation of the forecasts; because "highs" and "lows" were deemed too mysterious for comprehension; in the face of all this humiliating confusion, the forecasts, if they err, are criticized in a way that not only brings out all the old, but a new ignorance that is as invincible as it is hypercritical, and raises a popular prejudice against the Weather Bureau wholly unwarranted by the facts. . . . Forecasting is an art rather than a *science*. The art is based on science, and as the science improves so will the art.[9]

Watts did not stop with this spirited defense of the Weather Bureau but also published a kind of media guide for weathermen to use in dealing with the press, urging bureau officials to educate newspaper editors and reminding them that few newspapers could afford to assign the same reporter to weather stories week after week so each cub had to be educated anew. Bureau forecasters needed also to prepare written explanations of common types of weather, clear definitions of meteorological terms, and monthly summaries of norms and extremes. Although he rejected the use of a daily weather map because of its small size and complicated symbols, he urged forecasters to prepare special maps to tell the story of weather causation.

The use of maps to educate the public had the full support of Willis Luther Moore, chief of the Weather Bureau from 1895 to 1913. Although Moore's aggressive style and personal ambitions ultimately got him dismissed by President Woodrow Wilson, he was successful in selling the bureau to the nation's newspaper publishers, and the weather map was the key. As Mark Monmonier ably documents, Moore persistently lobbied Congress for more money to print maps in various parts of the country, simplifying the map and adding explanatory notes. He also prepared maps both as inked copies suitable for photographic engraving and as engraved chalk plates. By discontinuing individual station maps, he gave the newspapers exclusive rights to distribute the new simplified map. The *Minneapolis Journal* published the first new map on March 1, 1910. Whereas only a few dozen papers published maps before 1910, almost 150 were publishing daily maps by July 1, 1912. However, Moore's successor, Charles D. Marvin, was less interested in promoting the work of the bureau in newspapers, and ten years later the American Meteorological Society reported that the *St. Joseph News-Press* was one of the few newspapers continuing to publish a daily weather map.[10]

Moore was nothing if not a tireless self-promoter. A former associate described him as "a picturesque farmer-statesman who announced that 'my cows know more about the weather than the Weather Bureau forecasters.'"[11] His book *Moore's Meteorological Almanac and Weather Guide 1901,* as Sarah Strauss observes, was "a demonstration of the transformative progression from weather lore to meteorology." But it is also a demonstration of how weather lore and meteorology coexist and how the struggle for authority was less one of science over superstition in the abstract and more a contest among the various institutions involved in redefining concepts of space and time, such as Stephen Kern explores in *The Culture of Time and Space 1880–1918.* After boldly proclaiming that "any intelligent person, by studying the few simple principles on which the daily weather map is founded, can make an intelligent estimate of the general character of the weather for his region one, two, and, at times, three days in advance," Moore admonishes that "the great mass of intelligent people have no conception of the methods of preparing the weather map and of the many and diversified uses to which a study of the data would lead." He then goes on to explain how 200 weather stations across the United States telegraph their temperature, barometric, and wind data to Washington, D.C., where the map is prepared. For Moore, the regularity of eddies of air across the country is both proof of "the wonderful precision of [God's] laws" and the source of American mental and physical energy.[12]

Where Moore led, the bureau's staff followed. Henry Allen Hazen examined

"The Origin and Value of Weather Lore" in the *Journal of American Folklore* the same year as Moore's *Almanac.* After debunking many traditional weather signs, Hazen offers an explanation: "The origin of a good deal of our weather lore is dependent upon the climate of the country in which it began, and in many cases the weather of the country will be a valuable criterion by which to trace such sayings." Though he discounts the condition of plants and animals in the fall as predictors of the severity of the winter, some of his colleagues were studying the correlation of animal behavior and meteorological conditions. Phenology, the study of the relation of climate to periodic biological phenomena such as bird migrations and the flowering of plants, was popular in Canada and parts of Europe but seems to have been less so in the United States, although Smithsonian secretary Joseph Henry practiced it from the 1840s through the 1870s and President Franklin Roosevelt's Science Advisory Board recommended in 1935 a phenological survey of the country. Phenological techniques are still in use. A 2005 report from an organization called Clean Air–Cool Planet places the first day of "natural spring," when lake ice breaks up and lilacs and apple trees begin to bloom, on March 13, a week earlier than 150 years ago.[13]

Weather lore remains a popular topic, and most books on weather include some examples, but in the early years of professionalization the topic was pursued by government meteorologists for three reasons. First was scientific curiosity. To what extent could weather lore be verified by modern techniques? Second was establishing a bridge between citizens and scientists; if some weather lore was verifiable, the new science might seem less threatening. Finally, whether science debunked or upheld weather lore, it was asserting scientific authority over popular beliefs. As early as 1869, the British meteorologist Richard Inwards published *Weather Lore,* with the observation that "the state of the weather is almost the first subject about which people talk when they meet, and it is not surprising that a matter of such importance to comfort, health, prosperity, and even life itself should form the usual text and starting point for the conversation for daily life." The third edition, in 1898, was organized by types of weather where Aristotle, Bacon, and southern U.S. proverbs shared the pages.[14]

In the United States, a Signal Corps officer, Lieutenant H. H. C. Dunwoody, prepared an American version, *Weather Proverbs,* in 1883, with familiar examples such as, "When clouds are gathered toward the sun at setting, with rosy hue, they foretell rain," and odd ones like, "When wild geese fly to the southeast in the fall, in Kansas, expect a blizzard." He focuses on the prognostic value of weather proverbs, while conceding that "the ablest meteorologists of to-day, aided by the most perfect meteorological instruments and the results of years of accurate instrumental observations, are still unable to give reliable forecasts of the weather for a longer period than two or three days, and frequently not longer than twenty-four hours." Dunwoody also includes a paper by two British meteorologists arguing that many proverbs reflect an accurate understanding of cyclonic and anticyclonic weather patterns, as well as a considerable amount of data gathered by Frank Cushing on Zuni weather prognostics.[15] A third compilation of weather lore was published in 1902 in the *Weather Bureau Bulletin*. Edited by Edward B. Garriott, "Weather Folklore and Local Weather Signs" was specifically intended to separate accurate weather proverbs from the useless by matching them with the frequent weather patterns of 143 American cities and towns. For example, the entry for Vicksburg, Mississippi, notes: "Cirrus and cirro-stratus clouds move from the west, and when precipitation can be associated with their appearance it occurs two to three days after they appear. No other characteristics of cloud formation that presage rain have been noted by the observer," thus confirming the popular rhyme, "Mackrel clouds in the sky, / Expect more wet than dry."[16]

The coincidence of an interest in folk beliefs about the weather and an expanding American empire led to a focus on the vocabulary and lore of weather in "exotic" cultures. Even before Franz Boas remarked on multiple words for snow in Eskimo, William Churchill compiled a list of 267 words for clouds and 436 for winds in a dozen Polynesian languages. Churchill introduces his lexicon with an encomium to the Pacific Islanders, who are "observant of nature in all her moods. . . . To the islander, all the more by reason of his very insularism, the sky is everything." The *Monthly Weather Review* carried items on Native American and Philippine weather words and devoted long articles to the history of "Indian Summer."[17]

The upshot of all this, however, was more antiquarian than scientific. In the 1920s and 1930s, W. J. Humphreys, a Weather Bureau meteorologist and prolific popularizer of weather information, was fond of quoting the first stanza of an apparently anonymous doggerel:

> What is it moulds the life of man?
> The weather!
> What make some black and others tan?
> The weather!
> What make the Zulu live in trees,
> And Congo natives dress in leaves
> While others go in furs and freeze?
> The weather!

To which Humphreys adds, "The weather is always with us. It is the one thing about which anybody and everybody can and does talk. Yet, if any dozen people who, like the rest of us, had been talking about the weather from the days of their earliest prattling, should attempt to define this thing they know so well, it is likely that all would be perplexed and that no two of the definitions would be exactly alike." He then offers his own parody:

> What makes some glad and others sad?
> The Weather.
> What make the farmer hopping mad?
> The Weather.
> What puts a mortgage on your land,
> That makes you sweat to beat the band,
> Or takes it off before demand?
> The Weather.[18]

By removing the exclamation marks and focusing on the economic aspects of the weather, Humphreys loses some of the jaunty tone of the original but avoids its racism and environmental determinism. The verse nicely sums up the nature of weather talk, with its emphases on climate and culture, personal comfort, and the need to reduce weather to an artifact of the poet's pen. Mr. Dooley was wrong; there are more than two kinds of weather. "Human weather" is manifested in both the atmosphere and the noosphere—the place we live in, and a place we want to live in.

Two additional stanzas appear in a 1922 anthology, but the verse may be much older, and I suspect it has been parodied many times:

> What makes the summer warm and fair?
> The Weather!
> What causes winter underwear?
> The Weather!
> What makes us rush and build a fire,
> And shiver near the glowing pyre—
> And then on other days perspire?
> The Weather!
> What makes the Cost of Living high?
> The Weather!
> What make the Libyan Desert dry?
> The Weather!
> What is it men in ev'ry clime,
> Will talk about till end of time?
> What drove our honest pen to rhyme?
> The Weather![19]

Talking about weather, defining its terms, and claiming authority to speak on it became, in the early twentieth century, a perennial activity of journalists, meteorologists, novelists, and poets, and it remains so. Ambrose Bierce's definition in *The Devil's Dictionary* set a tone:

Weather, n. The climate of an hour. A permanent topic of conversation among persons whom it does not interest, but who have inherited the tendency to chatter about it from naked arboreal ancestors whom it keenly concerned. The setting up of official weather bureaus and their maintenance in mendacity prove that even governments are accessible to suasion by rude forefathers of the jungle.[20]

A less cynical assessment in 1913 by an anonymous writer for the magazine the *Independent* provided a more thorough hermeneutics of weather talk:

This is the season [summer] when people talk about why they talk about the weather. "'Tis a foolish custom," say some. "We will put a stop to it," say others. But both are wrong. A custom so ancient and universal is neither without reason nor easily abolished. A remark about the weather is like the move P to K4 in chess; it has become the established way of beginning a conversation and it holds its ground for the same reason, that no better opening has been discovered.

The author goes on to suggest that the utility of the weather gambit in conversation is that it is a great equalizer: "Everybody has an equal share in the weather and when two meet they have the same weather."[21] They have the same measurable amount of wind and same degree of temperature, perhaps, but how often have I heard my neighbor say, "Pretty cold, isn't it?" when I think the weather's fine.

The second reason weather is a safe topic, according to the writer, is that the alternative, "How are you?" too often leads to a lengthy response filled with personal medical history. Third, weather allows everyone to be an expert: "The Chief of the Weather Bureau and the ordinary farmer may comment and speculate with equal assurance and while each may despise the other for his ignorance and presumption and while this feeling may be justifiable on both sides, this does not detract from the pleasure of the conversation, but rather adds to it. In fact, many people converse for no other reason than to acquire this feeling." Finally, says the writer for the *Independent,* the chief value of weather talk is that it is useless. Nothing can be done about the weather, the writer maintains, so it does not require action. This, of course, was not quite true, since pluviculture was an established faith on the Great Plains and the reputable engineer Carroll Livingston Riker had just published his plan to divert the Labrador Current and warm the climate of eastern North America. Still, the point that "however much you may deplore the badness of the weather nobody is likely to hold you responsible for it" is valid, as is the conclusion that "being then both useless and useful the weather is unequalled as a topic for conversation in its formal and preliminary stages."[22]

This was not the last word, of course, nor could the bureau allow it to be. In 1920, a bureau meteorologist assured readers of the U.S. Department of Agriculture Yearbook that weather *was* worth "speaking of," mentioning aviation, blizzard, and hurricane forecasts; forest fire, frost, and flood warnings; and crop, highway, and shipping bulletins. Bureau forecasts are scientific, not superstition or guesswork, and he claimed 88.4 percent accuracy for the forecasts. The bureau, he concludes, saves farmers and shippers millions of dollars and costs taxpayers about a nickel a household annually. Eighty-two years later, forecasting and other Weather Service work cost the average U.S. household about thirteen dollars a year.[23]

Are we getting our money's worth? Are forecasts more accurate? These were difficult questions to answer in 1900, and they remain so today. Criticism of the Weather Bureau a century ago centered on (1) inaccurate forecasts and false claims of verification; (2) poor leadership and inefficiency in the Washington bureaucracy; (3) monopolistic control of meteorological research and forecasting; and (4) imprecise terminology. The popular novelist Emerson Hough presented these criticisms in *Everybody's Magazine* in 1909. Hough's outrage seems to have been fueled by what he perceived as the bureau's arrogance as much as anything. He quotes "one of the best known meteorologists in the world," who was skeptical of the bureau's claims of verification of forecasts:

> "If [a forecaster] attempted to forecast for each day on either a fair or foul basis, he might get quite a per cent. of verifications. But if he tried to tell what part of a day would be fair or foul, whether there would be a cloudy sky or one partly cloudy, whether there would be little rain or much, it might go hard with him. Each added detail would cut down the per cent. of verifications. *Yet sixty per cent. of this kind of verification might be of far greater value than one hundred per cent. of loose and general verifications!*" [24]

The Weather Bureau had been claiming 88 percent forecast accuracy since 1905, but on a scale so large that local weather conditions were often at variance with the forecast. Public confidence in the predictions waned, opening the way for ridicule. Nor was sarcasm limited to Americans. An article in the British journal *Nature* in 1914 compared the daily forecasts for London for all of 1913 with the actual weather. The author "concluded that someone following the rule 'Tomorrow will be like to-day' would have been right almost as often as the Meteorological Office was."[25] Part of the problem was the use of ambiguous forecasting terms. A more serious problem lay in the interpretation of forecasts by the public and the ways in which users acted on them. Finally, and most troublesome, were the limits on weather forecasts imposed by the chaotic nature of the atmosphere.

Naming the Weather

The bureau sought to address the first problem by trying to impose rigorous definitions on common terms and producing scientific taxonomies as empirical research produced new data. As articles in the *Monthly Weather Review* clearly indicate, Abbe kept abreast of international developments, and American weather terminology was constructed with transnational communication in mind. The July 1898 issue of the *Review* brought to the attention of American weather observers and forecasters the work of the International Meteorological Congresses of Vienna in 1873 and Munich in 1892 on definitions and symbols for twenty-two meteorological phenomena. Weather Bureau chief Mark Harrington had recommended their use by American observers in a circular dated January 1, 1894. The symbols were accompanied by brief, simple definitions: "1. Rainfall—Indicates that an appreciable quantity of rain (one hundredth of an inch or more) has fallen during the day or since last observation; also that the day is a rainy day as distinguished from snowy or clear days. 2. Snowfall—Indicates that an appreciable quantity of snow has fallen during the day."[26] The other twenty symbols were for hailstones, sleet, silver frost, glazed frost, ice needles, drifting snow, snow covering, fog, high haze, dew, hoarfrost, strong wind, thunderstorm, heat lightning, solar aureola or glory, lunar aureola, solar halo, lunar halo, rainbow, and auroral lights.

Concern for ambiguity in weather terminology continued in the revised edition of *Instructions for Voluntary Weather Observers* in 1901, but the definitions of the terms included did little to clear the air, due in part to the anonymous author's apparent desire to pack as much information as possible into a page and a half on "nomenclature." For example:

> The word "storm" will refer to a disturbance of the ordinary average condition or to unusual phenomena, and unless specifically qualified may include any or all meteorological disturbances, such as wind, rain, snow, hail, thunder, etc. This word may be qualified by some peculiarity, i.e., sandstorm or duststorm (such as the "simoon"), hot wind (such as the "khamsin" or "foehn" or "chinook"), cold windstorm (such as the "norther" and the "pamperos"), cold rainstorm and snowstorm (such as the "blizzard").

The Weather Bureau's world, the pamphlet implies, is divided by "peculiarities" of weather and united by "cyclones" too large to be "observed by one observer or at a single station."[27]

The Weather Bureau had inherited many of the voluntary observers, first organized by the Smithsonian in 1848, and relied on them both for "knowledge of the climate of a region" and for "improvement or rules and methods for making

Boy standing by cooperative observer weather station in Granger,
Utah, ca. 1900. National Oceanic and Atmospheric Administration.

weather predictions," the major duties of the bureau as set forth in the organic act
of 1890.[28] With only 546 full-time personnel and 173 weather stations in 1900, the
bureau was quite dependent on the more than 3,000 cooperative observers, as
they came to be called, yet there is little acknowledgment of them by the bureau
until after the organization of the American Meteorological Society in 1919. The
Bulletin of the American Meteorological Society provided a public forum for the
volunteers, and the chief of the Weather Bureau encouraged the AMS to provide
training in meteorology for them. In the 1920s, most volunteers confined their
observations to temperature, precipitation, and general atmospheric conditions,

working without barometers or anemometers. Faulty and broken instruments were and remain a constant problem.[29]

Volunteers continue to outnumber professional employees by almost 3 to 1. In 2005 there were more than 11,000 cooperative weather observers whose important work has been recognized by the NWS by upgrading both their instruments and their training, while the number of NWS employees is approximately 4,700 working in 122 field offices, thirteen river forecast centers, and nine national centers across the United States.[30]

Since volunteers frequently made unauthorized local forecasts, spoke at community meetings, wrote for small-town newspapers, and even supplied daily reports on radio, the language they used was important to the bureau, which was striving to standardize the meanings of weather words. Cooperative observers, who tended to be better educated than the average citizen, often expressed pride in their work and frustration over lack of recognition.[31] A small but vocal number of volunteers were women. When the Weather Bureau decided to honor the longest-serving volunteers in 1935, it discovered that three of the thirty-eight who had made continuous daily observations for more than forty-two years were women and featured them in photographs. Miss Louisa B. Knapp of Plymouth, Massachusetts, had put in forty-nine years; Miss Alice Scudder of Moxee, Washington, began making observations in 1892; and Miss Annette Koch of Pearlington, Mississippi, began a year later. When Elwood Kirkwood of Mauzy, Indiana, celebrated fifty-seven years of daily observation in 1938, his wife was acknowledged because she made the observations when her husband was absent. Husbands substituting for wives may be rarer, if the example of Betty Thompson of Glady, West Virginia, is typical. She began serving as a cooperative weather observer in 1973, but her retired schoolteacher husband prefers to garden and build model railroads.[32]

Mrs. Ross Woods of Palmetto, Tennessee, continued her father's observations upon his death in 1905. In 1928 she recalled for readers of the *Bulletin* her passion for weather observation:

In fancy I stand before the instrument, not at the time I set the thermometer and make my daily record, but this is the hour before bedtime and this is my observation; above me is the sky "that beautiful parchment on which the sun and moon keep their diary." I see it "sometimes gentle, sometimes capricious, sometimes awful, never the same for two moments together, almost human in its passions, almost spiritual in its tenderness, almost divine in its infinity," and I am glad I am numbered even though in a humble way among those who scan the sky.

Woods saw her weather observations as a service to her neighbors, who called to ask when they should drain their car radiators and put more cover on the potatoes; to her community, when her records were subpoenaed by the courts and used by insurance companies; to the nation, through frost warnings; and to the world, by uniting nations in the peaceful pursuit of meteorological knowledge. There were times, as a busy farm wife with small children, when she regretted her duties and hoped that a "cyclone" would blow her instrument shed away, but the pleasures outweighed the drudgery, and she wittily corrected her choice of words. "I should say 'hurricane' or 'tornado' to be strict in the choice of names."[33]

As Woods's comments suggest, conditions specific to the United States evoked discussions of proper terminology. In 1915, an unsigned article, "Weather Bureau Terms Used to Designate Storms," began the long and sometimes contentious process of defining cyclones, tornadoes, and hurricanes. The bureau was anxious to correct the popular use of "cyclone" as a synonym for "tornado." The former, according to Weather Bureau usage,

> is the name of any atmospheric system in which the barometric pressure diminishes progressively to a minimum value at the center, and towards which the winds blow spirally inward from all sides. The system overspreads an approximately circular or elliptical area at least 50 miles, generally several hundred miles, and often over one thousand miles in diameter. A cyclone is any such system of winds, except a tornado which is rarely greater than a mile in diameter, or a whirlwind which is seldom more than a few yards across.[34]

The article further urged classification based on the geographic origin of cyclones—tropical or extratropical. Origin was important because tropical cyclones were popularly known as typhoons in the Pacific and hurricanes in the West Indies. Further complicating clear definitions of storms was the use of "hurricane" in the Beaufort wind scale to indicate winds of 75 mph or more, irrespective of their geographic origin. The bureau continues to struggle to assert its authority over vernacular terms.

Some words were more difficult to control than others. In the same year, Eleanor Buynitzky of the Weather Bureau library traced the evolution of the meaning of "fair," arguing that traditional usage implied "partly cloudy," but that since 1888, the Weather Service restricted its use to mean absence of precipitation. Thus, a "fair" day could be completely overcast as long as it did not rain. Buynitzky quotes U.S. Signal Service and Weather Bureau instructions to observers from 1870 to 1915 to document the changing meanings and the attempts to reduce ambiguity, but she concedes that the bureau contributed to the problem by instructing voluntary observers in 1892 to report "fair skies" when the sky was four-tenths to seven-tenths obscured. In 2007 the matter was still unsettled. In *The Glossary of Weather and Climate* published by the AMS in 1996, "fair" is defined:

With respect to *weather,* generally descriptive of pleasant weather conditions, with due regard for location and time of year. It is subject to popular misinterpretation, for it is a purely subjective description. When this term is used in public weather forecasts (The National Weather Service), it is meant to imply (*a*) no precipitation, (*b*) less than 0.4 *sky cover* of low clouds, and (*c*) no other extreme conditions of cloudiness, visibility, or wind.

The NWS remains adamant that the word "fair" not be used to forecast sky conditions.[35]

A definition of atmospheric pressure also proved problematic. In February 1918, Charles Marvin, chief of the Weather Bureau, reviewed the "Nomenclature of the Unit of Absolute Pressure," reluctantly accepting Vilhelm Bjerknes's term "millibar" for measuring atmospheric pressure. The problem for Marvin was inconsistency and the impossibility of measuring pressure in absolute units. "Pressure," Marvin scolded, "can be conceived only with reference to some area over which it acts, and pressure multiplied by area is a force. Since the dyne is the standard unit of force, a pressure of 1 dyne per square centimeter seems to constitute the logical unit of pressure."[36] Marvin did not see the need for a new term for a unit of pressure, since pressure was meaningless without force.

Nevertheless, both British and French scientists had been using "barad" and "barye" (from the Greek, *barys*—weight or heavy) as a unit equal to 1 dyne per square centimeter since the 1880s. In 1906 the great Swedish meteorologist Vilhelm Bjerknes, in need of a simple measurement of atmospheric pressure to describe his concept of weather fronts, without resorting to cumbersome expressions such as "Pressure at that height where gravity potential is $x.10^5$ C.G.S.-units," began his campaign to establish the millibar as the universal standard.[37] By 1919 he had succeeded, as a note and humorous verse in the first issue of the *Bulletin of the American Meteorological Society* record:

> The Congress of Directors of Independent Meteorological services recently decided at Paris, by a majority, that the millibar should be adopted in preference to the millimeter of mercury for use in stating the pressure of the atmosphere.

> The Directors of Science in Congress assembled,
>   Agreed that in future no discord should mar
> The values of pressure so often dissembled
>   By units derived by the platinum bar.
> The inch and the meter and gravity trembled,
>   As into the Congress there tripped lightly skipping
> An innocent damsel who'd just 'scaped a whipping;
>   Her name in plain English was Miss Milly Barr,

Dear Milly Barr
Bjerknesian star;
A thousand of you
Shall be ever our cue,
As our standard of pressure wherever we are.—E.G.[38]

The verse by E.G. (probably Lieutenant Colonel E. Gold of the Meteorological Office, London) caught the exuberant mood of scientists at the beginning of the revolution in meteorology.

Marvin was happy to have the revolution start without him. But while he fussed over old definitions, events were taking place in Scandinavia that changed the terms of debate. In 1917 Bjerknes established the Geophysical Institute in Bergen, Norway, and by 1922, the Bergen school of meteorology had proposed a theory of polar and tropical "fronts," the interface between air masses of different temperatures, and was developing a mathematical approach to the workings of the atmosphere. Most histories of the U.S. Weather Bureau fault Willis Moore and Charles Marvin, who succeeded Moore as chief in 1913 and who served until 1934, for resisting the adoption of these new theories and methods. Marvin had risen from the instrument division, where he had invented a number of meteorological devices, but he failed to appreciate the potential of the Bergen approach.[39] In 1933, the crash of the dirigible *Akron* in a storm off the New Jersey coast led to the creation of the Committee on the Weather Bureau, headed by Nobel Prize–winning physicist Robert A. Millikan, that recommended the adoption of the Bergen school techniques, but the bureau's budgets were cut, and the implementation of the recommendation did not take place for another two years.[40]

Historians who dismiss the backwardness of the Weather Bureau fail to appreciate that the bureau was more than the Washington, D.C., headquarters. With more than a thousand employees and 200 stations, the bureau was diverse enough to permit challenges to Marvin's authority. A glimpse of the world of local weather stations in the late 1920s and early 1930s is provided in a brief reminiscence by Jack C. Thompson. Thompson, a recent high school graduate, obtained a job as a minor observer in the Sacramento, California, office of the bureau in the summer of 1928. The meteorologist-in-charge was Nathaniel Taylor, whose career stretched back to the Signal Corps weather service. While Thompson's description confirms Kristine Harper's observation on the lack of formal meteorological education in the Weather Bureau, it also reveals something of the place of the bureau in local civic life:

Mr. Taylor was a spry old gentleman, whose erect bearing, white hair, and neatly clipped mustache were a familiar sight on the streets of Sacramento. Indeed, "Weatherman Taylor" was well known throughout the entire

Sacramento Valley and, as a consequence of occasional newspaper articles, throughout northern California as well. Lacking any formal training in meteorology, he was nevertheless a competent, practical forecaster who commanded the respect and affection of the community he served.

This local recognition of the Weather Bureau's meteorologist-in-charge was characteristic of those earlier days. Throughout the United States, almost everyone in town knew who their "weatherman" was. Furthermore, despite its connection with the federal government, the weather office was looked upon as a purely local institution.[41]

Thompson's discussion of the daily routine in the Sacramento office and his introduction to the ideas of the Bergen school are also illuminating. Each morning at 7:00 the staff gathered to read the coded weather reports from stations throughout the country delivered by Western Union. The barometric readings were entered on a map and the isobars drawn, red for barometric pressures below 30 inches, blue for pressures of 30 inches and above. Thompson recalled,

> With the coming of the Bergen school's weather fronts, the central office ordered that all isobars be drawn in black pencil. The red color was then to be used for warm fronts, and the blue for cold. However, some intransigent old-timers at smaller or special-purpose stations refused to give up their red and blue isobars, claiming that the accuracy of their forecasts would suffer thereby. But fronts, air-mass analysis, and other modern ideas came anyway, and the colored isobars disappeared.[42]

After the maps were prepared and forecasts made, the staff printed the *Daily Weather Report,* containing a smaller version of the manuscript weather map, tables of temperature and rainfall for stations in the United States, the forecast, river stages, and flood warnings.

The printed *Daily Weather Report* was mailed to a substantial list of subscribers throughout the Sacramento Valley. In those days, mail delivery was more frequent than it is today, so that the *Weather Report* was always received by subscribers living in Sacramento or in nearby towns the same day it was mailed. At more distant locations, it arrived the next day. This practice of issuing a printed weather report was rather common then, especially at weather offices staffed by three or more people. However, with an increase in the number and popularity of radio stations and home receivers, this procedure was discontinued, and the locally printed weather report disappeared.[43]

Thompson went on to earn a degree in meteorology from UCLA, studying with Jacob Bjerknes, Vilhelm's son, and to work in the bureau's Fruit-Frost Service in Pomona, California. Later he worked in the Washington, D.C., office and at the World Meteorological Organization in Switzerland. Although they may not be typical, Thompson's experiences and career show that a competent cadre of career weathermen emerged in the 1930s despite problems with the leadership of the bureau. The U.S. Weather Bureau, dependent on Congress for funding, and bearing the marks of its military origins, chose the classic bureaucratic strategies of CYA and curry favor. The first approach involved using ambiguous forecast terms such as "fair" and avoiding potentially contentious terms such as "tornado." As a technique for establishing the bureau's authority, defining the weather had its limits; so, too, did wooing its customers.

Using Forecasts

In a political system in which the budgets of government bureaus are determined by elected representatives who are, in turn, responsible to voters, the consumers of the Weather Bureau's products are the ultimate judge of its worth. These farmers, shippers, and fellow citizens voted, and the bureau knew that, like post offices, weather stations were where Americans were most likely to see their federal government in action. Daily forecasts might be chancy, but the bureau was certain that it knew its customers and developed specialized services for dozens of weather-dependent occupations. The just-mentioned Fruit-Frost Service, established in 1917, is one example; others, such as the River Observations and Flood Warnings, were legacies of the Signal Corps era. Special weather forecast services for cotton and sugar growers were begun in 1881 and for grain crops in 1887. Daily and weekly bulletins were published containing temperature and rainfall records from special stations in agricultural states that enabled farmers to plan spraying for insects and fungi and for harvesting of hay and, in a few states, cranberries. Climatological data provided by the twice-daily records of the volunteer observers were used by farmers experimenting with new crops and hybrids, manufacturers determining the locations of plants, home seekers, and physicians advising patients. Fire weather warnings were issued by district forecasters in Washington, Oregon, Idaho, California, and Minnesota beginning in 1914, and mountain snowfall surveys to determine the amount of spring runoff were begun in 1910 in Utah, spreading to other western states in the next few years. The Highway Weather Service was authorized in 1919. Forecasting for military, private, and commercial flying began in 1916 but failed to provide enough coverage for the rapidly expanding air transportation industry, which began to hire its own meteorologists and set up private weather services in the 1930s.[44]

As it tried to fill every niche and meet the needs of diverse groups of weather consumers, the bureau was stretched thin, and frequent budget cuts disrupted some specialized forecasts. As will be seen in the following chapter, the rise of private weather services, consulting meteorologists, and fields of applied and industrial meteorology was a direct result of the bureau's inability to meet the needs of all the nation's weather-dependent activities. Although the question of whether the public understood forecasts and acted on them properly, and whether the bureau was providing the information that farmers, shippers, and travelers really wanted, came up occasionally, customer satisfaction with Weather Bureau products was the least of the obstacles it faced in fulfilling its mission.

A novel conjunction of public education and customer service occurred in 1931 when the *Daily News,* one of New York City's major tabloids, built a new office on Forty-second Street between Second and Third Avenues. Designed by Raymond Hood, architect of the *Chicago Tribune* building and Rockefeller Center, the lobby had glass walls 9 feet high. Six panels held the largest glass weather map ever made. Representing the Northern Hemisphere, the map was redrawn daily. Weather instruments on the roof provided temperature, wind direction and velocity, atmospheric pressure, humidity, and rainfall measurements to gauges in the lobby. The paper hired meteorologist J. Henry Weber from the Weather Bureau to maintain the apparatus and give twice-daily lectures.[45] Six years after the *Daily News* building opened, the trade journal *Printer's Ink* published a tongue-in-cheek review of the weather station and discovered that Weber was receiving more than 4,000 calls a month from dozens of businesses seeking weather information. Among the callers were individuals planning weekend activities, Good Humor Ice Cream preparing the next day's production, building superintendents planning heat requirements, contractors pouring concrete, department stores evaluating stock requirements and advertising campaigns, architects of skyscrapers seeking wind velocity data, operators of ballparks and racetracks, and, of course, "weather nuts."[46] The *Daily News*'s foray into the weather business might be a simple footnote in the history of American weather were it not for its influence on one of the most important theoretical meteorologists of the twentieth century, Joseph Smagorinsky, who recalls that he was initially inspired to study meteorology by frequent visits to the paper's lobby.[47]

By 1940 the meteorological community had grown, and the number of university and private-sector meteorologists was catching up with federal employees, creating two additional groups of knowledgeable weather information users. They were also becoming more critical of the bureau's communications with its audiences. Two articles in the October 1940 issue of the *Bulletin of the American Meteorological Society* addressed this issue. G. Emmons of New York University, which had developed programs in upper air research as well as forecasting under the leadership of the energetic South African–born geophysicist and oceanographer Athelstan Spilhaus, titled his paper "Suggestions for Improved Presentation of Weather Information to the Public." Emmons reviews two problems: terminology and newspaper weather charts. The second article, by Helmut Landsberg, a professor at Pennsylvania State College, reports on a test he gave to freshman

English composition students to determine their understanding of forecasting terms. Landsberg's survey, while not a rigorous study, offers the first empirical evidence of the problem.

Emmons was critical of the bureau for emphasizing barometric pressure because he felt it was one of the least important weather elements as far as ordinary human activity was concerned. He found the terms "area of high pressure" and "trough" unintelligible to the layman:

> Weather bulletins prepared for public dissemination ought to deal exclusively with the elements to which human life and activity are most sensitive. The "air-mass" concept, which is fundamental in modern methods of weather analysis, lends itself naturally even to highly simplified discussions of changes in these elements, for the reason that every intelligent layman is more or less familiar with the operation of the simple physical laws on which this concept is based. For example, he realized that wind is air in motion. He also has an elementary knowledge of certain climatic relationships, such as the existence of low temperatures at the poles and high temperatures at the equator.[48]

Emmons's suggestions for weather maps were complicated and included the location of "fronts" and isotherms in increments of 32°F from 0° to 96°F. "This is a logical scheme, since 64°F is close to the temperature of maximum comfort, while 96°F is approximately the temperature of the human body." His scheme may say more about an engineer's approach to weather than a layman's, but it raises an interesting point about concepts of comfort that I will explore in chapter 5. Emmons urged newspapers to print a box explaining forecast terms with the daily forecasts, to avoid the term "snow flurries," and to provide "the man in the street" with information that will allow him to "judge for himself whether today's weather will be good, bad or indifferent from his personal point of view."[49]

On nine different days, Landsberg asked classes of twenty to eighty-three students to underline terms describing the day's sky aspect (fair, cloudy, partly cloudy, clear, overcast), wind (calm, light, moderate, fresh, strong, gale), precipitation (light rain, moderate rain, light snow, heavy snow, snowflurry [sic], drizzle, mist, light shower, heavy shower, hail, sleet, thunderstorm, fog, haze), and temperature compared to the previous day (warmer, colder, not much change). He then compared their choices with those of the bureau's forecasters, finding that they were in agreement 69 percent of the time on temperature, 62 percent on sky aspect, but only 40 percent on wind. Landsberg was unable to explain this discrepancy on wind, other than to say that there had been no strong or gale force winds on the days the test was given, and that many students seemed to confuse "light" with "calm" and "moderate."[50]

His final conclusion, that "public opinion" seems favorable to the proposed terminology, is partially at odds with Emmons. Landsberg retains the terms "fair" and "snow flurry," although he is willing to add terms such as "possible" (1:2 chance), "probable" (2:1 chance), and "likely" (4:1 chance) to precipitation forecasts. He also identifies a fundamental problem with the current forecasts:

> The forecaster wants to see his forecast verified; the less specific his forecast or the more latitude a term according to his own definition permits the better his imagined score. The general public wants information. It is often unaware of the wide leeway that the forecaster permits himself in using certain terms. . . . It should also be stated that ambiguity in forecasts is partly due to the large territories that are covered by forecasts. . . . If the forecast is vague it will not serve a useful purpose; if it is specific it will for certain disappoint part of the customers in other localities. More decentralized forecasting is very necessary for specificity.[51]

Landsberg's candor about the personal and professional motives for ambiguous forecasts calls attention to the inherent conflict between producers and users of weather information. Landsberg probably already knew that the "general public" was aware of the bureau's credibility problem. In April 1940, *Fortune* magazine informed its readers about how the weather worked and the Weather Bureau didn't. The article explained the contribution of the Bergen school and praised the selection of Francis Reichelderfer as the new chief of the bureau. The magazine supported increased spending on research by both the bureau and private forecasters, pointing out that forecasts cost Americans only a few cents per capita:

> All of which leads to the question—exactly how good or bad are Weather Bureau forecasts? Statistically they look pretty good, but that statement needs copious footnotes. In the first place there is no generally accepted tolerance of error in rating forecasts. The Weather Bureau used to claim 85 to 90 per cent accuracy, but claimed correct ratings whenever a mere *trace* of rain fell, whether the forecast had predicted rain or not. . . . Then there are those good old weasel phrases like "probably showers," "generally unsettled," and "partly cloudy," which make impossible any realistic rating of accuracy. The Weather Bureau tries to discourage such looseness, and some forecasters have expunged "probably" from their vocabularies except when guessing the time of day that rain may start or stop; but it is not forbidden, and inexplicit clichés are still common in district forecasts.[52]

A decade later another Penn State professor, Hans Neuberger, tested more students, this time in an introductory meteorology course for non–science majors. Using Weather Bureau terminology, students were asked to describe the day's atmosphere. Although the students used many terms correctly, Neuberger found no significant improvement in their understanding after completing the course and concluded "that people in general fail to understand forecasts as now presented and that this, in turn, may adversely affect the attitudes of the general public toward weather forecasting. . . . Furthermore, our experience seems to show that education of the public to a more complex terminology is impossible. Simplified forecasts and thus more inclusive terminology would be more meaningful to the majority of the public, not to mention the improved verification scores that would result."[53] Was Neuberger simply more cynical than Landsberg, or was "blame the consumer" endemic in both the federal and academic weather communities? In either case, caveat emptor obviously applies to forecast users.

Although there seems to have been no further empirical studies of weather consumers, by the 1970s bureau officials were beginning to see the need for identifying new categories of weather-information users and addressing different forecasts to each. Shortly before the U.S. Weather Bureau was made part of the National Oceanic and Atmospheric Administration (NOAA) in 1970, the year of the first Earth Day, the newly renamed U.S. Weather Service joined with the American Meteorological Society to hold the symposium "A Century of Progress." George Cressman of the bureau reviewed "Public Forecasting—Present and Future," arguing that the primary problem was not forecast accuracy but timely dissemination to specific audiences. Cressman contends that, while large-scale events such as changes in temperature and precipitation can be accurately forecast twelve to twenty-four hours in advance, giving time for dissemination by newspaper as well as by radio and television, most severe weather occurs on a small horizontal scale, leaving less time for collecting and processing data and issuing predictions or warnings to the media. Citing a study of public users of weather forecasts in Baltimore and Denver conducted for the Weather Bureau by the National Bureau of Standards, Cressman writes that there was a consensus that eight of ten forecasts were correct (the venerable 80 percent), but that half the respondents said that they made their own estimates after hearing the forecast and looking outside. The majority wanted short-range (six-hour to twenty-four-hour) local forecasts.[54]

Vague terminology is just one obstacle to verifying forecasts; another is changes in forecast procedures and management practices. Several studies show significant fluctuations in verification over twenty-five-year periods and, in the case of heavy snow forecasts between 1962 and 1969, an actual decrease in forecast accuracy. Cressman attributes this to "the fact that these forecasts are prepared in the context of the larger-scale weather patterns, but the actual heavy snow events occur as smaller scale, short life-time events."[55] He predicts that advances in data collection and analysis will improve the forecasts of severe storms such as tornadoes and hurricanes but cautions, "I don't want to paint a picture of the citizen of the future obsessed with weather and bombarded from all sides with a barrage of the best possible weather forecasts. He should enjoy the weather, without being bored by overexposure to forecasts." In less than a decade, plans were under way for a twenty-four-hour weather channel, as if to test Cressman's gloomy forecast.

Cressman was followed by an official from the Department of the Interior, who emphasized the importance of weather forecasting for the burgeoning recreation sector of the economy. After polling various recreation and sports associations, he urged the NWS to consider issuing a "Camper's Weather Forecast" similar to the "General Aviation Weather Forecast."[56] As Americans increasingly found greater psychological satisfaction in their leisure pursuits than in their work, and as participation in boating, biking, and other forms of outdoor recreation increased, the National Weather Service was faced with a new and demanding set of weather consumers. Cressman recognized this the following year when he spoke at the annual AMS meeting on the uses of public weather services. Again lamenting the paucity of information, he turned to the thousands of letters and telephone calls received by the NWS, as well as an ongoing AT&T survey, to list the public's requirements for weather information. One surprised him. The AT&T survey indicated that people want the latest weather information for reassurance. They want to know that the "government is watching the weather and that bad weather will arrive and depart more or less on schedule."[57] Other new clients for weather information were pollution-control agencies, mothers sending their children to school, children returning from school, commuters, and, of course, spectators at outdoor sporting events.

Forecasting, which had been the primary purpose of the Weather Bureau for its first half century, had lost ground to research after World War II, with the advent of radar, computers, and satellites and the acceptance of Bergen school theories, but it would not go away. Researchers had, in some cases, oversold the prospects for improved and extended forecasts. It was an era of optimism in science and technology. The development of nuclear energy, "better living through chemistry," and the exploration of space seemed to support the claims of the weather modifiers, the snowpack enhancers, and the rainfall augmenters. Controlling the weather might make forecasting redundant. Of course, it was never that simple, but these developments and the rising expectations of the public based on the promises of some meteorologists working with computers brought forecasting back into focus.

From the start there were critics of the promises of numerically based forecasting and skeptics of greatly improved forecasting. C. S. Ramage, professor of meteorology at the University of Hawaii, cited studies that showed no improvement in NWS forecast accuracy from 1966 to 1975. He attributed this to two factors, atmospheric turbulence and overcentralization of the NWS. Turbulence, or sporadic and irregular fluctuations in the flow of the atmosphere, is a particular problem for numerical weather forecasting because of the difficulty of applying statistical smoothing procedures without distorting the physics. The second problem was not scientific but congressional law. Budget cuts in 1973 forced NOAA administrator Robert M. White to eliminate federally funded state climatologist positions. Climatological files were sent to the National Climatic Data Center in Ashville, North Carolina. Ramage felt their removal handicapped local forecasters in serving their constituencies.[58] Within a year, however, thirty-two states were providing some local climatological services, and by 1985, all but two states had restored research and public information programs.[59] Climate remained in the background of weather talk until medical research rediscovered the connections between climate and disease, and evidence for global warming raised awareness of rapid climate changes. Ramage continued his criticisms of NWS forecasting, concluding in 1993, "Like it or not, weather forecasters are viewed with some awe and some distrust, along with soothsayers, astrologers, geomancers, prophets, and oracles."[60]

This may be as close to the truth about the public's love-hate relationship with weather forecasters as we can come. As someone once remarked, weathermen are wrong often enough that we can make fun of them, but they are right often enough that we cannot ignore them. A *Time* magazine essay in 1978 tried to answer the question of why the weather is a compelling topic:

Since the weather is to man what the waters are to fish, his preoccupation with it serves a unique purpose, constituting a social phenomenon all its own. Far from arising merely to pass the time or bridge a silence, "weathertalk," as it might be called, is a sort of code by which people confirm and salute the sense of community they discover in the face of the weather's implacable influence. By dispensing a raging blizzard, a driving rainstorm or even a sunny day, the weather tends to ameliorate the estrangements inherent in cultural divisions and social stratification. . . . People everywhere, including the U.S., confront the weather with marvelously confused feelings and attitudes. They love it as an unrivaled spectacle and fear it as an unrivaled destroyer. One day they curse the rain, the next they dream of walking in it barefooted with a lover. They study meteorology in school, while clinging to the conviction that the weather can be forecast on the basis of the behavior of bugs, animals and vegetation. Groundhog day is still observed.[61]

Time's essayist also notes that Americans are perversely proud of their weather, no matter how abominable, and although they dream of controlling the weather, they fear that if it were predictable they would have nothing to talk about. Like the editorial in the *Independent,* sixty-five years earlier, this piece uses "weathertalk" as evidence of the nation's essential values of equality and individualism. Both writers celebrate weather as part of the natural environment that is presumed to make America great. Though each writer has his tongue firmly in his cheek, there lurks in both pieces a doubt about the ability of Americans to go beyond superficial chatter about heat and cold. We may not be able to predict the weather, but we can with certainty predict conversations about it.

As the ability to predict severe storms rapidly improved, the problem of communicating with the public in a timely and clear manner remained. The NWS continued to study public understanding and misunderstanding of terminology without much resolution of the problems. A spate of deadly and costly storms in the 1970s—the Lubbock, Texas, tornado of May 11, 1970; Hurricane Agnes on June 15–25, 1972; the New York City heat wave of July 22–29, 1972; and tornadoes in Alabama, Michigan, and Ohio on April 3–4, 1974—led to passage of the first Disaster Relief Act and, by the end of the decade, after more tornadoes and a severe winter freeze, to the creation of the Federal Emergency Management Agency (FEMA). Among the more troublesome terms were those used for hazardous weather, especially "watch" and "warning." Two NWS meteorologists with considerable experience in issuing storm warnings reviewed the situation in 1978, pointing out:

> There are two aspects of the terminology used in weather information that must be understood before any further discussion can be fruitful. To begin with, certain terms are used in a descriptive sense. These are the words normally used to describe a class of weather information releases (e.g., warnings, statements, outlooks, and forecasts). The adjectives that are used to further define the class of release are similarly used. As a result, the NWS has 44 separate product titles (e.g., Flash Flood Statement and Tornado Warning) with which to identify releases about dangerous weather parameters.
>
> The other aspect of weather terminology concerns the words that are actually used within these weather releases. These are what the public actually hears in a broadcast. Much of this terminology consists of hazardous weather terms . . . and associated action/response words.[62]

Among the former are potentially confusing terms such as "funnel cloud" and "tornado," "storm tide" and "storm surge," and "ice storm" and "sleet," while the latter include "advisory," "alert," "relocate," and "evacuate," as well as "watch" and "warning." The Army Signal Corps had introduced "warning" in 1870, primarily for hurricanes, but in 1943 the Weather Bureau began issuing "preliminary hurricane alerts" for the same circumstances as warnings, and then switched to "watch" in 1956. "Forecasts" of tornadoes, begun in 1952 after U.S. Air Force meteorologists demonstrated the usefulness of radar in tracking convective storms, were labeled "watches" after the deadly Palm Sunday tornado outbreak of April 11, 1965. "Watch" was later applied to flash floods and winter storms, and its use was greatly expanded in the 1970s. Polls showed that a majority of NWS employees believed that the public understood "warning," but they were less sure that it understood "watch." Some in the NWS wanted to replace "watch" with "alert" or "possibility," while others resisted, scornful of a public that still called the NWS the "Weather Bureau."[63] By 1980 the director of university affairs for NOAA was sanguine about the state of weather forecasting. "Today the general public has high respect for the forecaster and his work," he wrote, calling attention to the increasing number of articles on forecasting in the *Bulletin of the American Meteorological Society* and the founding in 1975 of the National Weather Association (NWA), which "moved forecasters and weather forecasting onto the center stage of discussion in meteorology."[64]

The American Meteorological Society also responded to the growing market for forecasting by creating another journal, *Forecasting and Weather,* in 1986. Its mission, according to its first editors, was "to be a medium for useful and productive dialogue among AMS members who are operational forecasters and those who are researchers. The broad goals are to increase the understanding of weather events that present significant operational forecast problems, provide a forum for new forecasting techniques, and assess progress in analysis and forecasting."[65] Stimulated by the need for more precise terminology, analysts and critics of forecasting became more sophisticated in their discussions. Addressing the meaning of "that was a good forecast," Allan H. Murphy, of the College of Oceanic and Atmospheric Sciences at Oregon State University, an early and frequent contributor to the new journal, begins with the obvious distinction between forecaster and user definitions of a good forecast. "From the forecaster's point of view, the goodness of a forecast is generally related—in one way or another—to the degree of similarity between the forecast conditions and the observed conditions. On the other hand, users are primarily concerned with whether or not a forecast leads to beneficial outcomes in the context of their respective decision-making problems."[66] This simple dichotomy masks a whole set of problems, however, as Murphy demonstrates. From the forecaster's perspective, a forecast is also a "good" forecast if it corresponds to the forecaster's best judgment derived from her knowledge base, regardless of the observed conditions after the forecast. Murphy defines these three types of forecasts as "consistency," "quality," and "value."[67]

Problems arising from a "consistent" forecast, beyond the fact that it may be flat-out wrong in predicting the actual weather, are that (1) judgments in the forecaster's mind are seldom available for evaluation, and (2) forecasts need to include all the information that potential users require. What Murphy is getting at is that a forecaster may feel that her judgment is sound and that it was based on the best information, but she may lack information needed by specific users. "To design such forecasts in a rational manner, it is necessary to obtain detailed information about the users and uses of the forecasts." This gap between a forecaster's judgment and a user's needs is compounded by forecasts that fail to communicate the uncertainty inherent in forecasters' judgments. Forecasters sometimes use probabilistic terms to indicate that their data are insufficient or that atmospheric conditions make it difficult to forecast with any confidence, but these terms may not be understood or be useful to the user. Goodness of the second type, "quality," is achieved when the forecast corresponds to the observed conditions of the weather. Murphy introduces statistical tests of a quality forecast that use ten variables rather than a simple verification of temperature and precipitation predictions. Measuring the "value" of a forecast, as defined by Murphy, is even trickier, since weather forecasts possess no intrinsic value and acquire value only when they influence decisions that save money or lives. A decision maker needs to estimate what might be saved if quality forecasts are available in the future, as well as how much more might have been lost if no forecast had been available. Murphy argues that forecasters must consider the kind of information users need and integrate all three types of goodness into their forecasts.

One of the more confusing concepts for consumers of forecasts is "probability of precipitation" (PoP). This phrase began to appear in the meteorological literature in the 1960s, and in 1965 the NWS initiated a nationwide program of PoP forecasts. The definition seems straightforward—"the likelihood of occurrence of a measurable amount of liquid precipitation (or water equivalent of frozen precipitation) during a specified time interval (e.g. 12 hours) at any given point in the forecast area; expressed as a percentage." Confusion arises, however, when the audience to whom the forecast is addressed does not understand the meaning of "measurable amount" and "point." PoP forecasts do not predict where or how much precipitation may fall in an area that may be several hundred square miles.[68] Many forecasters still hedge their PoPs with words such as "brief," "occasional," "intermittent," "frequent," "chance," and "likely," but one rationale for the PoP is that a user may more easily translate the risk of losing or making money because of rain. As Environment Canada puts it: "Suppose the probability of precipitation is 40%. The (concrete) contractor has to calculate the costs of redoing the job if it rains and, if he/she goes ahead, the expectation is that the risk is 40% of the cost. Compare this to the costs of delaying the job until the next day. If the costs of delay is [sic] less than the risk of going ahead, the best decision would be to delay."[69] Such calculations are, of course, improved if the forecaster also estimates the amount of precipitation expected if it falls.

In a number of studies, Murphy and his associates attempted to assess the problems related to PoP forecasting. In 1970 he polled almost 700 NWS forecasters nationwide on their attitudes toward and experience with PoP forecasting. There were a number of troubling findings. The forecasters surveyed did not think in probabilistic terms. They frequently interpreted the same data differently, hedged their forecasts, and believed that many users preferred categorical forecasts accompanied by verbal modifiers. About half those polled believed that the public interpreted PoPs correctly but used them improperly.[70] In May 1979, Murphy directed a survey of fifty female and twenty-nine male students in Eugene, Oregon, to learn what they knew about PoP forecasts. While most of the students understood numerical probabilities, a majority thought they applied to the whole forecast area, not to localized points within the forecast area. While urging further study and a NWS program to educate the public, Murphy felt that his respondents understood the problem of uncertainty inherent in all forecasts, verbal or numerical, probabilistic or categorical; this appreciation of the difficulties forecasters faced was at least as important as knowing what each percentage means.[71] Improving public understanding of forecasting is as important as improving the forecasts themselves.

An instance of improved forecasting, partially because forecasters understood their users' needs, occurred in the tornado-prone states of the Midwest and the South. The creation of the NWS Severe Local Storms Center (SELS) in Kansas City, Missouri, in 1954, and its expansion into the present Storm Prediction Center in Norman, Oklahoma, improved the capacity of the NWS to provide timely local severe storm forecasts. One study claimed an improvement in severe weather watches from 63 percent in 1975 to 90 percent in 1996. Of course this assertion requires some explanation: "A severe weather watch is considered verified if one or more tornadoes or severe thunderstorm events occur within the watch and during its valid time."[72] In the 1950s, a watch area might be as large as 25,000 square miles and the duration of the watch, two to seven hours. With improvements in computers and models, the addition of radar and satellite data, and better communication between forecasters and volunteer spotters, emergency managers, and the media, watches can be converted to warnings and lives saved—good forecasts by any standards. The red parallelogram "watch boxes" appearing on NWS maps and television screens today are still typically 24,000 to 30,000 square miles (120 by 200 to 250 miles on a side), but when the system works, local spotters (often sheriffs and highway patrol officers), radio and television stations, and emergency responders alert the affected area in a timely fashion.

Charles Doswell, an authority on severe storms, is decidedly less sanguine about the efficacy of the storm warnings for several reasons. With almost forty years' experience in severe storm research and forecasting, Doswell has earned a reputation as an outspoken critic of the NWS, but his criticisms generally confirm those that have been made for more than a century, by Dunne, Hough, and Ramage. These include mismanagement, bureaucratic inertia, and, above all, failure to understand what users want and to communicate honestly with them about what forecasters can and cannot do.

Doswell heaps scorn on the NWS and NOAA—first, for NOAA's taking credit for the weather radio system that went into service in 1963, before NOAA was created; second, for putting National Weather Radio on a frequency that was unreachable on normal radios and therefore required a special receiver; third, for the uneven quality of its broadcasts; and finally, for working for the commercial broadcast media, not the public.[73]

Doswell is the first weather commentator, I think, to pinpoint a major problem in public understanding of weather, that is, the definition of "normal." "Strange weather" is a major subtopic of weather chatter, and weird weather is often attributed, as Doswell notes, to global warming, El Niño, volcanic eruptions, and divine retribution. Moreover, these beliefs are inspired and sustained by Sunday supplements, tabloid newspapers, and television programs, whose primary purpose is to sell cars, beer, and fast food, not teach science. Since people often feel that if they have not personally experienced something it must be unusual, they are likely to define a new experience as strange. Human memory is also short, or at least highly selective; we often forget what the weather was yesterday, but think we know what it should have been and what it should be now. Any deviation is "strange." As Doswell, and every weatherman, insists, the most typical thing about the weather is its variability. Words such as "normal," "average," and "typical" have very little meaning outside of specific contexts and locations. The NWS has records for 140 years, more or less, for a few hundred weather stations in the United States and a few thousand more from the cooperative observers. Thus, averages of temperature, precipitation, and other weather conditions can be determined for any given day, month, or year, but these averages are merely statistics that do not describe real weather, which changes moment by moment.

Average weather is a snapshot, not a movie. In some places the weather is more variable than in others. This is one of the reasons we have 100-year events happening every few years. According to the media, the United States has experienced several "storms of the century" in the past 100 years. Where extremes are greatest, averages have less value in determining what is "normal." The NWS arbitrarily picks the preceding 30 years, currently 1970–1999, to define "normal." Normal temperatures for a specific locality will be different beginning in 2010. These 30-year periods are often used when referring to the climate of a place. Climatologists are currently experimenting with ways to calculate daily normal temperatures that incorporate daily temperature variability, providing a median of the distribution rather than a simple average of extremes. They hope that this will eliminate false assumptions about climate change that weathercasters and the public often make when a temperature is far above or below "average."[74] "'Normality' is a matter of definition," Doswell concludes. "In order to understand what 'normal' means, you have to know what was done to the data."[75]

Echoing Mr. Dooley's two kinds of weather, Doswell identifies two National Weather Services: "One in the field, doing the valuable and productive work of the organization. The other sits in offices and pumps out paperwork and rules, without doing anything even remotely productive itself."[76] What the administrators should be doing, Doswell thinks, is finding out what the value of NWS forecasts really may be. "Meteorology," Doswell believes, "finds it difficult to point to its value, mostly because its contributions lead to reduced costs and losses, rather than direct income generation. Meteorological efforts don't *put* cash in the bank, they *keep* cash in the bank."[77] The fault is not entirely the forecaster's, however, since once a forecast is made, it is the user's responsibility to take action based on it, even when the forecaster is uncertain. Educating users about the limits of forecasting is one answer, but NOAA's public relations office continues to promise a "No Surprise Weather Service," secure in the knowledge that federal agencies are protected from lawsuits resulting from faulty forecasts.[78]

The result of a century of debate over the relationship between the weather forecaster and the public is the emerging field of "weather policy." Largely the creation of Roger A. Pielke Jr., an energetic young political scientist at the University of Colorado who has an undergraduate degree in mathematics and whose father is a meteorologist, weather policy is a specialized branch of science policy as it has been practiced in the United States for the past 150 years. Following the standard distinction between policy for science and science for policy, Pielke sees weather policy as having two interrelated components:

> One is "policies for weather research and decision making." This includes government policies about weather research, forecast operations, and responses focused primarily on the National Weather Service, but more broadly constituted would include the Department of Agriculture, Federal Emergency Management Agency, Small Business Administration, and other agencies that deal with weather and its impacts. It also covers the private sector, notably including providers of weather information and the insurance industry. Of course, an important aspect of weather policy is the relationship between the public and private sectors.
>
> A second component of weather policy is "weather research for decision making" and refers to the connections between research and the actions taken in preparation for and response to weather. This aspect of weather policy is variously called "forecast use and value" and "connections of research and operations."[79]

Pielke's campaign to bring atmospheric scientists and social scientists together began in the early 1990s when he joined the staff of the Environmental and Societal Impacts Group at the National Center for Atmospheric Research (NCAR) in Boulder, Colorado. NCAR is itself a response to the growing awareness of the interrelation of atmosphere, climate, and the earth's environment. The mission of NCAR, which was established in 1960 with funding from the National Science Foundation, is to sponsor and coordinate research on meteorology, climate modeling, and the societal impacts of weather and climate. NCAR's headquarters, opened in 1967, were designed by I. M. Pei to symbolize for the public the center's lofty goals and self-image. Pei's two towers, five stories tall and constructed of reinforced concrete faced with red sandstone, are situated on a mesa and dramatically rise above the city of Boulder. The buildings are surrounded by a park with a "weather trail"—a nature trail with observation points from which to observe atmospheric conditions.[80]

Beginning in 1996, Pielke sought to enlighten scientists and policy makers on how to communicate better with each other. First, they should reach some consensus on the definition of weather-related problems; second, the NWS should be encouraged to study how forecasts actually save lives and property; and finally, policy makers must think about weather and climate policy in broader contexts than energy policy. Pielke uses the current confusion over increases in hurricane damage to illustrate the misdefinition of the problem. While some conclude that severe hurricanes are increasing in frequency, others claim that such storms pose less of a threat because of improved warnings. After the 2005 hurricane season and the damage inflicted by Katrina, the public has been swayed to the first position, but officials in FEMA, the Army Corps of Engineers, and the NWS have a vested interest in promoting the latter perspective. Both are in error in Pielke's view because neither considers the issue of population growth in hurricane-prone areas, so that "the increased damages seen in recent years, is [sic] largely a function of increased societal exposure to hurricanes rather than increased storm incidence."[81]

Even if it turns out that there is a long-term increase in severe hurricanes, Pielke's point is valid; people are getting in the way of environmental hazards and will suffer the consequences. As has been seen in New Orleans in the aftermath of Katrina, mitigation plans based on moving people away from the danger are probably not an option in a democratic society that values private enterprise and individual freedom. On the other hand, many Americans seem unwilling to accept responsibility for their risky behaviors, and the federal government has increasingly stepped in with massive aid for the victims of natural disasters.

Chaos and the Butterfly Effect

The third problem faced by weather forecasters is the incompleteness of weather observations. Small errors in initial observations are compounded as a weather system develops. Numerical weather forecasting holds the promise of great accuracy, but Edward Lorenz's discovery in 1961 that the dynamics of the atmosphere are inherently chaotic because of their "sensitive dependence on initial conditions" remains unchallenged. In 1972 Lorenz brought this new "uncertainty" principle to the attention of a broader scientific audience as the "butterfly effect."

Most nonscientists probably first learned about the butterfly effect with the publication of James Gleick's best-selling book, *Chaos: Making a New Science,* in 1987. In his prologue, Gleick introduces the term as "the notion that a butterfly stirring the air today in Peking can transform storm systems next month in New

York."[82] In the first chapter, titled "The Butterfly Effect," he tells the story of Edward Lorenz, a mathematician and meteorologist at MIT, who "saw that there must be a link between the unwillingness of the weather to repeat itself and the inability of forecasters to predict it."[83] The answer, Lorenz discovered, lay in small variances in initial conditions that eventually produced large differences from predicted outcomes.

Since he is interested in the larger topic of chaos theory, of which the butterfly effect is only a part, Gleick relegates the origins of the term to a footnote: "Lorenz originally used the image of a seagull; the more lasting name seems to have come from his paper, 'Predictability: Does the Flap of a Butterfly's Wings in Brazil Set Off a Tornado in Texas?' an address at the annual meeting of the American Association for the Advancement of Science (AAAS) in Washington, 29 December 1979 [sic]."[84] Gleick also calls attention to the resemblance of a butterfly's wings to Lorenz's illustration of a chaotic system on his computer.

The butterfly effect is a striking, memorable metaphor and will long remain one of the great contributions to the popular understanding of meteorology and chaos theory. For some, the image of the butterfly as the cause of subsequent great changes in nature recalled Ray Bradbury's short story "The Sound of Thunder" (1952), in which a time traveler to the prehistoric past steps off a designated path, crushing a butterfly. Returning to the present, he discovers that the outcome of a recent presidential election has been reversed and the undesirable candidate is in office, with dire consequences for American civil liberties. In his book *The Essence of Chaos* (1993), Lorenz recalls the origin of the phrase and attempts to clarify its significance. He notes that he avoided answering the question in the title but acknowledges that the similarity between a graphical representation of the random changes of variables he called the "strange attractor" might have been the origin of the term. In fact, he writes, the title of the paper was not his, but that of meteorologist Philip Merilees, the AAAS session's convener. Merilees has assured Lorenz that he did not know the Bradbury story.[85]

Lorenz recognizes that other symbols of chaos in meteorology have preceded his seagull and the butterfly. He calls attention to novelist and University of California English professor George R. Stewart's *Storm* (1941), in which a young meteorologist recalls his professor's remark that "a Chinaman sneezing in Shen-si may set men to shoveling snow in New York City."[86] Lorenz, who had received a copy of *Storm* from his sister while he was still a student, concludes, "Stewart's professor was simply echoing what some real-world meteorologists had been saying for many years, sometimes facetiously, sometimes seriously."[87]

Who were these meteorologists, and what metaphors did they use? When I wrote to Professor Lorenz for further information on the origins of the butterfly effect, he kindly responded with a letter and a copy of a paper he had just prepared for a conference in April 2003. In the letter he suggested that Stewart may have consulted with Jacob Bjerkness, who was teaching at UCLA in the 1930s when Stewart was working on the novel. Jacob and his father, Vilhelm, had long been engaged in the study of atmospheric instability. In his paper on the origins of numerical weather prediction, Lorenz calls attention to a paper by meteorologist Joseph Smagorinsky written four years earlier than his talk to the AAAS. Smagorinsky had written: "If we [satisfy certain conditions], then could we predict the atmospheric evolutions from the initial time with infinite precision infinitely distant into the future? Or would the flutter of a butterfly's wings ultimately amplify to the point where numerical simulation departs from reality?"[88] Lorenz is happy to attribute the origin of the phrase to Smagorinsky. Had Smagorinsky been inspired by Bradbury? Or was there another source?

Serendipity is not a branch of lepidoptery, but it can lead to wonderful discoveries. Reading through past issues of the *Bulletin of the American Meteorological Society* year by year and page by page to understand the history of the organization, my attention was caught by the following in the March 1941 issue:

"A condition called 'temperature inversion' could develop down in Texas, they told us, by which a mass of warm air balanced so precariously and unnaturally over a stratum of colder air that even so small a thing as the kick of a gnat's leg could create a roaring, rushing cyclonic disturbance that— traveling the usual storm path northeastward—could wipe out New York City." It had been ten years since I was exposed to such weather trivia in the Army Air Corps.[89]

The quotation, from a short story by Carey Worth Stevenson in *This Week Magazine,* April 13, 1941, drew a brief comment from Charles F. Brooks, who had been an instructor at a school for army weathermen established at Texas A&M University in 1918, before becoming a professor of meteorology at Harvard and one of the founders of the American Meteorological Society. Brooks remembered that William S. Franklin, professor of physics at MIT, had published a paper, "Much Needed Changes of Emphasis in Meteorological Research," in the *Monthly Weather Review* in October 1918, quoting the crucial passage:

Imagine a warm layer of air near the ground overlaid with cold air. Such a condition of the atmosphere is unstable, and any disturbance, however minute, may conceivably start a general collapse. Thus a grasshopper in Idaho might conceivably initiate a storm movement which would sweep across the continent and destroy New York City, or a fly in Arizona might initiate a storm movement which would sweep out harmlessly into the Gulf of Mexico.[90]

We still do not know how the grasshopper metamorphosed into a butterfly, or if the citizen of Shen si sneezed before Stewart's novel in 1941. All we can surmise is that the idea of small changes in initial conditions leading to enormous differences in final results had been "in the air" for almost a century.

Smagorinsky may well be the butterfly man, but grasshopper man William Suddards Franklin deserves a place in the history of chaos theory. His 1918 paper, possibly inspired by the ideas of Henri Poincaré but citing another French mathematician, M. J. Boussinesq, is both a challenge to the emerging science of meteorology to base its research on mathematical physics and a pioneering formulation of chaos theory. A few paragraphs before his insect example, he considers the wide departures from the average weather of a place, comparing the turbulence in the atmosphere to that of a raging fire caused by a tiny spark: "This possibility of the growth of tremendous consequences out of a cause that has the mathematical character of an infinitesimal is the remarkable thing; and this possibility is not only characteristic of fire, but is characteristic of impetuous processes in general."[91]

From the sketchy biographical information available, Franklin was himself impetuous, if not chaotic. His article is combative in tone, and he seems to have been especially agitated by educational reformers in the early twentieth century who idolized science and sought to make it part of the college curriculum without demanding high standards. "Science is *Finding Out* and *Learning How*," Franklin wrote in a self-published book on the scientific method, "but most men think of science in terms of results. These results have fascinated the crowd, and the great majority of men have adopted a scale of physical values for everything in life with a consequent neglect of quality and a denial of human value in everything. We have a philosophy of rectangular beatitudes and spherical benevolences, a theology of universal indulgence, a jurisprudence that will hang no rogues."[92]

Franklin was born in Kansas in 1863 and studied physics and electrical engineering at the University of Kansas, the University of Berlin, Harvard, and Cornell, from which he received a DSc in 1901. He taught at Lehigh University from 1897 to 1917, at MIT from 1917 to 1929, and at Rollins College in Florida in the year before his death in 1930. He was an advocate of playgrounds and outdoor education for children and wrote an account of two young men on a thirty-day hike from Loveland, Colorado, to Laramie, Wyoming, as well as a number of books on physics and electrical engineering.[93] His range of interests and passion for good science make Franklin an appealing figure.

Meteorologists and climatologists continue to struggle with chaos, chiefly by redefining it. Writing in the January 1989 issue of the *Bulletin of the American Meteorological Society,* A. A. Tsonis and J. B. Elsner, geoscientists from the University of Wisconsin–Milwaukee, simply deny chaos, arguing that "chaos theory, which mathematically defines randomness generated by simple deterministic dynamical systems, allows us to see order in processes that we thought to be completely random. (Apparently, the founders of chaos theory had a very good sense of humor, since chaos is the Greek word for the complete absence of order.)"[94] At the risk of ruining the joke by explaining it, this statement begs rebuttal. The English word "chaos" is usually traced to the Greek word for "abyss" and, by extension, "the first state of the universe" before the world was formed.[95] The world that was formed from chaos has been given order by myth and science, but neither poets nor physicists can be certain that some element of chaos does not lurk behind any apparent order. Moreover, as the great British meteorologist Napier Shaw wrote presciently in 1926, "Every theory of the course of events in nature is necessarily based on some process of simplification of the phenomena and is to some extent therefore a fair tale."[96] For Tsonis and Elsner, as for Henri Bergson, disorder is simply an order we cannot yet see. But this denies the meaning of chaos. It is perversely random and unpredictable. Today, computer models use grids of 20 miles instead of the 200-mile resolution used in the 1950s, but, as science writer Jeffrey Rosenfeld points out, "a butterfly as big as Manhattan can still elude detection."[97]

While the Lorenz-Merilees-Smagorinsky butterfly seems to have become almost extinct in scientific writing (it does not appear in the AMS's *Glossary of Weather and Climate* [1996] or its *Glossary of Meteorology* [2nd ed., 2000]), it has found an environmentally friendly niche in literature and popular culture. A recent sighting in Brad Leithauser's novel in verse, *Darlington's Fall,* confirms its survival. In the novel, Russel Darlington becomes fascinated with butterflies as a boy in late nineteenth-century Indiana and after graduating from college sails to Malaya via the South Pacific to collect and study rare lepidoptera. While pursuing a specimen on a mountainous tropical island, he falls and breaks his back. The remaining two-thirds of the book concern his adjustment to his injury and his dream of furthering Darwinian evolutionary theory by discovering the pattern and direction of life. Struck by the realization that his life would have been entirely different if he had not captured a rare moth when he was ten years old that brought the attention of a scientist at the university who encouraged young Russel's passion, he recoils from the possibility that life is random, that evolution is the result of unpredictable developments from small initial conditions. In Leithauser's deceptively simple verse:

> So Russel darkly foreshadows a notion
> Set to flourish decades after his death:
> The idea, dear to the chaos mathematician,
> Of the "butterfly effect"—Edward Lorenz's vision
> Of a world wherein momentous, unreckonable
> Consequences spin off from the tiniest motion
> And a flap of papery wings in the Andes may
> Portend a howling, deadly aftermath,
> Summoning an eventual coastal storm that will
> Kill thousands, thousands of miles away.[98]

Note that, with time, the feckless butterfly has gained range and strength, from the Amazon to the Andes, from a tornado to a hurricane. Computer programmers and economists also seem attracted to the buttery insects and are equally cavalier with their geographic range. Peter Bernstein, in his wonderful history of chance and risk, quotes a chaos-besotted computer scientist from Berkeley who "estimated that the gravitational pull of an electron, randomly shifting position at the edge of the Milky Way, can change the outcome of a billiard game on Earth." Bernstein goes on to assert that the most popular example of the concept of chaos is "the flutter of a butterfly's wings in Hawaii that is the ultimate cause of a hurricane in the Caribbean."[99] The capricious insect also turns up in an advertising flyer mailed in the summer of 2003 by the business newsmagazine the *Economist*: "When a butterfly flutters its wings in one part of the world, it can eventually cause a hurricane in another. . . . Try 4 issues RISK-FREE." The flyer extends the butterfly metaphor to include political climates affecting financial markets.

Religion, Politics, and the Weather

Not all talk about weather concerns definitions, forecasts, or chaos. Sometimes the presumably safe topic veers off into equally popular but conversationally taboo subjects such as religion and politics. Avoid politics and religion, stick with the weather, we are taught from childhood, but if the weather's bad, God or politicians often get blamed. As the founders of the Weather Bureau knew, meteorologists were in competition with shamans, wizards, and almanac makers. They debunked some examples of popular weather lore and incorporated others in their own writings. In creating the daily weather report and convincing the public to consult it, the "professors" invented a ceremony, secular perhaps, but as powerful as any in a religious creed. The daily weather report, whether consumed from newspapers, radio, television, or the Internet, is, I believe, a ritual of a belief in the unseen and ineffable power of nature; it is a ritual that confirms both our need to know and our need to be mystified, to feel in control while experiencing the thrill of the uncontrollable. In *Leaning into the Wind: A Memoir of Midwest Weather,* Susan Allen Toth, invoking William James's classic study of religion, believes that, for midwesterners, weather is "an immersion in the varieties of religious experience." Not all midwesterners claim to think about God, she continues, "but they definitely think a lot about the weather."[100]

"Mythology is a primitive meteorology," concludes the French philosopher Gaston Bachelard in his 1943 essay on air, dreams, and the imagination. Meteorology, though few in the field would care to admit it, may be sophisticated mythology. Napier Shaw's comparison of scientific theories to fairy tales was intended as a drop of humility in the deluge of science over religion in the early twentieth century. A critic, seeking to explain why the weather is so often anthropomorphized in the media, asserts that "meteorology performs the same structural task as superstition (a.k.a. religion), explaining forms of life that are outside the control of those experiencing them." The Romanian historian Lucian Boia, surveying centuries of writing about weather and climate by European philosophers, is somewhat more tentative when he writes, "by way of a conclusion," that "perhaps we are in the process of inventing a religion of climate."[101]

The long-standing tradition of calling weather and other natural disasters "acts of God" does not, of course, mean that a person using the phrase necessarily believes in an active deity. Legal dictionaries simply replace "God" with "forces of nature," a kind of pantheism that drives both fundamentalists and environmentalists crazy. Evangelical ministers find tornadoes and hurricanes (largely confined to the Bible Belt) convenient evidence of an angry God, natural forces being simply his tool, while the unreligious find nothing "natural" in such events, merely the failure of people to get out of the way.[102]

Disaster researchers have been interested in religious interpretations of natural disasters for some time. Russell Dynes and Daniel Yutzy, surveying the field in 1965, identified three types of explanation for disasters: "naturalistic," "fatalistic," and "supernatural." The first they equated with scientific laws. When victims attributed the disaster to neither natural laws nor God's will, they were identified as being fatalistic, believing in chance or bad luck. While both of these categories are obviously too broad to tell us much about how people understand tornadoes, hurricanes, floods, and heat waves, the authors break the third category down into more specific responses. "Supernatural explanations will vary," they write, "according to the dominant conception held of the nature of the supernatural power and of man's relation to this power." Dynes and Yutzy propose a model with four types of religious belief. Type A supposes a belief in a wrathful God and basically good humans. Type B assumes an angry God and evil humans. Type C believes in a benign God, but sinful men and women. Type D believes both God and humans are basically good.[103]

When a weather disaster strikes a Type A person, he feels no responsibility but knows that the angry God must be appeased. The appeasement may take various forms, from sacrifices thought to placate him to scapegoating. Curiously, Dynes and Yutzy think that humans with a Type B belief system simply despair and take no action, but since this conception of God and humankind is reflected in some fundamentalist sects, it seems likely that individuals would accept their responsibility for God's anger, act to reform themselves, but not expect quick-and-easy forgiveness. The jeremiads preached by Puritan ministers in colonial New England often responded to natural disasters in this way. Type C beliefs also cause people to acknowledge their sins and repent, but, believing in a benign God, they expect to be spared next time. Dynes and Yutzy cite Billy Graham's admonition to the citizens of Waco, Texas, after the May 11, 1953, tornado that killed more than 114 people, as an example of this belief. Graham was quoted as saying, "The storm

"Psalm after Storm." A flag, a cross, clouds (one of them a contrail), and a minister preaching in his tornado-ruined church symbolize the interplay of nation, nature, technology, and faith. *Tampa Tribune,* February 5, 2007.

shows what God can do if we do not repent. Out of the storm comes a message of warning to the 50 percent of Wacoans who do not attend church." While a God-fearing fundamentalist whose trailer was blown away might be expected to resent being punished for the behavior of his heathen neighbors, the subtext here seems to be the apostle Matthew's observation that "God maketh His sun to rise upon the good and the bad, and raineth upon the just and the unjust." Types B and C seem less distinct in practice than in theory. Dynes and Yutzy argue that the Type D situation, in which a benign God punishes a basically good people who take no action, but accept God's will, is characteristic of contemporary American society. They quote a tornado survivor who focused on the other survivors, saved by God's will, not on the victims. God and humans get off the hook in this scenario, although FEMA and the Weather Service may not.

Dynes and Yutzy cite some evidence that victims are more likely to choose supernatural over naturalist explanations of weather disasters when the death toll and property losses are high, but the standard used to determine high and medium is not clear. My purpose in critiquing their commendable paper is not to point out their shortcomings but to raise what I think is an important question. How has religious belief influenced the ways in which weather policy is discussed? If Americans are increasingly risk-adverse and increasingly willing to have state and federal governments pay for uninsured disasters, to what extent is this a reflection of religious beliefs? During the Depression years of the 1930s, the nondenominational Protestant journal *Christian Century* reported, with some concern, that meatpacking companies were subverting rural ministers by preaching that farmers who withheld their crops and livestock from market to force a better price were the cause of the drought: "God is angry because little pigs have been slaughtered and productive acres withheld from cultivation," according to these men. Break your contract with the Agricultural Adjustment Act, the farmers were told, and rain will fall.[104] Those who suggested that New Orleans was destroyed in 2005 because God is angry about abortion rights, gay marriages, or the war in Iraq were similarly manipulating religious belief systems for political ends.[105]

We all retain beliefs about weather control from our Stone Age ancestors. As a child visiting relatives in north Texas, I was advised to "kill a snake and hang it belly up on a fence to bring rain." Even at nine or ten I knew this prescription had a better chance of working in August in Texas than in my native Nevada. Praying for rain will probably never go out of existence among farmers, even the nonreligious.[106] Probably, too, our linking of weather and religion, in the twenty-first century, is simply what *Time* magazine essayist Lance Morrow has called "The Religion of Big Weather." "What we have," he writes, "is weather as electronic American Shintoism, a casual but almost mystic daily religion wherein nature is not inert but restless, stirring, alive with kinetic fronts and meanings and turbulent expectations (forecasts, variables, prophesies). We have installed an elaborate priesthood and technology of interpretation: acolytes and satellites preside over snow and circuses." Morrow goes on to argue that we welcome the uncontrollable in life because it brings ideological relief from the current belief that everything is political and manipulable: "The moral indifference of weather, even when destructive, is somehow stimulating. Why? The sheer leveling force is pleasing. It overrides routine and organizes people into a shared moment that will become a punctuating memory in their lives ('Lord, remember the blizzard in '96?')."[107] I suspect that Morrow is using Shintoism as a metaphor for nature worship, but his points about television weather as a kind of performance art, and as ritual, can hardly be denied.

Religion was an issue, too, in the heyday of weather modification, from 1950 to 1975. In polls conducted toward the end of that period, 34 percent of respondents in northern California believed that cloud seeding violated God's plans, while 38 percent in South Dakota and 46 percent in Colorado held this conviction.[108] The Dooleyian dichotomy of Weather Bureau weather and human weather took a new twist when control of weather seemed a possibility. The issues raised by cloud seeding were, however, more political than religious. If we can control the weather, whose weather do we control, and who gets to control it? Acts of man proved as hard to anticipate as acts of God.

Weather and religion have been intertwined since the beginnings of human history. The recent interest shown by some evangelical Christians in the problems of global warming has renewed the debate over the meanings of various passages in Genesis. Is man supposed to subdue the earth or be a good steward? The rhetoric of the Evangelical Environment Network and the Evangelical Climate Initiative borders, in the view of their opponents, on nature worship by putting the earth, one of God's creations, before God himself. Recognizing this dilemma, evangelical environmentalists are already shifting their emphasis from ecology to poverty, focusing on how climate change will affect the world's poor. This is a logical step toward linking weather and politics, which itself is chained to religion.[109]

There are several dimensions to the topic of politics and weather, all with mythologies of their own. Take the effect of weather on voters. In a well-publicized paper in 2004, two Princeton University political scientists, Christopher Achen and Larry Bartels, argue that "voters regularly punish governments for acts of God, including droughts, floods, and shark attacks. As long as responsibility for the event itself (or more commonly, for its amelioration) can somehow be attributed to the government in a story persuasive within the folk culture, the electorate will take out its frustrations on the incumbents and vote for out-parties."[110] As this statement suggests, their argument rests on three assumptions: first, that acts of God can be distinguished from acts of humans; second, that the "stories" about these events that fix blame on incumbent elected officials must be convincing within what they call "folk culture"; and third, and most radically, that few American voters are well informed about issues or able to make rational choices in the voting booth, acting instead on "blind retrospection," blaming incumbents when unforeseen disasters occur. Of their four somewhat whimsical examples, two are weather-related. The other two are the killer shark attacks on the New Jersey shore in July 1916 that appear to have cost Woodrow Wilson votes in the November presidential election, and the influenza pandemic of 1918, which did not seem to have had a negative impact on the Democratic candidates, but which Achen and Bartles use to make a point about voters' misunderstanding of the difference between causation and responsibility: "As long as no one supplied a convincing argument that the government did control or should have controlled the spread of the pandemic or its horrific consequences, the pain of millions failed to have any force in the political process."[111]

Achen and Bartels correlate the results of twenty-seven presidential elections from 1896 to 2000 with the Palmer Hydrological Drought Index (PHDI), an index of long-term moisture supply. Employing a variety of statistical tools, the authors conclude that "wet or dry conditions in a typical state and year . . . cost the incumbent party seven-tenths of a percentage point, while 'extreme' droughts or wet spells . . . cost incumbents about 1.5 percentage points."[112] They defend their statistical methods that show that floods and droughts in general have a negative effect on electoral support for the incumbent president's party. "That negative effect is not coincidental; nor is it simply a matter of voters rationally punishing particular presidents for failing to prepare adequately for or respond adequately to particular threats. It is a pervasive risk to the reelection chances of every incumbent party, and no more controllable than the rain."[113]

In the election of 2000, there was a severe drought in parts of the South and West, and excessive wetness in the Dakotas, New York, and Vermont. Achen and Bartels estimate that 2.8 million people voted against Al Gore because their states were either too wet or too dry. They are convinced that climate cost Gore the states of Arizona, Louisiana, Nevada, Florida, New Hampshire, Tennessee, and Missouri, "almost three times as many electoral votes as Florida's infamous 'butterfly ballot.'"[114] If there is validity to this analysis, incumbents will need to hire anthropologists to understand the "folk culture" of their constituents so their PR flacks can spin "stories" about natural disasters that deflect blame from them. The Achen and Bartels view of the American voter is scornful, and they reject both the Jeffersonian belief in an enlightened voter and the contemporary notion that, even if voters are largely ignorant of the key issues, they at least recognize good and bad government performances. Their position goes far beyond the studies that show that bad weather conditions on election day keep some voters from the polls, but it is not unrelated. Whether the weather affects elections through "blind retrospection" or through mild inconvenience, or both, it introduces an unwanted element of irrationality into a sacred ritual of democracy. Chance is no more welcome in politics than in science.[115]

Weather Humor and Play

As if a century of weather talk about forecasting, chaos, religion, and politics were not enough, weather jokes, weather clowns, and weather comedy fill even more of our days and nights. Weather and its many manifestations—snow, clouds, rain, wind, heat, cold—are manifestly paradoxical, a quality it shares with humor and play. From the simple, wry observation "Climate is what you expect and weather is what you get," to the numerous satires of television weathercasters, weather jokes highlight the unpredictability of weather and the predictable outrage of people who expect weather to be predictable. My personal favorite expression of the paradoxes of weather is the visual joke of the "Melted Snowman," aka the "California Snowman." A parody of the souvenir snow globe, it is a plastic dome about 4 inches high, half filled with water, with a floating black hat and miniature lumps of coal and a carrot resting on the bottom. It is an absurdist commentary on regional climate stereotypes, nostalgia for traditional winter amusements, and a meditation on human and global mortality.[116]

Weather, humor, and play are linked in several ways, most obviously in their constant tension between order and disorder, the expected and the unexpected, concealment and revelation, the trivial and the serious. They are also ubiquitous in human life. Outdoor play and games were, for centuries, determined by seasonal weather. Technological developments in architecture, heating, and cooling began to deseasonalize sports in the twentieth century, and by the end of the century weather was regarded as either a minor annoyance or an element to be manipulated to an advantage. Managers, fans, and players of major-league baseball teams believe meteorological factors—humidity, air density, wind—affect hitting and pitching. Coors Field in Denver, elevation 5,280 feet, led all major-league ballparks in home runs during six of the seven years from 1995 to 2002, causing speculation that Denver's thinner air was responsible. Three scientists from the University of Colorado in Denver reevaluated the statistics and studied the weather patterns in the ballpark, concluding that the high number of home runs was due to team emphasis on power hitting and to the nervousness of pitchers on both teams who convinced themselves that Coors Park was difficult to pitch in because of the altitude. In fact, the meteorological data indicated that local afternoon winds blowing in from center field more than offset the Mile-High City's expected 10 percent advantage in long-ball hitting. To counter the shrinking effect of low humidity on the ball, the Colorado Rockies baseball team uses a humidor to keep balls at 90°F and 40 percent humidity, or 70°F and 50 percent humidity in 2006, according to a story in the *New York Times*. [117] Trying to micromanage the weather is just one more indication of the power of weather over leisure activities. Enjoying many of the post–World War II outdoor recreational pursuits—backcountry skiing, surfing, hot-air ballooning, rock climbing, ocean kayaking—requires fairly sophisticated knowledge of local weather conditions and the physics of the atmosphere.

Weather is a component of some indoor games. In "Smog: The Air Pollution Game," marketed by Urban Systems, a Cambridge, Massachusetts, consulting firm in the early 1970s, players try to manipulate and plan their city's environment but are dealt "Outrageous Fortune" cards that introduce weather-related problems such as a change in the wind direction that sends clouds of pollution over housing projects, causing a loss of political support. Another board game, "The Weather," was sold in museum shops in the early 1980s. The game consisted of a map of the United States on which players placed manufacturing and retail businesses that were subjected to weather and other natural hazards including volcanic eruptions. Skill and luck in selecting sites unaffected by disasters determined the winner. Computer games such as "Eastern Front (1941)" by Atari and "Second Front: Germany Turns East" by Strategic Simulations use winter weather as an obstacle that the players must take into account. "Operation: Weather Disaster," developed by Discovery Communications in a CD-ROM format, pits the malevolent "Dr. Rainwater," a former TV weatherman in Arizona who went berserk from forecasting nothing but sunny skies, against players who use "Millibar," a robot with a database of meteorological information, to thwart "Rainwater's" attempts to destroy the world with severe storms. In 2005, self-described "weathergeek" Joshua Kelly created a "fantasy weather league," in cooperation with Wunderground. Participants submit their five-day forecast for an airport location chosen by Kelly. The forecast, submitted the day before the week begins, consists of high and low temperatures for each day of the week, daily wind direction and speed, and precipitation amounts. Participants are paired, and Kelly has devised a point system for correct and near-correct forecasts, with a total of thirty-two points possible each day. Like fantasy baseball leagues, such games allow the truly weather-obsessed to form cohorts in cyberspace.[118]

A much simpler level of play consists of weather toys found in most inexpensive gift shops. Among such items are a tornado in a bottle, some with references to *The Wizard of Oz*, and variations on the "old Indian weather rock," a small stone that, if hung on a string outside, tells you it is raining if the rock is wet, that it's windy if it's swinging, that it's sunny if it's casting a shadow, and so forth. These items often have regional variants; the weather rock is a piece of wood in Nova Scotia and a piece of string attached to a drawing of a mule in Ohio. Slightly more complicated weather toys were described by Weather Bureau librarian Charles Talman in 1926. These "toys" included the familiar "weather house" (often a Swiss cottage), a simple hygroscope operated by a piece of twisted catgut that tightened and loosened with changing humidity. Typically, when the air is dry, a woman emerges from the house; when it is moist, a man comes out, often with an umbrella.[119] Weather humor in the 1920s included attempts to invent a rainmaker in the mold of logger Paul Bunyan. Wayne Carroll, a lumber dealer in Gothenburg, Nebraska, is credited with creating Febold Febolson, a Swedish immigrant of many talents, including rainmaking by building bonfires around all the lakes in the region, which caused the lakes to evaporate and form clouds that brought precipitation. Bunyan and Feboldson are both examples of what folklorists call "fakelore," an imitation of oral tradition created by individuals.[120]

Gender differences in weather watching are the subject of a humorous essay in *Parents* magazine in 1998. When the pregnant author goes to the doctor for her routine sonogram, her husband and the technician point out the fetal head, spine, and legs, but she cannot recognize a single human feature. "It's because you don't understand The Weather Channel," her husband admonishes. Returning home, she dutifully watches storms scud over the map, but all she can see is "the curve of a baby's head and belly sweeping across the country."[121] Although part of the humor in this piece rests on the dubious assumption that men are more weather-obsessed than women, it also plays with the mixing of messages in an electronically mediated world, something with which earlier weather humorists did not have to contend. The *New Yorker,* ever the bellwether of the cultural fads of America, has published a series of cartoons on weather phobias and manias. On March 19, 2001, a cartoon by Robert Mankoff shows a woman on the telephone saying, "We'd love to come, but the weather mongers have paralyzed us with fear," while her husband stares at the TV set. Two years later a cartoon by

"Cloud Chart." Roz Chast. *The New Yorker,* April 14, 2003, 59. ©The New Yorker Collection 2003 Roz Chast from cartoonbank.com. All Rights Reserved.

B. Smaller shows a woman looking out a picture window and saying to her husband, who is reading a newspaper, "It's so nice out. Let's do something weather-related." A week later Roz Chast offered a full-page cartoon, "Cloud Chart," with eight types of clouds never envisioned by Luke Howard, including "Sigmunds," "Clouds with an uncanny ability to make you feel anxious or depressed," and "Duhs," "No-name generic clouds having no meteorological significance whatsoever."[122]

Earlier weather cartoons seem to have been more politically motivated. In his "Alphabet of Joyous Trusts" (1902), F. Opper illustrated the twenty-sixth letter with a bloated capitalist charging a beleaguered citizen for breathing, and the

"Old Man Weather. That Not So Comic Character: No Time for Comedy." Henry C. Barrow. *Omaha World-Herald,* 1953. Library of Congress.

rhyme: "Z is the Zephyr Trust. Prophets declare. / We must soon pay for breath, when the Trusts own the air." In the same Progressive Era, Joseph Keppler commented on the presidential race of 1912 with a cartoon of five heads—Theodore Roosevelt, William Howard Taft, "Radicalism," "Insurgency," and "Standpatism"— blowing at a GOP elephant weather vane. Weather disasters always bring out cartoonists' sarcasm. After costly floods in Rochester, New York, in 1913, John Scott Clubb mocked a local Weather Bureau pronouncement, "If we don't get the big rains that we had last Spring there will be no danger of a flood," with a panel of five cartoons showing a man standing in a puddle reading about the city's flood preparations as tiny spiderlike creatures labeled "Germs" cling to his clothes, and another in which two men with signs announcing the sale of their homes discuss the "water proof" cellars. By the more libidinous 1920s, humor magazines such as *Life* satirized the attitudes of the rich in cartoons such as John C. Conacher's drawing of a plump matron in a lounge chair on the patio of a tropical resort asking her young companion to "Order me a dry martini, Elaine—and, in a low even voice, read me the weather reports from Chicago." In some regions weather was no joke. A cartoon by Henry C. Barrow in the *Omaha World-Herald* in 1953 showed "Old Man Weather" carrying a watering can, fan, thermometer, weather balloon, and wind gauge, while in the background a tornado levels several buildings. The caption reads: "That Not So Comic Character: No Time for Comedy."[123]

For humorists, weather is always comic. Mark Twain wrote several sketches on weather topics, one of the best being his monologue on the "sumptuous variety [of] New England weather." "In the spring," Twain writes, "I have counted one hundred and thirty-six different kinds of weather inside of four-and-twenty hours." After commenting on the variety and unpredictability of New England weather, he concludes with praise for the ice storm:

When a leafless tree is clothed with ice from the bottom to the top—ice that is as bright and clear as crystal; when every bough and twig is strung with ice beads, frozen dewdrops, and the whole tree sparkles cold and white like the Shah of Persia's diamond plume. Then the wind waves the branches and the sun comes out and turns all those myriads of beads and drops to prisms that glow and burn and flash with all manner of colored fires, which change and change again with inconceivable rapidity from blue to red, from red to green, and green to gold—the tree becomes a spraying fountain, a very explosion of dazzling jewels; and it stands there the acme, the climax, the supremest possibility in art or nature, of bewildering, intoxicating, intolerable magnificence. One cannot make the words too strong.[124]

Too strong? Try to hear these words with Twain's low-key delivery, concealing the excesses of florid rhetoric to strike just the right note of genuine appreciation for the ice storm and ironic comment on those who would make it an icon of the region. Twain's other great weather joke, the novel *The American Claimant,* offers Colonel Mulberry Sellers as the epitome of American free enterprise, who at the novel's conclusion is seeking to control the world's climates by manipulating sunspots. To complicate the joke, Twain announces, "No weather will be found in this book. This is an attempt to pull a book through without weather. It being the first attempt of the kind in fictitious literature, it may prove a failure, but it seemed worth the while of some dare-devil person to try it, and the author was in just the mood." Claiming "to borrow such weather as is necessary for the book from qualified and recognized experts," he provides an appendix with seven excerpts from popular writers of the day whose lurid depictions of weather he spoofed in his New England piece. This excerpt from fellow humorist "M. Quad" is a good example:

> Now the rain falls—now the wind is let loose with a terrible shriek—now the lightning is so constant that the eyes burn, and the thunder-claps merge into an awful roar, as did the 800 cannon at Gettysburg. Crash! Crash! Crash! It is the cottonwood trees falling to earth. Shriek! Shriek! Shriek! It is the Demon racing along the plain and uprooting even the blades of grass. Shock! Shock! Shock! It is the Fury flinging his fiery bolts into the bosom of the earth.[125]

The dozens of Internet weather humor sites, blogs, and e-mail attachments are part of Twain's legacy in the twenty-first century. Mississippi State University maintains a home page that includes two pages of weather jokes, the best being the story about the chief of an Indian tribe whose duty it was to advise his people about preparations for the winter. When they asked him if it was going to be a cold winter, he prudently replied that it was and they should start gathering wood to heat their homes. After a few weeks of wood gathering, the tribe asked the chief if they could stop. This time the chief called the local meteorologist and asked for the forecast. "According to our satellite data and computer models, it will be a colder than average winter," was the reply. So the chief encouraged his people to gather more wood. After several more weeks of gathering they were tired and asked the chief if he was sure it would be a hard winter. Again the chief checked with the Weather Bureau. "Are you sure your data are correct?" he pleaded. "We are really sure, it's going to be the coldest winter in recorded history," came the reply, "because the Indians are gathering wood like crazy."[126]

Although some jokes, like this one, ridicule the Weather Bureau while acknowledging our dependence on it, most are of the "how cold/hot was it?" variety. Cities, states, and regions lay claim to the hottest, coldest, windiest, wettest weather conditions as a matter of local pride. Most of these are of the tall-tale type of humor descended from Crockett Almanacs and Mark Twain. An enterprising writer created a book of Arizona weather humor, exaggerating the effects of the state's heat, dryness, and wind. Weather disasters stimulate comedic responses such as "30 Things Hurricanes Teach Us," an e-mail I received from a former student living in the Virgin Islands in October 2004, with such observations as "1. An oak tree on the ground looks four times bigger than it did standing. . . . 10. You can use your washing machine as a cooler. . . . 14. Downed power lines make excellent security systems. . . . 18. The life blood of any disaster recovery is COFFEE. . . . 24. No matter how hard the wind blows, roadside campaign signs will survive." During a period of bitter cold and snow in the East, a sympathy e-mail arrived from Southern California with a "photo attached illustrating the excessive damage caused to a home from a West Coast storm that passed through the Los Angeles area a couple of days ago. It really makes you cherish what you have, and reminds us not to take life for granted!!!" The photo shows one plastic deck chair upended.[127]

Weather disaster humor is often a kind of emotional relief. The *Onion,* the popular contemporary humor magazine, published "Tornado Safety" measures as recommended by the "National Weather Service." These included "Stay calm. This will be difficult, since you are almost certainly going to die," and "Live a little, for once: Strap yourself to the roof of your house and rage at the heavens." Another Internet site describes the "Moojita Scale," a parody of the Fujita Scale for tornado severity (see chapter 5) that also pokes fun at the cliché of flying cows in the movie *Twister:*

> Mo Tornado—Cows in an open field are spun around parallel to the wind and become mildly annoyed.
> M1 Tornado—Cows are tipped over and can't get up.
> M2 Tornado—Cows begin rolling with the wind.
> M3 Tornado—Cows tumble and bounce.
> M4 Tornado—Cows are airborne.
> M5 Tornado—S T E A K!!![128]

Weather disaster humor can also be an attack on the purveyors of weather disaster in the media. Another humor magazine, the *National Lampoon,* parodied both television weather reports and *Time* magazine in the piece "Shitstorm, U.S.A." (1984) about a sophisticated forecast computer, "GROUNDHOG [*Greatest, Rightest, Original, Ultimate, Never Defective Handful Of Guesses*]," which goes haywire causing downpours of cement, vaginal deodorant, carbon tetrachloride, as well as excrement; pennies fall on Wall Street, killing hundreds of bankers. The classic comic routine on television weather remains Don Knotts's performance on *The Steve Allen Show* in the 1950s. Playing a weatherman forced to make up a forecast without any information, Knotts stumbles along using bits of meteorological jargon while marking a weather map, his "highs" and "lows" eventually spelling, "H-E-L-P."[129]

Professional meteorologists, like all scientists, develop their own brand of weather humor. Keith C. Heidorn, whose Web site introduces him as "the Weather Doctor," collects humorous quotations from a variety of sources, from Mark Twain to George Carlin. He prefers pithy comments on the power of weather to humble human pride. An example comes from the early twentieth-century Indiana humorist Kin Hubbard: "Don't knock the weather, nine-tenths of the people couldn't start a conversation if it didn't change once in a while." Other meteorologists kid themselves: "You might be a weather nut if . . . Your favorite holiday is

Groundhog Day; [or if] . . . you hated the movie 'Twister' because of its unrealistic portrayal of storm chasing, but you've seen it at least six times." Similarly, another meteorological comedian suggests, you might be suffering from "Supercell Deprivation Syndrome" (SDS) if you exhibit any of thirty symptoms, including "Innate fascination with the 'little tornado' in the bathtub drain" or "Purposely move into a mobile home."

Weather Talk: Wind in Perpetual Motion

If weather talk could be measured on a kind of Beaufort Scale of verbosity, the preceding pages could be rated from breezes to hurricanes. The popularity of weather as a hobby, a business, and a political issue has grown exponentially in the past few years. A century ago, those curious about the weather might have picked up William Ferrel's *Popular Treatise on the Winds,* first published in 1889, which survived for a second edition in 1911. At more than 500 pages of technical information on the atmosphere, cyclones, tornadoes, and thunderstorms, it was not for the casual reader. Perhaps the first attempt at a truly popular book on weather was *The Story of the Earth's Atmosphere,* by the British meteorologist Douglas Archibald, whose brief, well-illustrated book was published in the United States in 1897 as part of the Library of Useful Stories series. Mark Harrington, first chief of the new civilian Weather Bureau, attempted an American equivalent in *About the Weather* in 1899. Willis Moore, Harrington's successor, followed in 1900 with his *Moore's Meteorological Almanac,* mentioned earlier in this chapter. Popular weather books appeared with increasing frequency. Among the best were Edwin C. Martin's nationalistic *Our Own Weather* in 1913, and Alexander McAdie's fact-filled *Man and Weather* in 1926. Moore tried to maintain his authority after his dismissal from the bureau by writing *The New Air World: The Science of Meteorology Simplified* in 1922, but it was little more than a rehash of his *Almanac.* Charles F. Brooks of Harvard and the Blue Hill Observatory followed with his lively *Why the Weather?* in 1924.[130]

"Don't Apologize for Talking about the Weather!" was the title of a five-part series on weather in the muckraking *American Magazine.* Running from February through June 1926, the articles emphasize the extremes and variety of weather in the United States and Canada and their salubrious influence on American character. After describing the rigors of winter and the power of rain, the author, Stuart Mackenzie, grows even more ecstatic over "Wind! A Star Performer in the Drama of American Life" and concludes proudly that "Our Tornadoes Are the Fiercest of All Storms." Beginning in 1931, when Weather Bureau librarian Charles Fitzhugh Talman collected his miscellaneous newspaper and magazine pieces into *The Realm of the Air: A Book about Weather,* weather books were coming a little more frequently. Talman even makes a jocular "author's apology" for contributing another book on weather. Bureau employee W. J. Humphreys continued crowding the "QC" section of library shelves with *Weather Rambles* in 1937 and *Ways of the Weather* in 1942.[131]

World War II revolutionized meteorology and expanded the number of weather enthusiasts. The acceptance of the Scandinavian forecasting system by navy and then Weather Bureau meteorologists, the development of radar, the training of thousands of young men and women as weather observers for the military services, and above all the advances in electronic computing encouraged a boom in weather-book publishing that continues to the present. The titles of the spate of books in the 1940s and 1950s reflect the widening scope of meteorology, from storm warnings to environmental science. The respected Penn State professors Hans Neuberger and F. Briscoe Stephens emphasized the importance of weather on society in their book *Weather and Man* (1948), while the colorful Caltech meteorologist Irving Krick cast an equally wide net in *Sun, Sea, and Sky: Weather in Our World and in Our Lives* (1954). The founding of the popular journal *Weatherwise* in 1948, and the entertaining books by Eric Sloane and David Ludlum raised the level of weather talk. Many meteorologists working today recall being turned on to weather by Paul Lehr, Will Burnett, and Herbert Zinn's *Weather: Air Masses, Clouds, Rainfall, Weather Maps, Climate*. It was part of the Golden Nature Guides series for children first published in 1957 and frequently revised. *Weather,* by Philip Thompson, Robert O'Brien, and the editors of Time-Life Books, published in several editions from 1965 through 1980, is another frequently mentioned inspiration.[132]

Today's equivalents of the books by Moore, Martin, Brooks, Talman, and Humphreys are John D. Cox, *Weather for Dummies;* Mel Goldstein, *The Complete Idiot's Guide to the Weather;* Robert Henson, *The Rough Guide to Weather;* and Jack Williams, *USA Today Weather Almanac.* Meteorologists or science writers, these authors, like their predecessors, are well educated and able to communicate easily with the public, but as some of the titles suggest, the tone has changed. Competing with television and the Internet, popular weather books at the beginning of the twenty-first century are packaged to catch the eye of a new generation of Americans for whom The Weather Channel (TWC) is old news, a generation for whom extreme sports often include extreme weather.[133] These and similar books also demonstrate the extent to which the weather is being recognized as fundamental to ecology. There is money to be made in utilizing, not ignoring, weather and climate. Many Americans in regions famous for good or bad weather have known this for years, but it is becoming part of a new environmental awareness.

Ford Carpenter's *Climate and Weather of San Diego, California,* published by the San Diego Chamber of Commerce in 1913, may be the first of the books on local weather, and the genre flourishes today, with more than thirty books (by my cursory count) on the weather of individual cities, states, and regions—most of them published in the past twenty years. Some, like Ben Gelber's *Pennsylvania Weather Book* and Jon Nese and Glenn Schwartz's *Philadelphia Area Weather Book,* are outstanding local histories covering not only unusual weather events and seasonal regularities but the institutional histories of local weather forecasting and media dissemination of weather information. Nese and Schwartz include some of the clearest and best-illustrated explanations of how weather works available anywhere, and their bibliography and Web sources are extensive. While some of the local weather books are tainted with boosterism, they add to the conversation on weather in interesting ways. San Diego's climate has long been regarded as ideal, and Carpenter, who was the local forecaster, begins his book with a history of weather observations in the area from its discovery by Juan Rodriguez Cabrillo in 1542 to the present. Attention to the city's Spanish heritage provides the transition into his explanation of its perfect weather, the clouds that usually block the morning sun from March through October, keeping the city cool in the summer. These clouds are called *velo* in Spanish, and Carpenter explains at length that the word, like the cloud, has no equivalent in English. It is not "high fog" but a unique creation of Southern California topography and, by extension, an element of its distinctive culture.[134]

By one estimate there are 30,000 to 35,000 men and women working in the atmospheric sciences. The National Weather Service has almost 5,000 employees. It relies on more than 11,000 cooperative observers. Since 1997, another 8,000 volunteers have participated in the Community Collaborative Rain, Hail, and Snow network (CoCoRaHS), created by Colorado state climatologist Nolan Doesken to measure the extent of precipitation neighborhood by neighborhood, first in Fort Collins, now in nineteen states and the District of Columbia. The American Meteorological Society has more than 12,000 members, roughly divided in thirds among government, private, and academic meteorologists and other scientists. *Weatherwise* magazine has more than 13,000 subscribers. Web sites such as the one maintained by Alan Steremberg, president of The Weather Underground Inc., receive thousands of comments daily on its wunderground blog. WeatherMatrix, created in 2000 by a young meteorologist now employed by AccuWeather in State College, Pennsylvania, has more than 13,500 members in its online commonwealth of weather enthusiasts. Weather bloggers claim 5,000 to 10,000 hits on their Internet sites daily.[135] These figures only scratch the surface of the weather community; it is safe to say that there are millions of Americans who are, in the terms used by The Weather Channel, "weather engaged." In April 2007, according to Nielson/NetRatings, weather.com, TWC's Web site, had 39 million unique visitors. In late 2007, web designer Ben Tesch launched "Cumul.us," an Internet site that invites everyone to predict the weather. Their predictions are compared to those of the National Weather Service and other professional forecasters and averaged. Each individual's forecast is rated by how well it matched the aggregate. Based loosely on the "wisdon of crowds" theory that group decisions are more accurate than those of experts, "Cumul.us" is the ultimate democratization of weather. Everyone gets to vote. This is a weather community that no one dreamed of a century ago.[136]

There were, however, speculations of a different nature. In the summer of 1938, Willis Gregg, chief of the Weather Bureau, sent letters to his fellow meteorologists asking them to speculate on what the profession might look like in fifty years. In a brief report compiled by W. C. Devereaux and published the following year in the *Bulletin of the American Meteorological Society,* thirty of the responses were excerpted. Most of the predictions focused on scientific advances in forecasting and control of weather. Charles Brooks and William Humphreys, always attentive to popularization, foresaw a more rapid dissemination of meteorological studies to libraries and "robot reporters—instruments that not only keep a continuous record of the weather elements, but which, at the touch of a button, or automatically at regular intervals, also tell all about the weather there at the time."[137] Only two of the contributors allowed themselves wild fantasies on the social implications of weather. The irrepressible Major E. H. Bowie of the San Francisco office facetiously suggested that the only way to end the dust bowl was a Works Progress Administration project to lower the height of the Sierra Nevada and Rocky Mountains, while T. A. Blair of Lincoln, Nebraska, perhaps inspired by personal experience with the dust bowl, issued this ominous forecast for a full century hence:

> In the year 2038 an American meteorologist discovers how to control the weather. . . . But difficulties arise. This control involves a shifting of the air masses and means that while the one area is getting the kind of weather it wants, another region is subject to unfavorable weather. . . . Political parties develop on the basis of these differences, and "pressure groups" attempt to control the WDA (Weather Distributing Administration). Soon other nations attempt to manipulate the weather. International complications begin and the human masses of the world are plunged into a war for the control of the air masses.[138]

Unthinkable? The following chapters reveal how close we are to 2038 in the continuing story of human efforts to shape and misshape the weather.

2

Managing Weather

It's an Ill Wind That Blows No Good

My neighbors in Pendleton County, West Virginia, like the way the wind blows on Spruce Knob, Goshen Ridge, and Jack Mountain, scudding clouds over the upper forks of the Potomac River and scuffing the red oaks, spruce, and sycamores along the ridgelines. They don't like mountaintop removal, and they especially don't like to be treated like hillbillies—the way they feel they were treated by US Wind Force, "a developer of renewable energy projects." US Wind Force wanted to build fifty wind turbines, each more than 400 feet tall, on Jack Mountain and string 17 miles of power lines across the county. The story of the thirty-month fight to prevent the selling of the wind (and the land needed to tie the wind down) touches on the central issues of this chapter—marketing and managing weather, and the expansion of the institutions that mediate our perceptions of weather.

Wind, the Weather Service tells us, is air in motion relative to the surface of the earth. Wind is caused by the planet's rotation and uneven heating of the earth by the sun. Wind is one of the most basic elements of weather. In West Virginia the prevailing winds come from the west and northwest, blowing over the narrow ridges of the Allegheny Front, but swirling to all points on the compass in the gaps and hollows that cleave the mountains. The 7,700 citizens of Pendleton County live in a thousand microclimates, their weather as personal as their religion and almost as important. The young man who supplies me with wood for my outdoor boiler complains that he has to use twice the amount I do because he lives another few hundred feet up the mountain. A farm on one side of a mountain gets more rain than one on the other side. My neighbors lean into their personal winds, the winds that have shaped the land, the forests, and the houses that dot the valleys and hills. Every year some of these houses are blown down by the wind.

A dozen or so names dominate the phone book and the memorial to the fallen in the nation's wars—Bowman, Evick, Harman, Judy, Kimble, Lambert, Moats, Pitsenberger, Probst, Puffenbarger, Raines, Rexrode, Smith, Sponaugle, and Waggy—the rest of us are "from aways," even those who have lived here for more than thirty years. Those who never left and those who wandered in agree on the importance of place. Where US Wind Force and its subsidiary Liberty Gap Wind Force saw numbers on a map indicating wind power forces of 4 and 5, residents saw watersheds that keep their wells full, ground cover for the game they hunt, and almost forgotten cemeteries where their ancestors are buried. US Wind Force looked at a color-coded map developed by the Department of Energy (DOE) and saw Jack Mountain in pink and purple, the colors indicating "good" and "excellent" wind power; the Pendleton County commissioners saw green in the money the company promised the county and, on November 16, 2004, signed an agreement without an opportunity for public comment.

Outrage over the proposal and the procedure began a week later when residents saw the application for the project published in the *Pendleton Times* as required by law. US Wind Force asked for an expedited review by the state Public Service Commission (PSC), leaving opponents just thirty days to request that the PSC delay approval. A citizens group, Friends of Beautiful Pendleton County, was organized. Community meetings were held in December, and citizens deluged the PSC and elected officials with letters opposing the wind turbines because of their environmental, aesthetic, and cultural impacts. Place and its meanings became more important as the process of gaining approval from the PSC moved forward. The battle over the construction of wind turbines and transmission lines on Jack Mountain compelled a community to think about the relation of land and sky, to learn the language of landscape planners and managers, to talk about "siting rules" and "wind power classes."

As wind farms spread across the nation, eddies of opposition have increased. Wind turbines, with enormous propeller blades, now dwarf the world's tallest tree (378 feet) and the Statue of Liberty (305 feet). When located in mountains, the tops of the ridges are stripped and leveled to provide a base for the turbines. Noise from the wind-turned blades and from the generators is audible for as far as 2 miles, and aircraft warning lights are visible even farther away. Birds and bats fall victim to the blades in numbers that alarm biologists and the public. Construction of roads and power transmission lines to and from the wind generators impacts water tables, causes soil erosion, and adds to bird flyway hazards. While advocates of wind power cite improvement in the "airshed," the supply of air in a given region, opponents decry destruction of the "viewshed," an area of land and sky visible from a fixed point. The wind stirs the mind's eye.[1]

Citizens of Pendleton County expressed alarm over the effect of the construction and operation of wind turbines on birds and bats, on watersheds and scenic vistas, on property values and cultural heritage. The announcement in March 2007 by two Canadian researchers that wind farms may have unintended local climate impacts, such as changes in heat and moisture fluxes, added to the concerns of the West Virginians.[2] The argument that wind energy may help reduce carbon and other emissions nationally was insufficient for people whose concerns were local. "It was interesting," one protester wrote, "that a Wind Force official was quoted that the electricity produced in Pendleton was leaving the State to be sold. This sounds familiar, like the past history of West Virginia—extractive minerals, other natural resources, and profits leaving the State and the environmental residues (stripped wastelands, slag piles and tailings, clear cuts) staying." Another distraught resident scrawled a note in pencil: "People who want noise and lights should go to a carnival." Opposition to the turbines came not from Luddites but from people like Harmon Moates of Moatestown, a mixed-race community that has existed in a gap on Jack Mountain since before the Civil War, who summed up the paradoxes of technology and wilderness by pointing to the forested mountain and telling a reporter, "I get on my four-wheeler . . . and I just get lost in there."[3]

Maps are supposed to keep us from getting lost. Like the daily weather map, maps for wind energy are an attempt to visualize a complex and dynamic reality. The *Wind Energy Resource Atlas* was compiled more than twenty years ago by the Department of Energy from wind speed measurements taken by the Weather Bureau, the Forest Service, private power companies, and the Pacific Northwest Laboratory of DOE. Under its "Siting Rules," the PSC requires another kind of map, a 5-mile radius map on a scale of 1 inch to 4,800 feet or greater (U.S. Geological Survey topographic maps are usually 1:2000 feet) showing (1) major population centers; (2) transportation routes; (3) bodies of water; (4) topographic contours; (5) major institutions; (6) incorporated communities; public and private recreational areas; historic areas; religious places; archaeological sites; places of cultural significance that are recognized by, registered with, or identified as eligible for registration by the National Register of Historic Places or any state agency; and (7) land use classifications. Liberty Gap Wind Force's failure to submit a map with all this information and its omission of viewshed renderings from several historic overlooks and trails, including the Confederate Prayer Service Site, were cited by the PSC as one of the principal reasons its application was ultimately rejected.[4]

The PSC decision came just days after the Committee on Environmental Impacts of Wind Energy Projects of the National Research Council (NRC) published its report, which gives further support to those who think that natural beauty and human heritage may be more endangered than bats and birds by wind turbines. The authors of the report—engineers, ecologists, landscape architects, sociologists, and wildlife managers—unanimously conclude that decisions on the siting of future wind turbines and transmission lines must consider the opinions of people who live, work, and recreate in the area because "landscape consists of layers of meaning that may not be understood by an outside professional conducting a visual assessment." The committee recommends that a computer-generated map with a 10-mile radius be created to help evaluate the visual impacts of wind farms. Significantly, the committee recognizes the importance of weather conditions in evaluating visual impacts. "Generally, projects are evaluated using 'worst-case conditions,' e.g., leaf-off visibility and clear skies. An abundance of clear skies makes aesthetic impacts in that area no worse or better than visual impacts in a region that has more cloudy skies. Indeed, a scenic view that is only rarely visible may be even more highly valued than one that usually can be seen."[5]

The eastern side of the Allegheny Front is one of the foggiest in the nation; fog shrouds the gaps and valleys of Pendleton County more than sixty days a year, a little less than on the coasts of Washington and Maine. From the summit of Shenandoah Mountain, before the morning sun warms the air, the landscape is one of dark islands rising from a white sea. No towering propellers break the surface of the clouds, sucking the breezes. The people of Pendleton County manage their weather for different reasons and map their winds in krummholz, the flights of birds, and their own imaginations.[6]

Weather Wealth and Illth

Who owns the sky? This question is asked in the title of a book by the journalist and entrepreneur Peter Barnes. His concern is not directly with the weather but with all the things the sky, the atmosphere, provides for life on earth, some of which we call weather. The sky shields us from asteroids and meteors; it filters harmful ultraviolet rays; it maintains the earth's temperature; it replenishes the earth's water; it provides oxygen to our lungs; it absorbs our exhausts and moves them somewhere else; it powers boats and windmills; it lifts airplanes; and it carries sound. "There is nothing more fundamental to us than the sky," Barnes writes, "our sky, our *unique* sky. We're sky animals. We live on the land, but *in* the sky. We inhale from and exhale into it about fifteen thousand times a day."[7] Prompted by the threat of global warming and the increase in carbon dioxide and other pollutants in the air, Barnes proposes what he calls a "Sky Trust," a private corporation with elected trustees that would manage the atmosphere for the common good, charging polluters the market price of their environmental impact and paying citizens dividends.

The plan is utopian; Barnes makes no apology for that. The alternative is unbreathable air and a steadily diminishing quality of life for our posterity. Inspired by Garrett Hardin's seminal essay "The Tragedy of the Commons" (1968), Barnes argues that the sky is a commons that belongs to all the people on earth, but he is realistic enough to see that the problem can only be solved one nation at a time. And, since the United States is one of the biggest polluters in the world, an American Sky Trust is a good place to begin. The mission of the Sky Trust would be to preserve the mix of gases in the sky by issuing permits for burning carbon-based fuels up to a limit established by Congress and charging market prices to individuals and corporations. As those costs rise, there would be greater incentive for conservation.

The idea that ecosystems have value beyond the agricultural and mineral products we extract from them is relatively recent among economists. Barnes cites Robert Costanza's 1997 estimate that the value of seventeen ecosystem services, such as purifying the air, detoxifying wastewater, pollinating crops, and regulating climate, is $33 trillion a year. The sky, Barnes reminds us, is a sink into which we dump airborne waste. And just like a landfill or an ocean, the sky is a sink with a finite capacity. We are running out of sinks faster than we are running out of the Earth's resources from which we produce toxic wastes. Using the cap-and-trade model that has become part of our existing clean air legislation, Barnes sees the Sky Trust as a giant mutual fund. Every American would be a member with one nontransferable share. If you burn less than your assigned share of carbon, you receive a yearly dividend; if more, you pay for the excess.

Barnes discusses the ways in which the Sky Trust would work in more detail than I can go into here, but his attempt to quantify the value of the atmosphere is directly relevant to my belief that the selling of weather information and products needs to be seen in the larger context of market-driven environmental policy. Reviving the concept of "illth" as proposed by the British philosopher John Ruskin, Barnes adds another $4.372 trillion to the cost of our current gross domestic product (GDP). Illth is simply the devastation and waste that are the by-products of wealth. The costs of ozone depletion, air pollution, commuting, and job-related stress are all illth. Adding the costs of weather disasters and the costs of natural-hazard mitigation makes the figure considerable higher. If we subtract illth from the GDP, Barnes demonstrates, our national wealth has actually fallen since the 1970s.[8] Weather produces wealth and illth. Specific examples of this can easily be seen in attempts to market the weather's products—climate, sun, wind, and snow.

Climate, defined as the weather typical for a place or region over a number of years for which there are records, is often modified by adjectives such as "salubrious," "ideal," and "delightful," or "harsh" and "inhospitable." Until recently, climates were assumed in popular belief to be relatively unchanging. Given past weather data, settlers, vacationers, and health seekers felt assured they would find the weather they sought. Real estate developers, boosters, and resort owners sold local climates that were often more imaginary than real, however. Thanks to the prodigious research of historian Karen Ordahl Kupperman, we know that colonial Americans were puzzled by the extremes of temperature and weather they experienced because they believed that latitude was the chief determinant of climate; New England, due west of Spain, should have been milder. Hot climates were both feared and desired, since they were associated with both disease and wealth. By the end of the nineteenth century, physicians had created an extensive literature on medical geography that helped, in the words of historian Conevery Bolton Valenčius, American settlers to understand themselves and their land.[9]

By 1889, medical geography had specialized into medical climatology. Dozens of books on climate and health were available to wealthy sufferers of various diseases. Dr. Bushrod Washington James, a prolific medical writer, published a summary of the field in *American Resorts; with Notes upon Their Climate,* with chapters on seaside resorts, freshwater resorts, mountain resorts, ocean voyages, mineral springs, and summer and winter resorts from Alaska to South America, a grand tour that a more radical historian might call invalid imperialism. Not that there was general agreement on what constituted climatic utopia. In 1904, Willis Moore asserted that the nearest to an ideal climate, possessing dryness and an equable temperature, was "the great southwest." Geographer Ellsworth Huntington preferred the Pacific Northwest, since he felt that diurnal means should not exceed 75°F and lows in the winter should not fall below 30°F. Huntington's desirable annual mean temperature was 51°F; Moore's was 59°F.[10]

As the United States became an urban nation, and as improved heating systems and air-conditioning allowed the control of indoor weather, Americans grew accustomed to paying for their weather and climate. The American Meteorological Society (AMS) *Bulletin* kept its readers abreast of the latest theories of the effect of weather and climate on health. A brief review of A. C. Rollier's *Heliotherapy,* in 1925, promoted the value of sunshine at high altitudes because mountain air is rich in ultraviolet rays. Two years earlier, however, Ellsworth Huntington and Charles F. Brooks had advised a reader "that light as intense as that of Colorado is harmful to blonds." Up to a certain point sunlight is beneficial, they write, then point out,

> if a choice were to be made between living in Denver and Boston for example, the adverse features of the climate of Boston, e.g., the extremely sudden weather changes, frequent slushy periods in winter, and hot, muggy spells, would have to be balanced against the harmful effect of intense sunlight and other unfavorable features in the climate of Denver. The person concerned would have to consider these, especially in relation to his type of work.[11]

Increasing attention was paid to microclimates. At an AMS meeting in 1921, Brooks reported on work by city planners in Worcester, Massachusetts, who wanted wind, temperature, and sunshine data to assist in the location of industrial zones, parks, and residences. Brooks advised that parks occupy hilltops, "where people can get the purest air and the greatest stimulation from the winds and sun." The best residential areas adjoin the parks on the southern and southeastern slopes. Brooks's Harvard mentor, climatologist Robert DeCourcy Ward, jokingly commented that "some people who own real estate in Worcester will give Dr. Brooks large retainers and others will lay for him with a large club."[12] Selling climate could be as profitable and as dangerous as selling weather. Anticipating a postwar boom in medical climatology, the 1946 *Bulletin* published a translation of an Austrian report on health-resort climatology that outlined a number of requirements and concluded with a quotation from a French climatologist: "Nice vend du soleil, l'hiver, comme Manchester vend de la cotonnade."[13]

World War II had an unanticipated effect on sunshine sales in Tucson, Arizona, according to an article by Cleveland Amory in the *Saturday Evening Post* in 1944. Although state and private organizations like the Sunshine Climate Club had earlier spent hundreds of thousands of dollars advertising in magazines ranging from *Vogue* to *National Geographic* to the *Journal of the American Medical Association* to attract health seekers and sun worshipers to Tucson, they now were trying to discourage further population growth because there was not enough housing for war workers at aircraft and other war-industry plants. Amory's lighthearted article points out the irony of having too much of a good thing, but it obscures the darker side of the Sunbelt boom that resulted in overbuilding, water shortages, and, ironically, air pollution. Seeking, as every journalist does, a scientific authority to confirm the value of his topic, Amory quotes Dr. Bernard Wyatt, an arthritis specialist, "who once proved by radiometer that Tucson had 11 per cent more ultraviolet in its sunshine than Pasadena, California." Every percentage point meant big bucks in the sunshine rating game.[14]

By the 1970s, the value of sunshine as a source of energy rivaled its value for human health. Solar energy in the United States, as Donald Beattie, former solar energy program administrator for the National Science Foundation (NSF) and acting assistant secretary for conservation and solar applications in the Department of Energy, 1977–1978, explains, dates back to the 1930s, when thousands of solar water heaters in Florida and Southern California were in operation. The dream of solar power extends even further back into the nineteenth century, and, as early as 1903, an ostrich farm in Pasadena, California, was operating a 10 horsepower steam engine to pump water for irrigation using a 33-foot mirrored dish to focus the sun's rays. A series of university-sponsored symposia led to the establishment of the Association for Applied Solar Energy in 1955. In 1971, this became the International Solar Energy Society. The rapid growth of the nuclear power industry partially eclipsed the sun's potential, but in 1968 Congress authorized the National Science Foundation to begin a program on national problems, including energy. The NSF set up a small program that by 1971 was called Research Applied to National Needs (RANN) to encourage the private sector to develop new energy sources. In the same year, President Richard Nixon, riding the crest of the environmentalist wave, called for a program to ensure "clean" energy. NSF and the National Aeronautics and Space Administration (NASA) produced a report recommending a $3.5 billion program that would result in a solar energy infrastructure that would provide 20 percent of America's electric power and 35 percent of the total building heating and cooling load. A power struggle between NASA and NSF, similar to the rivalry between the National Oceanic and Atmospheric Administration (NOAA) and NASA in the atmospheric sciences, delayed the development of solar energy until mid-1973. On June 29, 1973, the president ordered Dixy Lee Ray, chairman of the Atomic Energy Commission (AEC), to prepare a report on the nation's energy future. In October of that year the OPEC oil embargo began, and Nixon proclaimed "Project Independence," with the goal of making the United States independent of foreign energy sources by 1980. A brighter day for solar power was about to begin.[15]

What did this mean for the atmospheric sciences? To answer that question, the National Center for Atmospheric Research (NCAR) held a workshop in October 1974, chaired by meteorologist Verner Suomi of the Space Science and Engineering Center at the University of Wisconsin–Madison. Participants came from NCAR and from private industry, including James R. Mahoney, a meteorologist who had founded the consulting firm Environmental Research and Technology and later became assistant secretary of commerce for oceans and atmosphere in the George W. Bush administration. Reporting to the AMS, Suomi noted the recommendation for basic research in the Ray report and stated that the mission of the NCAR panel was "to identify the areas where improved weather science and technology offer potentials for a particularly large benefit to the national energy picture." "Weather influences all energy delivery systems," his report continues, because of the sensitivity of components of those systems to disruption by severe weather, and because of the variation in demand caused by fluctuations in the weather. As a dump for chemical waste products, the atmosphere was linked to current energy production and consumption. As Suomi candidly put it, "Quite literally, atmospheric variabilities are a valuable waste disposal *resource*."[16]

The NCAR report also mentioned meteorology in agriculture, but unlike a century earlier, this did not mean a few regional frost warnings or precipitation forecasts but biolayer microclimate analysis, from the soil to a few feet above, where "direct measurement of CO_2 eddy flux down onto the crop would slow the photosynthesis rate and the plant growth rate. Such a measurement is feasible and, employed as part of a larger program probing the microclimate and heat and water fluxes of the growing crop, would be a worthwhile study."[17] Within eight years, the U.S. Bureau of Reclamation, in partnership with the Bonneville Power Administration, began to install the first of a network of AgriMet (*Agri*cultural *Met*eorology) weather stations on farms in Idaho, Oregon, Washington, and Montana. Each AgriMet station is equipped with sensors for air temperature, precipitation, solar radiation, wind speed and direction, and relative humidity. Some stations also measure soil temperature, solar radiation, crop canopy temperature, and evaporation. Data are transmitted via Geostationary Operational Environmental Satellite (GOES) every fifteen minutes and then forwarded to farmers and power companies via the Internet. As Suomi points out, "By knowing the water-holding capacity of the soil and the root depth of the crops, irrigators can use ET (evapotranspiration) information to apply the right amount of water to their crops at the right time. Benefits of scientific irrigation scheduling include water conservation, reduced energy and fertilizer costs, reduced soil erosion, and protection of water quality by reducing runoff and groundwater infiltration of pesticide- and fertilizer-laden water." AgriMet is just one of the many microenvironmental sensor systems now collecting data from treetops to ocean beds. We are discovering that weather can literally be a tempest in something as small as a teapot.[18]

Suomi's panel also considered weather modification and energy. By 1974, public opposition and scientific skepticism had cooled earlier enthusiasm about rainmaking and snowmaking and fog and hail suppression, but the NCAR report held out hope for continued research and specifically mentioned cloud reduction to enhance solar power. Inadvertent climate and weather modification is mentioned, but not global warming. In conclusion, the report expressed the kind of hubris that had for nearly a century made weathermen the object of ridicule: "There is under way a substantial effort to improve the understanding and ability to predict and control the atmosphere." Even when it acknowledged that "the production of energy requires dumping of combustion products and waste heat into the atmosphere and hydrosphere," and that "man is faced with an environmental energy dilemma," the panel was sanguine about the future and saw no need to recommend conservation.[19]

A solar power bonanza was beginning, but, as noted, meteorologists were preoccupied with their own problems in the 1970s and 1980s. Energy policy and climate and weather policy were going their separate ways. Beattie calls the years 1974–1977 the buildup years, with abolition of the AEC and its replacement by the Energy Research and Development Administration (ERDA), the Nuclear Regulatory Commission, and the Energy Resources Council. ERDA replaced the NSF as the chief funding agency for solar research. NSF and ERDA public relations offices oversold the progress of solar power, but progress was being made, and the Carter administration had ambitious plans for both nuclear and solar power. Creating a Department of Energy in 1977 was the first step. President Carter and his assistant secretary for conservation and solar applications, Omi Walden, pushed for commercialization of solar energy. On June 30, 1979, a solar hot-water heater went into service at the White House, and the president announced a goal of 20 percent solar power for the nation by 2000. Less than a month later, President Carter fired several cabinet members, including DOE secretary James Schlesinger, who had opposed the installation of solar energy equipment in federal buildings because they did not think the technology was reliable. Walden was removed from her position and given a new position as adviser for conservation and solar marketing. By the end of 1980, a lame-duck Carter administration left a legacy of good intentions and some significant legislation on solar energy, but little real progress. Subsequent administrations had little commitment to alternative energy in any form. President Clinton's proposed Btu tax was gutted by Congress, and R & D funds for solar power declined.[20]

The destabilization of the Middle East and the consequent increases in the cost of gasoline and other fuels have undoubtedly raised public awareness of the nation's dependence on foreign oil, but the sun has dimmed, literally, as a source for power. Studies reveal that global pollution has resulted in "global dimming," and less sunlight is reaching Earth than in the recent past. Solar power advocates continue to develop projects from single buildings to grid-connected generators, and there is no doubt that the percentage of American energy generated by solar technologies will grow, perhaps even faster with the benign neglect by Congress. The Energy Policy Act of 2005 (PL 109-58), signed into law August 8, 2005, has little to say about solar power, concentrating instead on biofuels and hydrogen fuels (the "Freedom Car"), cleaner coal, and new nuclear power plants. A measure of congressional desperation is the provision in the act for the extension of daylight savings time by three weeks in the spring and a week in the fall, purportedly to reduce the consumption of electricity.[21]

Wind, as the Pendleton County episode illustrates, makes more noise than the sun, and proponents of wind-generated energy seem louder in their sales pitch. Like solar power, wind power has been utilized for a long time, emerging during the 1973 gasoline shortage as the next big thing. The fortunes of solar and wind power have risen and fallen in more or less the same cycle over the past thirty-five years, but their environmental and cultural impacts have been significantly different. Twice as much public money has been spent on solar energy development as on wind energy, but neither sun nor wind produces more than 1 percent of the electricity generated in the United States. Wind power is more familiar, since windmills for pumping water were common in nineteenth-century America. Electricity from wind generators was demonstrated by Palmer Putnam, an engineer, in Vermont in 1941, and by the mid-1970s federally sponsored demonstration wind turbines dotted the Midwest, the Southwest, and the East Coast from Rhode Island to North Carolina. Technical problems plagued the big turbines, however, and a split developed between those who saw wind as "soft" energy primarily suited to individual homes and those who dreamed of providing 20 percent of America's energy from large-scale wind farms. The American Wind Energy Association (AWEA), founded in 1974, vigorously promotes the latter view and provides the public face of the wind industry.[22]

The National Weather Service has sunshine and wind direction and velocity records from the late nineteenth century forward, which solar and wind energy promoters have used to locate power plants, but many of these observations were made at airports or other sites that were neither especially sunny nor windy. Thus, a need arose for meteorologists who specialized in the measurement of sun and wind. E. Wendell Hewson of Oregon State University was one of the first to focus his research on wind power. Working with four Oregon public utility districts, he began, in 1971, to test various configurations of airflow around terrain features in a wind tunnel and studied various models of wind turbines from Germany, Russia, France, Great Britain, Denmark, and the United States. He also set up meteorological stations at more than a dozen sites along the Oregon coast from the Columbia River to Winchester Bay. Hewson explored every aspect of wind power from site location and terrain modification (deepening the cut in a saddleback to augment the Venturi effect of wind channeling) to aerogenerator design for coastal breezes and mountain gusts. He recognized the problem of visual pollution, but not the impact of the turbine blades on endangered birds and bats, negative impacts that would cause divisions in the environmental movement a generation later. Moreover, even proponents of wind turbines admit that "wider use of wind requires the invention of a new kind of weather forecasting, according to the Electric Power Research Institute, a nonprofit consortium based in Palo Alto, Calif., sponsored by the utility industry and its suppliers. Rather than forecasting from temperature or rainfall, what is needed is a focus on almost minute-by-minute predictions of wind in small areas where the turbines are."[23]

Recognized as a possibility from the beginning of wind power promotion, opposition to the aesthetics of wind turbines was nevertheless dismissed until wind farms with thousands of towers began appearing in populated areas. In 1987, two professors from the Department of Environmental Design at the University of California, Davis, sent questionnaires to people living near the Altamont Pass wind farm in California's Central Valley asking for their opinions of the project and enclosing six color photographs of the turbines from various distances. Of the usable responses, 55 were from women and 142 from men. Most of the respondents were between the ages of thirty and forty-four. Most agreed that the wind turbines were interesting and futuristic, symbolizing progress. Those who expressed a strong dislike of the machines, finding them ugly, cluttered, conspicuous, and out of place in the natural landscape, also questioned whether they were necessary or merely tax shelters. Well-educated men were the most critical. The researchers were surprised to find that those subjects most familiar with the specific environment of Altamont responded most negatively to the wind turbines, but, as Martin Pasqualetti, a geographer who studied the response to the wind farm on San Gorgonio Pass near Palm Springs, California, points out, people expect "permanence" in their landscape and, in the West, expect buffers between the places of energy production and consumption. For those who would sell the power of sun and wind, these are important lessons.[24]

Not everyone finds wind farms unsightly. Some see profit in marketing the image rather than the actual wind. Righter mentions that both Lexus and US Air ads have featured a "Tehachapi Wind Farm Tour." The blades of wind generators in advertising are usually framed by clear sky, detached from any topographic features that might suggest that the towering structures and whirling propellers mar the landscape.

The use of nature to sell technology is not, of course, new. One measure of the growing support for environmental protection is the effort by businesses to appeal to guilt-ridden consumers by promoting their products as "green" or identifying themselves with environmental causes. These days, almost all carmakers appeal to fuel efficiency (Honda's little white cloud of 30.1 miles per gallon floats above an identical cloud labeled 24.8 "Industry Average") or all-weather safety (a Lexus television commercial in 2002 featured the car in mortal combat with angry winds blowing dangerous obstacles in its path). Subaru goes beyond symbolism, financially supporting the Mount Washington weather observatory, among other environmental causes. Shell Oil recently ran a series of double-page ads in major magazines under the headline "Cloud the Issue . . . or Clear the Air." On the first page an outline of the Shell logo appears on a cloud; on the second page the familiar yellow and red shell appears against a clearing blue sky. Text and image make it evident that the company wants you to know that it is working to reduce pollution, although why what looks like an ordinary cumulus should represent global warming remains a mystery.[25]

Some call this eco-porn. My favorite corporate appropriation of weather, and one of the most elaborate, is by Air France, which in 2002 hired film director Michel Gondry (*Eternal Sunshine of the Spotless Mind* and *The Science of Sleep*) and model Audrey Marnay for a promotional film, *The Cloud,* depicting the pleasure of travel. "The poetry of the sky," according to an article in an in-flight magazine, "stems from the complicity between the young woman and the cloud, which becomes in turn the collar of her coat, the pillow on which she rests, and the playground of the pillow she passes . . . reminding viewers of Air France's commitment to its customers to 'Make the sky the most beautiful place on earth.'"[26]

Weather sells all sorts of outdoor products, as a casual glance at the ads on The Weather Channel reveals. If outdoor pollution is too bad, IonicBreeze promises to clean the air in your house. If the sun is too strong, Sun Setter awnings help you "outsmart the weather." "Wind & Weather," a mail-order business that offers weather vanes, lawn fauna in the shape of farcical frogs and comical cats, as well as home weather stations, has been supplying weather-conscious gardeners and hobbyists since 1976. One of its weather gadgets "features a 'Weather Boy' sporting one of 15 outfits that change according to changing barometric pressure"—a high-tech version of the old Swiss cottage with just two figures, one with an umbrella and one without.[27]

Freeing the Slaves of the Demons of Weather

Dreams of harnessing weather for the benefit of humanity may have been inspired by the boys' books available to the first generation of solar and wind power engineers. The author of one of the most interesting books, Francis Rolt-Wheeler, immigrated to the United States from England in 1893 and worked for newspapers in North Dakota, Minnesota, and Illinois. In 1904 he was ordained a Catholic priest and soon after began publishing his U.S. Service Series, novels about young men who work with the U.S. Geological Survey, Forest Service, Census Bureau,

Fisheries Bureau, Bureau of Indian Affairs, and Postal Service, among others. *The Boy with the U.S. Weather Men,* the story of a youth named Ross Planford who lives in western Mississippi, north of Vicksburg, and has many exciting weather-related adventures, appeared in 1917.[28]

Ross's experiences with floods, frosts, tornadoes, hurricanes, and lightning lead him to organize the Mississippi League of the Weather and to publish the *Issaquena County Weather Herald. The Boy with the U.S. Weather Men* is illustrated with seventy-two photographs of weather phenomena, most supplied by the Weather Bureau. Rolt-Wheeler clearly intended his book to instruct readers in meteorology. His preface begins: "The savage fury of the tempest and the burning splendor of the sun in all ages have stirred the human race to fear and wonder." He concludes with these thoughts:

> We are not slaves to the demons of the Weather, now—not as we once were. The United States Weather Bureau, day by day, draws closer and closer the chains which bind the untrammeled violence of sun and storm. High, high in the atmosphere is a world all unexplored, where no man can dwell; where, as yet, no human-made instrument has reached. This unknown world calls for explorers, it calls for adventure, it calls for daring and patient work. It is for Man to tame the forces of the sky, and tame them he must and will. To show how much the Weather Bureau is accomplishing, to depict the marvels of its work, to portray the ruthless ferocity of the forces as yet uncontrolled and to reveal the gripping fascination of this work, in which every American boy may join, is the aim and purpose of The Author.[29]

The message is that science is opening new and limitless frontiers to boys who are willing to acquire the necessary skills. Rolt-Wheeler was enthusiastic about forecasting and weather modification, but he goes beyond the confines of science to such social problems as racial discrimination and prejudices against the physically disabled. Regrettably, like other boys' books authors of the time, Rolt-Wheeler completely ignores young women.[30]

In the course of the story, Ross saves his lame friend Anton twice, once from a flood and again from a tornado. Becoming a cooperative observer makes Anton feel useful and happy. At the end of the book Anton saves Ross's life when Ross is stuck by lightning. When the weatherman, Mr. Levin, teaches the boys to make a sundial and to keep records of daily temperatures and weather conditions, he impresses them with the importance and magnitude of the task. Anticipating chaos theory, Rolt-Wheeler describes the complexity of the weather in terms boys of the time would understand. "Suppose you had a thousand marbles of different colors," he has Mr. Levin tell the boys, "and you dropped them from the top of a house to the hard ground below, a rough and rocky piece of ground, could you ever figure out what kind of a pattern they would make?"[31] Small differences in the way the marbles were dropped and landed preclude a precise forecast. Rolt-Wheeler's insight into the difficulties of weather prediction because of the influence of past conditions enables him to understand the complexities of social as well as atmospheric conditions.

Two of Rolt-Wheeler's characters are African American—Dan'l, who predicts weather from the behavior of animals and birds, not "fermometers," and Caesar, a boy who draws beautiful and accurate pictures of sunsets, illustrating the weatherman's points about the effects of volcanic ash in the atmosphere. With encouragement from the weatherman, Ross, Anton, and some other boys decide to integrate their club despite local segregation. "I think it's the absolute duty of every American boy to help every other American boy when he gets a chance," the weatherman tells them, "whether his skin is black or white."[32] The boys later save Dan'l when he is falsely accused of shooting a man, by showing the sheriff that Dan'l's shoes are dry even though it was raining at the time of the crime. Rolt-Wheeler then expounds on forensic meteorology. Dan'l shows his gratitude by buying the boys a crystal so they can make a sunshine recorder. The weatherman also encourages Dan'l to collect weather lore, which is published in a folklore magazine. As if to underscore the violence of weather that needs to be tamed, Dan'l is killed by a tornado. Caesar fares better; he is admitted to the Mississippi League of the Weather, and a "negro college" [sic] lithographs his drawings, which become a source of pride in Issaquena County.

The boys also photograph clouds and produce their own cloud atlas. Their most spectacular achievement, however, is upper-air exploration by kites. Rolt-Wheeler attributes scientific kite flying to William A. Eddy and recommends a tetrahedral kite invented by Alexander Graham Bell, illustrated on the cover of the book. Eddy, of Bayonne, New Jersey, was the acknowledged originator of meteorological kite flying. Beginning about 1890, Eddy lifted thermometers into the sky and in 1894 took his kites to Blue Hill Observatory, where, with the assistance of S. P. Fergusson, he flew an aluminum thermograph 1,500 feet.[33] The boys get their kite to an altitude of 3 miles (15,840 feet)—an amateur world record, announces the weatherman—but the wire breaks and the boys recover the runaway kite only after a long chase. The book ends with a discussion of the importance of weather forecasts to businesses and residents of areas subject to severe storms.

Rolt-Wheeler was one of a number of writers who tapped the market for youth adventure books in the early twentieth century, but he is unique, I think, in focusing on boys working with federal government agencies, and this may be why his books are less known than those about Tom Swift, the Rover Boys, and the Hardy Boys. Like the heroes of those books, Ross Planford and the other members of the Mississippi League of the Weather are plucky, honest, and patriotic, but they were also, thanks to Rolt-Wheeler's sense that the new science and technology require state sponsorship and organization, budding bureaucrats. Progress in an age of science and technology would not come only from heroic geniuses; it required

Cover of *The Boy with the U.S. Weather Men,* 1917, showing a kite used to carry weather instruments.

finding a place for everyone in the intricate network of specialized knowledge. The Weather Leaguers were also avid consumers and managers of information. While Rolt-Wheeler holds out the promise of freedom from the demons of the weather, he simultaneously celebrates the tightening of chains to control the violence of sun and storm. Enslaving nature to free humans is an old paradox.

The Boy with the U.S. Weather Men anticipates the expansion of weather education programs by the Weather Bureau, the American Meteorological Society, and later by NASA and NOAA.

The founding of the AMS in December 1919 initiated a new era in the marketing of weather information. The first AMS constitution set forth its objectives: "The advancement and diffusion of knowledge of meteorology, including climatology, and the development of its application to public health, agriculture, engineering, transportation by land and inland waterways, navigation in the air and the oceans, and other forms of industry and commerce."[34] In contrast to the Weather Bureau, which is responsible to the public and Congress, the AMS is a private, nonprofit organization dedicated to the creation of a professional identity for meteorologists. Its broad mandate emphasizes education and public information. Both the AMS and the Weather Bureau recognize the need to "sell" themselves and their products, and by the 1920s it was clear that weather was a valuable commodity.[35]

Weather Scouts and S'COOL Kids

Among the many consumers of Weather Bureau information, the Boy Scouts of America (BSA) and the Girl Scouts of the USA, as enduring symbols of American values, demand special notice. Moreover, the Scout weather pamphlets serve as a microcosm of the changes in public information and education disseminated by the Weather Bureau and the AMS over the century.

Organized in 1910, the BSA grew rapidly in both membership and organizational complexity. It instituted a system of merit badges to encourage Scouts to develop skills in camping, map reading, woodworking, and, in 1928, weather. William J. Humphreys, a Weather Bureau physicist, and Charles Fitzhugh Talman, the bureau librarian, consulted on the weather merit badge pamphlet.[36]

It is a seventy-three-page manual covering "Constituents of the Atmosphere"— water vapor, oxygen, nitrogen, carbon dioxide, argon and other gases, fog, rain, snow, sleet, glaze, and hail—with additional sections on halos, looming, lightning, Saint Elmo's fire, and the aurora. There are sections on instruments, the value of weather forecasts, and the interpretation of weather. This section is credited to Humphreys. Talman offers "An Outline of Meteorology," which is slightly more technical and provides a chart profiling the atmosphere in layers from the bottom 7 miles, the region of clouds and dust, to a fifth layer 127 to 187 miles above the earth, the region of auroral arcs. Talman notes that the atmosphere has been divided into layers since 1902 and that the branch of meteorology dealing with the study of all portions of the air not close to the earth is called "aerology." His section concludes with a discussion of the general circulation of the atmosphere, including the "polar front hypothesis," and the nature of tornadoes. There are photos of twelve types of clouds, and storm and weather flags and signals are illustrated. A bibliography includes all the serial publications of the Weather Bureau. Merit badge pamphlets were prepared and sold to Scouts to provide basic information on the topic. Boys studied the pamphlet and then contacted a "merit badge counselor," who confirmed that they had met the requirements. The system has remained basically unchanged for more than eighty years, but the contents of the pamphlets have been updated and the requirements have changed in ways that suggest changes in American attitudes toward the weather.[37]

The 1928 edition of the BSA Weather pamphlet was revised in 1943, with the assistance of Charles F. Brooks of Harvard University, a colleague of Humphreys and Talman. The contents are similar, but much of the material on the upper atmosphere is missing, replaced by twenty-three pages condensed from George R. Stewart's novel *Storm* and illustrations by Donald A. Smith. These are not insignificant additions. Stewart's account of a winter storm that develops over the Pacific Ocean and brings heavy snows and flooding to California (discussed in more detail in chapter 4) is told through the actions of a young meteorologist who plots the course of the storm as he receives data from Weather Bureau stations and instruments, and others who must deal with power outages, road maintenance, and air traffic. The novel both humanizes the abstractions of meteorology and naturalizes the nameless humans who struggle with the consequences of the storm. Some of this comes through even in the brief extract, aided by Smith's sketches, which range from the lurid (hand-to-hand combat by anthropomorphic polar and tropical fronts battling with lightning bolts and icicles) to the humorous (a cartoon explaining the thermometer in which a figure in a lab coat claims that centigrade "is more scientific," while a simian-faced figure points to Fahrenheit as "mentioned more often in weather reports").

The BSA excerpt from *Storm* is introduced by a paragraph urging Scouts to read the novel first as "entertainment," then to reread it "to see how the facts fit into place." The instruction that follows the story begins with the composition of the atmosphere and continues with a description and function of instruments—thermometer, barometer, anemometer, psychrometer, rain gauge. The authors then discuss cloud types and special phenomena—storms, lightning, thunder, rainbows, halos, coronas, mirages, looming, aurora borealis, and Saint Elmo's fire—and conclude with sections on climate, the Weather Bureau, the Airway Weather Service, map reading, and gliding. A reading list includes Brooks's *Why the Weather?*, Eric Sloane's *Weather Book,* and many government publications. The tone of the manual is dry, but occasionally Brooks's humor emerges. In describing the dangers of lightning the pamphlet advises: "Like a great many humans, lightning is lazy. It doesn't like to jump any farther than it must, and it prefers to flow through the substance which offers it the least resistance." Thus, you are safe in a house with a lightning rod because "the lightning doesn't have to smash its way through the house to reach the ground—which pleases both the lightning and everyone inside the house."[38] Contrast this with the 2002 merit badge pamphlet, which at seventy-six pages is a dozen pages longer than the 1943 volume and eschews any "entertainment." Lightning in this manual is a killer, and the Scout is told: "A steel-framed building or motor vehicle often is a safe place because the charge stays within the frame, which conducts it safely to the ground without danger to the occupants. Do not use the telephone or hold objects connected to electrical power (like hair dryers) during a thunderstorm." The new pamphlet, largely the work of Charles A. Doswell III, then of the National Severe Storms Laboratory, and Mike Branick, of the National Weather Forecast Office, both in Norman, Oklahoma, reflects changes not only in meteorology since the 1940s but also in the ways in which weather permeates American life in 2002. This is most obvious in the warnings of impending environmental disasters. In discussing the hydrologic cycle, a subject sketched very briefly in earlier handbooks, Doswell explains aquifers and concludes: "In many parts of the world, including the United States, the water in aquifers is tapped for human use, especially for irrigation. Once the aquifers are drained, it may be thousands of years before they fill again." The treatment of climate in the 1943 volume is purely descriptive, a map of the United States with ten zones and a page of record temperatures and rainfalls, whereas in the 2002 pamphlet climate is explained at length and problems caused by deforestation, careless agricultural methods, industrial pollution, and automobile emissions are mentioned. The potential harm of acid rain, greenhouse gases, and ozone depletion is explained, and the need for conservation of natural resources is emphasized. "It is up to you," Scouts are told, "to continue improving upon conservation methods that will protect the Earth's temperate and beneficial climate."

Meteorology is the cornerstone of environmental science, as Charles Brooks and the other founders of the American Meteorological Society knew in 1919, but weather was so pervasive, so ordinary, that the public had difficulty making the connection. The conservation movement of the late nineteenth century and the environmental movement of the late twentieth century focused on the immediate and the highly visible—disappearing trees, eroding soil, air pollution, endangered species, silent birds—while all around the atmosphere was absorbing the invisible chemicals that threaten life. We saw the forest but not the skies.[39]

That Doswell begins the 2002 weather pamphlet with the "ocean of air" metaphor (discussed in chapter 3) is further recognition of the impact of satellite perspectives on the earth and the success of NOAA's educational programs since 1970. The linking of the two "oceans" in one government agency is one of the most important developments in the history of environmental management. Doswell's BSA weather pamphlet draws attention to the intimate connections between oceans and sky, placing weather at the center of the ecosphere.

A final comparison of the 1928, 1943, and 2002 BSA weather pamphlets yields additional insights into the changes in American attitudes toward weather in the twentieth century. All three pamphlets list the "requirements" for a merit badge in weather. Many of them are the same or very similar—construct a weather vane or another instrument, keep a daily weather log (for a month in 1928 and 1943, a week in 2002), name dangerous weather conditions, be able to read a weather map (the 1928 manual has one small weather map, 1943 has maps for two days, the 2002 has a week's worth)—but there are significant changes over the seventy-five years. Earlier requirements such as "explain the value of weather prediction," "write a brief account of the United States Weather Bureau . . . [and] be able to interpret the charts and graphs contained in their publications," and "explain points connected with electrical and optical phenomena in the air, i.e. . . . rainbows, mirages, looming, halos, northern lights, St. Elmo's fire, lightning and thunder" have been replaced with "tell what causes wind, why it rains, and how lightning and hail are formed" and "define acid rain, identify which human activities pollute the atmosphere and the effects such pollution can have on people." Scouts are now given options to interview local Weather Service officials, radio and television weathercasters, and private meteorologists to learn about severe weather conditions in their communities, and they are urged to give five-minute talks on weather safety or acid rain.

The Girl Scouts of the USA introduced a weather proficiency badge in 1947 but do not rely on a special handbook. The Girl Scout Handbooks are age-graded. The 1947 Intermediate Handbook weather activities were written by Margaret Chapman and Marie Gaudette, a nature specialist with the Girl Scouts. The activities are similar to those required of boys—keep a daily record of wind, temperature, and cloud formations for a month; know the significance of at least three cloud types; know first aid for sunstroke, heat exhaustion, shock, and frostbite; be able to interpret a weather map—but are broader in approach. For example, the Intermediate Girl Scout is encouraged to "select and share with your troop some poems or stories about weather, wind, rain, snow, clouds, sky." Collecting and exhibiting pictures of rain, snow, clouds, and frost and their effects are also activities that count toward the proficiency badge. These activities were revised and updated in 1990 and now include finding out about weather modification, including asking the question, "What are some of the reasons we would want to change it?" Aspirants for the current weather proficiency badge are also encouraged to find out about and prepare for weather emergencies and to "keep a chart for one month that shows how weather affects your activities." The Girl Scout approach to weather is more holistic than that of the Boy Scouts, with its emphasis on the mechanics of the atmosphere.[40]

Seventh graders at Tucker Valley Elementary Middle School, Hambleton, West Virginia, making observations of clouds for a S'COOL project, 2007. Photo courtesy Eileen Poling.

Weather as a personal concern, a hobby like stamp collecting with an aesthetic as well as an educational dimension, has been replaced in the latest Boy Scout weather handbook with an emphasis on weather as social phenomena, with impacts on communities, nations, and the whole earth. Where Humphreys and Talman required a Scout to write an outline describing the climate of the United States and the Scout's own state, Doswell introduces climate change and challenges Scouts to think about the future. Weather prediction is now the product of satellites, Doppler radar, and computer models. In the bibliography the Scout is directed to merit badge pamphlets on emergency preparedness, environmental science, and soil and water conservation, as well as to Weather Service publications on flood, severe storm, tornado, hurricane, and lightning hazards. Thus, Scouts are initiated into the risk society, prepared to become future natural-disaster managers.

From tentative steps to dispel "popular errors concerning the weather" to extensive programs for students from kindergarten through college, the Weather Bureau and the AMS have created generations of weather-smart Americans. Where bureau scientists once feared that teaching meteorology in high schools would mislead students into thinking that weather processes are simple, today's AMS and the National Weather Service sponsor annual workshops for K–12 science teachers and projects that involve students in cloud observation coordinated with satellite data. With funding from the National Science Foundation, NASA, and NOAA, hundreds of teachers and thousands of students participate annually in the study of the atmosphere, water resources, and severe weather.[41]

S'COOL, a science education program involving the study of clouds, was developed not by the National Weather Service or NOAA but by NASA, a formidable rival barely mentioned in discussions of the weather community. S'COOL was conceived in 1996 by Lin H. Chambers of the Radiation and Aerosols Branch of NASA's Langley Research Center in Hampton, Virginia, just prior to the launch of three low Earth orbit satellites carrying instruments for the Clouds and Earth's Radiant Energy System (CERES), designed to monitor the impact of clouds on Earth's radiation budget. Radiation from the sun, only about half of which reaches Earth, is absorbed, scattered, and reflected by clouds. This process helps create weather and climates. Working with elementary school science teachers, Chambers coordinates the observations of clouds by children from more than 500 schools worldwide. As the CERES satellites pass over them, the children record cloud and surface properties on a one-page form that is e-mailed, faxed, or mailed to the Langley Center and made available online after processing. The observations are necessary because the satellites record only the top layer of clouds and may miss significant aspects of the cloud cover.[42]

Their effort, NASA tells the students, is called "ground truth" because,

in order to "anchor" the satellite measurements, we need to compare them to something we know. One way to do this is by what we call "ground truth," which is one part of the calibration process. This is where a person on the ground (or sometimes in an airplane) makes a measurement of the same thing the satellite is trying to measure, at the same time the satellite is measuring it. The two answers are then compared to help evaluate how well the satellite instrument is performing. Usually we believe the ground truth more than the satellite, because we have more experience making measurements on the ground and sometimes we can see what we are measuring with the naked eye.[43]

With this simple and unpretentious explanation, NASA introduces students to the scientific method and makes it accessible. Chambers has encouraged the teachers to be creative and incorporate the CERES observations into the curriculum. Teachers have built on the S'COOL program to inspire students to observe the sky and learn the cloud classification; to write poems and essays about the sky; to learn English, if it is not their native language, or to learn another language by communication with S'COOL participants outside the United States; to learn

Internet and computer skills; to teach responsibility and punctuality; and to develop scientific curiosity. Since S'COOL is just one of several NASA and NOAA education projects that involve the study of the atmosphere, it is fair to assume that they will promote as much interest in the weather as the Boy Scout pamphlets have. NASA and NOAA have independently developed and marketed cloud charts, games, books, videos, and lightning safety brochures, sometimes in cooperation and other times in competition with commercial publishers, triggering more tempests in Weathertown, U.S.A.[44]

From Weather Bureau to Weather Service

Selling weather knowledge to Scouts was easier than getting the attention of a skeptical public. The years between writing the first Scout weather pamphlet and creating the multiplicity of Internet sites and school projects were often turbulent because of internal and external politics, scientific and technological change, and national priorities. "Popularize Meteorology: Why Not?" read the title of an article in the *Sunday Times* of New Brunswick, New Jersey, reprinted in the *Bulletin of the American Meteorological Society* in December 1928. The anonymous author observed the popularity of astronomy clubs and summer courses in astronomy at Rutgers University and lamented the lack of such clubs and courses in meteorology. Apparently unaware of the many public lectures by Weather Bureau employees and the popular books by Edwin Martin, Alexander McAdie, and Charles Brooks, the writer mentions just one meteorological hobbyist, S. K. Pearson of Plainfield, and a column, "Look Aloft for the Weather," by S. Partridge in the *New York Evening Post*.[45] This, of course, is a perennial problem that might be called the "pearls before swine" conundrum. The Weather Bureau and the AMS cast thousands of educational programs before the public, but until someone catches sight of what he thinks is news, the efforts are generally ignored.

A sign that the Weather Bureau was attracting some new attention appeared in June 1932, when an award-winning architecture student, Robert Allen Ward, submitted his thesis for a bachelor of fine arts in architecture at Yale University. His project, "A National Center of Meteorology," was summarized in the *Bulletin of the American Meteorological Society* the following year. Naively ambitious, considering the state of the economy and the federal budget, Ward's plan was for an eight-square-block campus located from D to H Streets, NW, between Twenty-sixth Street and the Potomac River, where the Kennedy Center and the Watergate apartment complex are today. Going beyond a mere office for the bureau, Ward proposed "an Institute of Meteorology, to serve as the Central Office of the United States Weather Bureau, as a means of education and the dissemination of knowledge to the general public, and as a haven of study for scholars from all over the world."

At the north end of the quadrangle, facing the phenological gardens, Ward placed the two-story Administration Building, with a 120-foot tower containing meteorological instruments. Explaining his severely functional design, Ward wrote: "It is fitting that meteorology, the youngest of sciences, should be enshrined in architecture of the most modern type. There is no place here for unnecessary ornamentation, for the fetters of tradition to cramp the buildings into a mould unsuited for their purpose. . . . yet, by their lack of ostentation and their functional qualities, [the buildings will be able] to express something of the 'clear, cold light of science.'"[46] Other buildings, lining the gardens, included a library, auditorium and museum, climatological offices, a printing and instrument building, and eight small laboratories for visiting scholars. If the center had been built, it would certainly have raised the profile of the Weather Bureau and meteorology in general. Not until the construction of I. M. Pei's NCAR building in 1967 were the atmospheric sciences provided with an architectural symbol to match their ambitions.

The cold, clear light of science that Ward dreamed of shining from his tower was temporarily dimmed. In 1933, the Weather Bureau came under attack, first by the American Society of Civil Engineers, then by President Franklin Roosevelt's Science Advisory Board. Marketing weather took a backseat to circling wagons. The civil engineers had become increasingly unhappy with forecasts issued by bureau employees because they lacked the specificity needed to plan major outdoor construction projects such as highways, dams, and bridges. Their report recommended sweeping changes in the Weather Bureau from the top down. The engineers strongly implied that Charles Marvin, the seventy-four-year-old chief, was incompetent; that the bureau operated too few stations; that many of the stations were in the wrong location; that since only 22.5 percent of the forecasters in those stations had college degrees, the majority were unfit; that the cooperative observers were poorly trained and supervised; and that what useful data were available were distributed slowly and inefficiently. In March, Oliver Fassig replied for the bureau, denying the validity of most of the criticisms and placing the blame on inadequate budgets for salaries and the legacy of the Signal Corps past. The following month, Charles Brooks of the AMS offered a longer and more tempered response, praising the quality of the engineers' committee and calling their report constructive.[47]

Brooks saw the crisis in the bureau as the most important since its transfer to the Department of Agriculture in 1891 and welcomed the engineers' report as an opportunity to begin making much-needed reforms. He vigorously defended the work of the cooperative observers but acknowledged the need for a greater emphasis on research and the need to accept new ideas about forecasting. It is significant that he did not mention these new ideas by name, since discussion of the Scandinavian methods was taking place in the international meteorological community, but he knew that Marvin had abruptly dismissed the brilliant young

Swedish meteorologist Carl-Gustaf Rossby, who came to the United States on a fellowship from the Sweden-America Foundation and worked for the Weather Bureau in 1926.

Rossby went on to establish air routes in California under a grant from the Daniel Guggenheim Fund for the Promotion of Aeronautics. In 1928 he became associate professor of meteorology at MIT, leaving in 1939 to serve as assistant chief for research at the Weather Bureau at the invitation of its new chief, Francis Reichelderfer, who had adopted the Scandinavian methods while in the navy's Office of Aerology. In 1941 Rossby left to head the new meteorology department at the University of Chicago, where he remained until returning to Sweden in 1950. Restless and energetic, Rossby served as president of the AMS, which he had once derisively called "the National Geographic Society of meteorology," in the crucial years 1944 and 1945, setting new goals and helping to raise its standards for professional membership. Although Rossby, a dynamic and colorful personality, communicated mostly with the scientific elites of weather, he did not neglect popular education and encouraged the development of private meteorology. Rossby's ability to sell the new concepts of atmospheric science has become part of the mythology of the field.[48]

The engineer's report and Rossby's conflict with Marvin were merely a prelude to bigger troubles for the Weather Bureau. Just as Fassig and Brooks were replying, the dirigible *Akron* went down in a storm off the New Jersey coast, killing more than seventy men, including some prominent naval officers. Critics promptly blamed the bureau for failing to forecast the hazardous conditions, as they had when the navy airship *Shenandoah* crashed in Ohio in 1925. Many scientists, fearing that the bureau's failures were adding to what they perceived as a revolt against science, lobbied for the creation of a Science Advisory Board. On July 31, 1933, President Roosevelt appointed nine men, including three—Isaiah Bowman, director of the American Geographical Society; Karl T. Compton, president of MIT; and Robert A. Millikan from Cal Tech—who formed a special committee on the Weather Bureau. The following month, the committee added Charles D. Reed of the Des Moines Weather Bureau station as a fourth member, probably at the suggestion of Secretary of Agriculture Henry Wallace, who was sensitive to the internal politics of the Weather Bureau, which clearly felt threatened by this external review by a geographer and two physicists.[49]

Millikan, who won a Nobel Prize in 1923 for measuring the charge of electrons, had organized training programs for army meteorologists during World War I and was familiar with some of the personnel problems in the Weather Bureau. He and Compton were eager to push the Roosevelt administration toward greater support for high-quality science and used their report on the Weather Bureau to make several larger points. The preliminary report was ready by mid-November and began with an acknowledgment of the importance of the work of the Weather Bureau and the broad scope of its activities. It rejected the criticisms of the civil engineers' report as too narrowly focused but repeated most of them and recommended fundamental changes in the priorities and organization of the Weather Bureau.

First and foremost, the committee recommended "that provision be made at once for extending the so-called air-mass analysis method over the United States, through the cooperation of the Weather Bureau, the Army, and the Navy." Next Millikan and the committee recommended that all meteorological data in aid of forecasting be consolidated under the Weather Bureau, but they also recommended decentralizing forecast work and the creation of more district forecast centers. Recognizing the growing importance of air transportation and the limited value of downtown weather stations where temperature and wind data were subject to "radiation and turbulence effects," the committee advised locating all stations at airports. Finally, going well beyond the engineers' report, the Science Advisory Board committee urged the Weather Bureau to recruit better-educated meteorologists, support further training for current employees, and encourage research. Moreover, the bureau was advised to "obtain the cooperation of other countries in the northern hemisphere, particularly Canada, Mexico and Russia (Siberia), in securing appropriate meteorological data which will disclose the movements of major air-masses over all these areas, in the interest of increasing the time range of weather forecasting."[50]

Needless to say, it took several years, interrupted by World War II, to complete the recommended reforms. Marvin retired in 1934 and was replaced by Willis R. Gregg, another career Weather Bureau meteorologist, and the following year the bureau was able to hire three Rossby protégés with Ph.D.s from MIT—Horace Byers, Harry Wexler, and Stephen Lichtblau—who had some familiarity with the Bergen School air-mass theory. When Gregg died in September 1938, navy commander Francis Reichelderfer was approved to replace him and led the Weather Bureau for the next twenty-five years, improving its image, raising morale, and overseeing its transformation into the age of radar, computers, satellites, and weather modification. Public education and selling the weather were de-emphasized in favor of raising professional standards, while the AMS dealt with the issues raised by television weather and the accelerating growth of private meteorology. By the 1960s, however, the time had come for yet another shift in the weather community and its image.[51]

The Weather Bureau emerged from World War II with new ideas, new tools, and some new personnel, but the relationships among the government, science, and the public were changing so rapidly that another shift in the marketing of weather was certain. On the surface, things seemed to be going well for the weathermen. Despite budget cuts in the mid-1950s, the bureau grew, establishing the Severe Local Storms Forecasting Unit (SELS) in 1953, the National Severe Storms Project (NSSP) for research on thunderstorms and tornadoes in 1955, and the National Hurricane Research Project in 1956. The NSSP became the National

U.S. Weather Bureau employee examining anemometer at National Airport, Washington, D.C., 1943. Fred Driscoll photographer. Library of Congress, LC-USW3-038496-D.

Severe Storms Laboratory (NSSL) in 1962, and SELS evolved into the Storm Prediction Center of the National Center for Environmental Prediction in Norman, Oklahoma, in 1997. These reorganizations and name changes were partly the result of strong personalities and internal politics, but they also reflect the emergence of environmentalism and the recognition in the atmospheric sciences that weather and climate were, in fact, the ultimate environmental factors without which the biosphere, the earth, would not exist. Ironically, Sky Awareness Day trailed Earth Day by twenty years.[52]

The butterfly whose flapping wings ultimately caused a storm within the Weather Bureau was the National Science Foundation, created by Congress in 1950 to promote basic science and engineering. According to George Mazuzan, a historian of that bureaucracy, "During an early time in the NSF operation, one discipline provided a microcosmic study of all the roles the federal agency has sought to fulfill. From 1958 through 1962, the NSF greatly increased its support of meteorology (later broadened to the atmospheric sciences), which had been up to then a generally neglected discipline with a pronounced inferiority complex compared to other sciences." Mazuzan argues that the low esteem in which meteorologists were held was the result of "impressions of the discipline from newspapers, radio, and particularly the new medium of television on which a weatherman, often depicted as a meteorologist, attempted, between commercials, to forecast the weather. That made it difficult to recruit scientifically-oriented students."[53]

In 1956, taking advantage of NSF funding, the Weather Bureau asked the National Academy of Sciences to appoint a meteorological research committee to advise the Department of Commerce in its planning. Once again, Carl Rossby came to the rescue of the Weather Bureau when he was appointed cochair of the committee, whose members included some of the most illustrious scientists and scientific administrators of the postwar period, among them Horace Byers of the University of Chicago, Hugh Dryden of the National Advisory Committee on Aeronautics, Thomas Malone of the Travelers Weather Research Center, John von Neumann of the Atomic Energy Commission, and Edward Teller of the University of California. When Rossby and von Neumann died in 1957, they were replaced by Henry G. Booker of Cornell University and Jule G. Charney of MIT.

The committee focused on the long-term development of meteorology. In its 1958 report it recommended that financial support for basic research be increased by 50 to 100 percent; that a National Institute of Atmospheric Research be created to coordinate interdisciplinary research in related physical sciences; that the NSF sponsor a university consortium to facilitate cooperation among academic and government scientists; and that the AMS take the lead in stimulating an interest in the study of weather. The Committee on Meteorology became the Committee on Atmospheric Sciences to indicate the broader scope of research, and a series of meetings led to the creation of the University Committee (later Corporation) on Atmospheric Research (UCAR) in 1959. Beginning with fourteen universities that had graduate programs in the atmospheric sciences, UCAR today is a consortium of more than sixty universities in the United States and more than sixty international institutions. As noted in the previous chapter, the *Preliminary Plans for a National Institute for Atmospheric Research,* prepared by the NSF and NCAR in 1959, called for the building of an institute with offices, computers, and a library. This became the National Center for Atmospheric Research in 1960. Because the first director of UCAR, Walter Orr Roberts, a Harvard astronomer, refused to leave his Rocky Mountain observatory, both UCAR and NCAR are located in Boulder, Colorado, supported by the NSF and other federal agencies, including NOAA (of which the National Weather Service is a component), NASA, and the Environmental Protection Agency (EPA).[54]

Sverre Petterssen, a Norwegian-born meteorologist teaching at the University of Chicago, was president of the American Meteorological Society in 1958–1959 and responded to the NSF committee challenge by developing, with Kenneth Spengler, the very able executive secretary of the AMS, an ambitious new educational program. With modest grants from the NSF, the AMS continued its Visiting Scientist Program that provided lecturers to colleges and schools; prepared booklets on careers in meteorology; laid plans for a series of "monographs" on meteorology for high school students; began to develop motion pictures and "lantern slides" for popular presentations; and held a summer science program for high school "boys" at a private school in Connecticut. That these initiatives seem a bit quaint and unprogressive underscores how little had changed in the selling of weather knowledge to the public since the 1900s. The AMS was eager to contribute to the marketing of meteorology, and Spengler and the AMS committees charged with the task worked hard and fast.

The first films, *Above the Horizon* and *Formation of Raindrops,* were ready by 1964. The first monographs, Louis Battan's *Unclean Sky* and George Ohring's *Weather on the Planets,* published as inexpensive paperbacks, appeared in 1966, followed by Duncan Blanchard's *From Raindrops to Volcanoes,* Elmar Reiter's *Jet Streams,* and James G. Edinger's lucid and poetic *Watching for the Wind.* Private initiatives had already taken the lead. *Weatherwise* was still a small-circulation magazine, but it had added a column on audiovisual materials by the mid-1960s. Bell Labs hired the Academy Award–winning director Frank Capra in 1958 to produce *Unchained Goddess,* a lesson on weather animated by Shamus Culhane, who had worked on *Snow White* and *Pinocchio.*[55]

In addition to spreading the gospel of the new meteorology, the AMS was struggling to reorganize itself in order to provide higher professional standards. Rapid scientific and technological advances inevitably led to exaggerated claims for long-range forecasting and weather modification. The popular press was eager to report unsubstantiated claims of thirty-day forecasts and of rain on demand. The AMS responded with a code of ethics and disciplinary actions against members who violated them. Dealing with unwarranted statements by nonmembers was, of course, more difficult. The AMS vacillated between ignoring them and answering with its own public relations campaign that "will emphasize the positive aspects, i.e., the things meteorology can do and hopes to do, while also making clear the present limitations."[56]

By 1963, weather satellites had been in space for three years, and cooperation among the atmospheric sciences was not an option. The Kennedy administration was remaking the federal bureaucracy, sending men into space, and preaching the gospel of environmentalism. In Robert M. White they found a man ready to take on what was often held up to be the classic example of institutional inertia, the Weather Bureau. White, an ambitious meteorological entrepreneur with a B.A. in geology from Harvard and an M.S. and D.Sc. in meteorology from MIT who, after seven years with the Air Force Cambridge Research Center, had founded his own environmental consulting firm, became chief that year. In a speech to a joint American Meteorological Society–American Geophysical Union banquet in April 1964, he outlined his ambitious plans for reorganization, merging the Weather Bureau with the Coast and Geodetic Survey into the Environmental Science Services Administration (ESSA). His remarks were prescient, the tone a bit apocalyptic: "We are becoming concerned with the capacity of our water resources, our earth, and our air to sustain us," he began.

And as our demands increase, so also does the rate at which we spoil and use up our environmental resources. . . . We are also more sensitive to environmental hazards, both old and new. Our cities, our economic organizations, and our governmental agencies have become highly concentrated and very complex organisms. Under these circumstances, a natural catastrophe can be more than a local event. . . . Hurricanes, tornadoes, tidal waves, and earthquakes have increasingly become an Achilles' heel of our modern industrial society, and nature's arrows occur with such frequency and are distributed over so wide an area that the need for an exclusive warning system is evident.[57]

White also pointed out that the environmental sciences are global and interdisciplinary, requiring interagency cooperation. The tune was vaguely familiar, but now someone was listening. Within two years White, with the assistance of Herbert Holloman, assistant secretary for science and technology in the Department of Commerce, and Admiral H. Arnold Karo, director of the Coast and Geodetic Survey, had created ESSA, with White as director. The Weather Bureau lost independence and status within the Department of Commerce, but White was replaced by an able administrator who saw opportunities as well as problems.[58]

George Cressman was a meteorologist with a Ph.D. from Chicago, a Rossby student, who had been the first director of the Joint Numerical Weather Prediction Unit (JNWPU), created in 1953 to coordinate the computerization of weather forecasting. This organization struggled unsuccessfully through the 1950s to become operational, stymied by strong bureaucracies and struggling technologies, but numerical prediction went forward in the navy, air force, and, in 1961, the National Meteorological Center of the Weather Bureau. Recalling the reorganization in an oral history interview many years later, Cressman rejected the notion that ESSA was bad for the Weather Bureau. "I think Bob White believed in what he was doing," Cressman said diplomatically.

I think he still believes that it was right. We just have a difference of opinion on that. Later on, it became not so relevant because they kept organizing themselves into grander and grander things. You know, ESSA became NOAA and eventually the Weather Service got back pretty much to where it was, organizationally. . . . I'll say this in Bob White's favor, that I believe that the research activities got a lot better support once they were put in a separate part of the organization.[59]

Cressman's first priority as head of the reorganized agency was to deal with the "really rotten morale" of its employees. Weather Bureau meteorologists had been kept in lower grades than other federal employees doing similar work, their pay was lower, and opportunities for advanced training were fewer. Perhaps because of his own background, Cressman understood the personnel problems and addressed them. He was born in West Chester, Pennsylvania, in 1919, the son of schoolteachers. A high school field trip to the Weather Bureau office in Philadelphia led to his becoming a cooperative observer, and he was hooked on weather. He studied physics at Penn State because the meteorology department had not yet been established. Upon graduation he enlisted in the air force and was sent to New York University for nine months of training. The air force made him an instructor, and he was able to teach in the program at the University of Chicago, which provided him with contacts for his later doctoral work. Sent to Idaho to do forecasting for bomber training flights, he ended the war in Homestead, Florida, teaching meteorology to pilots. Rossby arranged for him to do research in Puerto Rico before returning to Chicago. Cressman also did consulting work, including a brief stint in 1951 at the atomic bomb test site in southern Nevada.

With his unusually broad experience, Cressman appreciated the grassroots base of the Weather Bureau and the selling of its services. Two stories in his oral history illustrate this point. One was in response to a question about the frequent attempts by the Office of Management and Budget (OMB) to cut funding and close small weather offices and stations. The other is about international cooperation in the face of isolationist politics. Both reveal why Cressman's leadership was crucial to the transition from Weather Bureau to Weather Service in the 1970s. When faced with the threat of a station closing, Cressman advised:

> What you have to do is first go see the Congressman, you explain the situation, and you explain you're going to visit his town and you offer him an opportunity to show up and be assured that the press will probably be there. . . . It works like a charm. So then if I had a real problem, a nasty problem, I'd go down and see the Congressman. He used to give us hell in Congress, you know, but in the meantime he's got some beautiful press coverage in our forecast office. I'm not going to mention one because there have been a number of them. So I go and say, "Look. We've got this real problem and I need some assistance and I need to consult with you about what we're going to do about this station in your district. . . . There's a little station in Nebraska, I don't know if it's still there or not, but it was not too long ago. It was called Valentine, Nebraska. That was a one-person station, a lady MIC [meteorologist-in-charge] in there for a long time. It doesn't look like much of a station and it stands out in a balance book, you know. Budget people always had fits over it. But this lady was there and she would furnish a really first-class service in this remote state with severe weather problems. She

would work with the local people. She had her networks all set up and if there was going to be a storm, she would have everybody alerted and people who had their cattle out in some God-forsaken place would see to it that they got them to shelter. . . .

And so the local meteorologist in charge, or official in charge, has got to keep his eye on the local pulse. That's what he's there for, to provide local service. We could put a machine in to take the observations, but we can't put a machine in to do local service. . . . So our meteorologists or officials in charge in these country-wide stations are very important people in their communities and I think almost without exception, they understand their position and their responsibilities very well.

I guess that's what has made the Weather Service really loved out in the heartland.[60]

Cressman's second anecdote comes from his experience with the World Meteorological Organization (WMO) and the need for the United States to secure international cooperation in atmospheric research. In 1963 efforts began to create the World Weather Watch (WWW), a coordinated international system for collecting, analyzing, and distributing weather information. Implementation began in 1967. As Cressman recalls:

They came along with this concept of "World Weather Watch." I was not an author of this term, but I think it has been a useful way to look at activities in the World Meteorological Organization. I think it was, in effect, a useful political term. It enabled governments around the world to get support for a new program, and more support for an old program and it enabled the U.S. to lay on new programs of support for areas where there were few data, for example, in developing countries and oceanic areas and so by and large, it was a useful concept. . . . you go to Congress and tell them this is for World Weather Watch and it's going to benefit the United States and we're going to get all this help from other countries and we couldn't possibly afford to do this ourselves. That's a selling point. In the meantime, we could sell the idea of the aid programs. They provided observing facilities for countries who would then operate them.

"Voluntary Assistance program" was its name. It was a very useful program. It involved some of the smaller countries and poorer countries in providing data for meteorological services and, in turn, certainly encouraged them in meteorology in general. But that was really under the surface a self-centered program for the U.S.[61]

WMO emerged in 1950 as part of the newly organized United Nations, but it had a predecessor, the International Meteorological Organization (IMO), which was begun in Vienna, Austria, in 1873. WMO is a microcosm for the internationalization of science in an era of American political, economic, and technological dominance. As Clark Miller points out, creating a worldwide science of meteorology in the post–World War II era was a deeply political as well as technical exercise. An international organization with an ostensibly apolitical scientific mission served the United States in numerous ways. It encouraged intergovernmental harmony, promoted U.S. technology, and gained the nation a measure of security by linking meteorology to a liberal world order. By helping to establish a common vocabulary for viewing the earth's weather, the WWW was an agent of what might be called scientific globalization. The United States was unsuccessful in trying to control membership in the WMO, but it has dominated the scientific agenda. Miller concludes that the most significant result of the WMO is the growing perception of the atmosphere as a global commons, while the WMO congratulates itself, on its Web site, for being "a unique achievement in international cooperation: in few other fields of human endeavor, and particularly in science and technology, is there—or has there ever been—such a truly world-wide operational system to which virtually every country in the world contributes, every day of every year, for the common benefit of mankind."[62]

Environmentalism and internationalism were twin engines driving the creation of the National Oceanic and Atmospheric Administration in 1970. The Weather Bureau became the National Weather Service, its chief given the titles assistant administrator of weather services in NOAA and director of NOAA's National Weather Service. As Shelia Jasanoff convincingly argues, a global environmental consciousness was actively promoted by astronomer Carl Sagan, among others, using the photo of Earth taken by the *Apollo 17* astronauts, which became an icon of interconnectedness. An earlier image of Earth from space appeared on the cover of the first *Whole Earth Catalog* in 1968, and the architect Buckminster Fuller coined the phrase "spaceship Earth" the following year. NOAA expanded in many areas, but satellite launches kept the focus on Earth from above, prompting the economist Kenneth Boulding to remark that the global perspective had become the new icon of the American frontier.[63] Things seem to be going well for the Weather Service and atmospheric science, numerical weather prediction had all but replaced older methods of forecasting, research on severe weather was yielding good data, and the AMS had been successful in using its seal of approval to bring higher standards to television weather reports, though the public often still confused the science with the forecasting. Nevertheless, another small dark cloud of reorganization was on the horizon.

In 1990 NOAA asked the National Research Council, a division of the National Academy of Sciences, to provide a review for the modernization and restructuring of the National Weather Service. The National Weather Service Modernization Committee was formed and made recommendations throughout the decade, issuing its final report, *A Vision for the National Weather Service: Road Map for the Fu-*

ture, in 1999. Issued in conjunction with another NRC report by a different committee, the "road map" reviewed the recent history of the Weather Service and looked toward the year 2025. Like Weather Bureau chief Willis Gregg's fifty-year projection in 1938, the NRC committee was bullish on the future. It predicted weather forecasts with "greatly improved spatial and temporal density, accuracy, and timeliness of weather and climate data information." To accomplish this, however, the National Weather Service had to be an active partner in research, broaden the scope of its information for "consumers [who] will use the information to meet their commercial needs and personal interests by integrating traditional weather data . . . with other data on the environment and on specific economic sectors," and "participate with other public institutions, professional societies, and the private sector in educating the general public and specialized users about the causes and consequences of weather-related environmental phenomena."[64]

Nature stimulated the reorganization effort by starting a reign of destruction in August 1992, with Hurricane Andrew causing at least 26 deaths and more than $30 billion in damage in southern Florida and Louisiana; in September, Hurricane Iniki struck Hawaii, killing another 6 and adding $1.8 billion in property losses; in November, tornadoes in the southeastern states killed 26; the "Blizzard of the Century" in March 1993 added 270 fatalities; and the upper Mississippi and Missouri floods in the summer of 1993 drowned more than 40 persons and caused another $15 to 23 billion in losses. Naturally, the Weather Service was criticized for the bad weather, even if its forecasts were accurate. Congress held hearings in 1993 that were published under the title "Has Nature Gone Mad?" (an attention-grabber borrowed from *Life* magazine). Testimony from Elbert W. "Joe" Friday, assistant administrator for weather services of NOAA, and Robert C. Sheets, director, National Hurricane Center, assured the House Committee on Science, Space, and Technology that nature was sane, if erratic. If the congressmen had listened, they might have understood what Friday and Sheets were telling them: that coastal and floodplain real estate development was crazy, not nature, and that most evacuation plans were non compos mentis.[65]

Friday, National Weather Service director from 1988 to 1998, began to make changes shortly after his appointment. Friday came to NOAA from a career in the air force, where he had commanded weather detachments in Southeast Asia. He had a Ph.D. in meteorology from the University of Oklahoma. "Modernization," as defined in his 1994 report, chiefly meant expanded use of Doppler radar, the replacing of human weather observers with Automated Surface Observing System (ASOS), implementation of more sophisticated numerical models relying on upper-air observations, better integration of data from diverse observing systems, and a reorganization of the weather forecast offices, river forecast centers, and data collection centers.

Radar (radio detection and ranging) was developed in Germany, Britain, and the United States just before World War II, primarily to track aircraft. Although its usefulness in mapping the position of thunderstorms was recognized, weather elements—rain, snow, sleet—were considered "noise," not information. The bureau modified military radar in 1946 and began creating a network for the detection of severe storms in Texas, Oklahoma, Arkansas, and Louisiana in the late 1940s. This was largely replaced with equipment better suited to meteorology, and the network expanded in the mid-1950s. Doppler radar, which detects wind speed as well as precipitation, clouds, and lightning, was available from the navy by 1956, but the NWS was not able to purchase its own until 1988, when Weather Surveillance Radar (WSR-88D) was ordered from Unisys Corporation. The company had difficulty producing acceptable units until 1993, but by the mid-1990s NEXRAD (Next Generation Weather Radar) covered most of the United States, though not without many operational problems.[66]

Other aspects of the 1990s modernization program—automation, numerical modeling, integration of data, and reorganization of offices—were also ongoing efforts by the NWS to be a star on NOAA's ambitious team. While the purposes of radar networks and computer systems for the timely integration of data are relatively easy for the public to understand, the alphabet soup of WFO (weather forecast office), DCO (data collection center), NMC (National Meteorological Center), IFPS (Interactive Forecast Preparation System), and NDFD (National Digital Forecast Database) was not. Friday began his explanation of the new structure with the WFOs (116 in 1994, 122 in 2006), which would provide watches and warnings for local storms and floods, aviation forecasts, marine forecasts for the Great Lakes and coastal areas, and hydrological services such as identifying flood-prone areas. These functions were not radically different from those performed by the traditional weather stations that have varied in number from 173 in 1900, to 443 in 1947, to 242 in 1989. Consolidation of city and airport stations in the 1950s and budget reductions lowered the number over the years. The DCOs were created for Alaska and the Hawaiian Islands because of the large geographic areas they needed to cover.[67]

More controversial was the Automated Surface Observing System, unmanned weather stations. More than 850 were being installed in 1994, with more than 400 planned for the near future. In 2006 there were about 1,590. The array of sensors on an ASOS, and on the earlier Automated Weather Observing System (AWOS), records temperature, visibility, precipitation types and amounts, wind direction and speed, humidity and dew point, barometric pressure, sky cover and ceiling, and thunder, twenty-four hours a day. Some report every hour, but newer systems report every minute. Though touted on the current NWS Web site as more reliable than human observers who make subjective decisions about cloud cover and who need to rest, the ASOS instruments measure only the weather that passes through the sensor array, not the horizon-to-horizon perspective of a human observer. Although Friday asserted that the decision to automate "was made not to try to replicate the manual observing process," it was clearly the intent of the first Bush administration to downsize the NWS. The 115 WFOs replaced 142 weather stations, and 800 jobs were eliminated over the next six years. The lack of human observation has left climatologists without the data they need to accurately determine whether the cloud cover over North America is increasing or decreasing, an important variable in climate change. According to scientists at NCAR and the National Climatic Data Center, data from ASOS laser beam ceilometers and from satellites miss much of the cloudiness in the sky, meaning that "climate researchers can no longer make full use of the historical records of the last 100–150 years to study the climate."[68]

President Reagan tried and failed to sell NOAA/NWS satellites to private companies in 1983, but that defeat did not discourage the ideologues who wanted to privatize the NWS. In 1990 the NWS began to close its telephone service offering weather information to the public, referring these inquiries to private weather services. Local offices offering fruit, cotton, and other specialized forecasts for growers were shut down. As former director Cressman predicted, there was grower and congressional outrage, but much of the shift to privatization continued. Under Friday, the NWS paid the Contel Corporation of Atlanta to electronically distribute weather information gathered at taxpayer expense. Some of the contracting to private companies was forced on the NWS because of repeated failures in its Advanced Weather Interactive Processing System (AWIPS), which was supposed to link radar, satellite, ASOS, and human observer systems to provide real-time weather information. The Clinton administration did little to improve the situation, waiting until 1997 to reassign Joe Friday.

The NWS remains committed to replacing written forecasts with digital (i.e., numerical) data sent directly to users. If you have looked recently on the Internet at the NWS forecast for your hometown, you have seen a screen filled with information: a five-day forecast of day and night temperatures in graphic form, a "detailed" seven-day forecast in written form, a "hazardous weather outlook" (often with the information that "no hazardous weather is expected"), and links to a plethora of digital forecast databases that provide radar and satellite images, maps with temperature, precipitation, wind, and sky cover information, and hundreds of bits of information. While the IFPS helps to standardize local forecasts and provide more forecast information to commercial meteorologists, broadcasters, and risk managers, there is a potential for confusion. These users—AccuWeather, The Weather Channel, and others—tweak the data in various ways, create their own forecasts, and use a wide variety of terms and graphics to communicate with their audiences. Many meteorologists, including Bob Ryan, chief meteorologist for WRC-TV in Washington, D.C., have expressed concern that too much faith is being placed in the digital format, which is too deterministic. "Put simply," Ryan writes, "the real atmosphere is not a digital system. . . . The unavoidable uncertainties that we know are inherent in meteorological systems will forever tie us to problematic/probabilistic approaches to the forecasting process."[69]

Opinions about the modernization effort range from rating it moderately successful to condemning it as a complete disaster. Adding to the uncertainty, NWS directors have come and gone quickly since Joe Friday. David L. Johnson, who served as director from January 2004 to June 2007, was, like many NOAA administrators, a retired air force officer with no meteorological background. On June 12, 2007, Mary M. Glackin, a career Weather Service employee with meteorological training, was named acting assistant administrator for weather services and director of the National Weather Service. Whether she will be replaced by a political appointee or become the first woman to head the NWS remains to be seen.[70]

The *crux meteorologium* in the twenty-first century, it appears, is that weathermen and weatherwomen must live in two cultures, science and the marketplace. Representatives of the former goad the NWS to expand research and, incidentally, provide useful information upon which consumers may make informed decisions. Agents of the latter are content to use the information provided by government scientists but keep them out of marketing. Roger A. Pielke Jr. and William H. Hooke have cogently argued that the NWS has a clear mission to protect lives and property and that the accuracy of its forecasts is quantifiable, but little effort has been made to link the two. Is the investment in improving weather and climate information worth the cost? Is there a point of diminishing returns?[71]

Their answer to both questions is that the NWS needs to justify its requests for future research with evidence of societal benefits. The generally shared assumption, following an NRC estimate in 1998, is that in an economy in which $1 trillion of the total $7 trillion is directly weather-sensitive, benefits will demonstrably outweigh cost. A trillion bucks is more than enough to plant the seeds of avarice in both the scientific and commercial communities. The real concern of those in the scientific community is not that the NWS has failed to market itself and its products properly but that it has neglected its core scientific mission. In an essay in the May 2006 issue of the *Bulletin of the American Meteorological Society,* Clifford Mass, of the Department of Atmospheric Sciences of the University of Washington, expounds on "The Uncoordinated Giant: Why U.S. Weather Research and Prediction Are Not Achieving Their Potential."[72] Significantly, Mass, an organizer of the Northwest Modeling Consortium, a federal-state-local partnership of ten agencies providing real-time data on weather, hydrology, and air pollution, does not single out the NWS but refers to something he calls "the U.S. meteorological community" and "the large American weather prediction enterprise," composed of public, private, and academic atmospheric scientists and science administrators.[73]

The recent rapid growth of this meteorological community is one cause of the problems. According to Mass, "The private sector has grown most rapidly, now standing as an equal to government and academic sectors in its number of members, resources, and range of activities."[74] This is an overstatement. Assuming that most representatives of the "community/enterprise" Mass refers to belong to the

AMS, which has roughly 11,000 members, many of the 3,700 private-sector meteorologists work as independent consultants or as forecasters for various energy, transportation, and recreational businesses. Only a few of these consulting meteorologists have access to the kinds of facilities available to government meteorologists, or do the kind of sophisticated research done in the federal and academic sectors. But Mass's point about duplication of effort in the weather prediction community is valid. Moreover, the U.S. forecasters have become a smaller community in the global context. Consequently, Mass asserts, "The U.S. has lost leadership in global weather prediction." Historically, of course, the U.S. only gained leadership after World War II with the help of émigré scientists, by which time the meteorological enterprise was so complex that no single nation was likely to "lead" in all aspects of data collection, analysis, modeling, and dissemination of information.

Mass's focus is on the past fifteen years, specifically the NRC "modernization" plan, which, as I have suggested earlier, was vague enough that its "road map" could lead just about anywhere. The slogans that emerged from NOAA's public relations office in the late 1990s—"weatherproofing the nation" and "no surprise weather service"—express a profound misunderstanding of the atmospheric environment, an arrogance born of big science, and the intentionally meaningless clichés of salesmanship.

The public sees little of these internal conflicts, of course, and the image of the weather community remains a curious mix of wizards playing with vast assays of numbers on their computers and on-air weatherpersons in front of maps with kaleidoscopic patterns of colorful radar images. The weather-engaged, to use The Weather Channel's term, about 6 million Americans by its estimate, know enough to enjoy the infighting, choose sides, and log on to Internet sites to express their opinions.

"Enter: The Consulting Meteorologist"

Nineteen thirty-four is the year frequently mentioned as the beginning of private-sector meteorology, but the pages of the *Bulletin of the American Meteorological Society* chart an earlier origin. In the fifth issue, from May 1920, Charles F. Brooks, the twenty-nine-year-old director of Harvard's Blue Hill Meteorological Observatory and secretary of the nascent organization, surveyed the future prospects for the field under the title "Enter: The Consulting Meteorologist." Brooks quotes a speech given at the Association of American Geographers meeting about the value to business of having such meteorological information as the monsoon season in India. "But could a consulting *meteorologist* make a living, that is, would there be enough problems to keep him busy?" Brooks asks. "One answer," he writes, "is that there are now at least three successful consulting meteorologists in the United States who are thriving in spite of the fact that each of the 200-odd regular Weather Bureau stations is a free information bureau."[75]

Brooks continues by citing examples from several businesses requiring meteorological data—a power company in Utah needing rainfall records, a crop insurance company, a midwestern distributing firm using climate statistics to predict the sales of dry- and wet-weather goods, and an "airship corporation" currently searching for a consulting meteorologist. Brooks ended his piece on a triumphant note: "'Exit: The Fakir,' for with the rising appreciation of sound meteorological knowledge the field for the professional rainmakers . . . and for weather forecasters, 'by rule of thumb,' will become narrower. However, just as astrologers still hang on in spite of the present advanced stage reached by astronomy, it will be but natural for a peripheral cloud of weather-makers and storm-proclaimers to hover about the meteorological world."[76]

This is the rhetoric of a man trying to convince himself that meteorology and mammon can coexist. Weather Bureau meteorologists fought to distinguish themselves from the charlatans who promised rain for money. With that legacy, even honest consulting meteorologists were tainted by the role they would play vis-à-vis the *selfless* academics and public servants. Twenty-seven years later, business analyst Lawrence Drake told AMS members why he thought many meteorologists were reluctant to enter private industry:

> The meteorologist's training leads him to look at the world through the exigencies of purely scientific concepts and methods. When he says research he means in a meteorologist's heaven, a world of pure meteorological phenomena. A factory, a department store, a chain store system, a transportation system, with all the many problems they have involving meteorological elements, simply do not fit into the picture. . . .
> [Meteorologists] don't want to set up a meteorologist's booth alongside that of the palmist, the spiritualist, the numerologist, and the horoscope caster—a few of the established "long-range forecasters" some businessmen patronize.[77]

Less than a year after Brooks's welcome to the consulting meteorologist, Eugene Van Cleef, a geographer with Rand McNally, presented a paper at the AMS meeting titled "Weather, Climate and Advertising." Van Cleef had received his bachelor's degree in climatology from the University of Chicago when that science was primarily part of the discipline of geography. He worked for many private businesses both before and after becoming a professor of geography at Ohio State University. "The purpose of advertising," Van Cleef begins, "is to encourage the consumer to buy. The consumer, being only human, is temperamental, and as such responds sensitively to atmospheric changes." He goes on to give examples of the use of weather information by retailers, jobbers, manufacturers, and municipal agencies. "It is our policy to study the weather forecasts quite closely," says one Chicago department store advertising manager quoted by Van Cleef. "Thus in scheduling the advertising that is to appear the following day, the latest weather forecast determines to quite an extent our course of action."[78] Sixty years later the creators of The Weather Channel could not put it any better. Jobbers, distributing goods across the nation, need to think of continental weather patterns, manufacturers need to know global climates, and, Van Cleef argues, cities need to attract workers and tourists through proper promotion of their beneficial climates. The potential market for consulting meteorologists seems endless.

California chambers of commerce had been marketing weather and climate benefits for years, as Ford Carpenter's book on San Diego illustrates, but even in Southern California wind and rain could be capricious. In the early 1920s, William Randolph Hearst, building his mansion at San Simeon, hired private consultants to install a number of meteorological stations on his 157,000-acre estate to determine the appropriate plantings for his gardens.[79] At about the same time, Irving Krick, the colorful and controversial Caltech meteorologist, began selling his weather predictions to Hollywood studios. When Pacific Gas and Electric employed meteorologist Charles Pennypacker Smith in 1934, the profession became more visible. As David B. Spiegler's useful outline on the growth of private-sector meteorology shows, jobs for meteorologists outside the government and universities increased with the expansion of air transportation, experiments in weather modification, computerized weather modeling, new federal and state laws requiring closer monitoring of the environment, and the popularity of television weather forecasting, to which we can now add Internet weather information systems.[80]

The immediate post–World War II period saw rapid expansion of the private weather sector, the beginning of professionalization, and attempts to gain recognition from the public. The AMS, recently reorganized and reenergized by Rossby's presidency and Kenneth Spengler's appointment as full-time executive director, organized the Committee on Industrial Meteorology and Climatology in 1947, with Reichelderfer, Drake, and others, that came up with a "six-point program," calling for the Weather Bureau "to advise all field offices that industrial meteorology was a legitimate field of endeavor," and that issuing special forecasts to commercial users should be considered the province of consulting meteorology. Unfortunately for federal–private sector relations, the directive issued was equivocal. A major problem was the Weather Bureau's restriction of access to its Teletype and facsimile networks that carried the daily atmospheric data, a situation not remedied until after 1953 and not fully implemented until the Internet replaced the Teletype in the 1990s.[81]

In 1948 the National Association of Industrial Meteorologists was organized. The association convinced the AMS to form the Special Committee on Industrial,

Business, and Agricultural Meteorology with three members from the private sector, one from the Weather Bureau, and one, Van Cleef, from academia. In June of that year, the *Bulletin of the American Meteorological Society* began a professional directory in its back pages in which each company or individual consulting meteorologist was allowed a space about the size of a business card. Among the first six participants were A. H. Glenn and Associates of New Orleans and Washington, D.C.; Murray and Trettel of Chicago; Northeast Weather Service and Weather Advisors of Boston; Weathercasts of America of St. Louis; and Irving P. Krick of Pasadena, California.[82] In 2006 the *Bulletin* regularly carried between fifty and sixty entries, including A. H. Glenn, and Murray and Trettel, celebrating sixty years of weather consulting.

The climate for private meteorology was improving. Dwight Eisenhower's election in 1952 returned a pro-business Republican Party to office after twenty years of government activism. One of the first acts of his undersecretary of commerce for transportation was to appoint the Advisory Committee on Weather Services, chaired by Joseph J. George, former air force brigadier general and superintendent of meteorology for Eastern Air Lines. Five other committee members, including Charles Pennypacker Smith, came from the private sector. Spengler represented the AMS, and Athelstan Spilhaus, then at the University of Minnesota, academia. By the end of the year this committee turned in a fifty-nine-page report, "Weather Is the Nation's Business," a somewhat ambiguous title depending on which word is stressed.[83] There was no ambiguity in the committee's sweeping recommendations, which included rewriting the basic law establishing the Weather Bureau, decentralizing its organization, replacing its aging staff with better-educated meteorologists, placing greater emphasis on basic research, canceling the present Teletype restrictions, and actively encouraging private meteorology. Chief Reichelderfer dutifully issued a circular letter to all stations ordering the implementation of some of these recommendations, specifically those dealing with the encouragement of private weather services. Four years after the report, the AMS established criteria for certifying consulting meteorologists and a "seal of approval" for radio and television meteorologists.

The AMS strove to define the consulting meteorologist as more than a mere technician. "Fundamentally, the consulting meteorologist is a practicing professional meteorologist whose practice is founded upon an understanding of the existing knowledge of the atmosphere and its behavior and upon the abilities he has acquired in applying this understanding to the vital affairs of man." Qualifications for certification included (1) knowledge, established by a college degree, or fifteen years' experience in the field, and an examination administered by a board; (2) experience, at least five years in the profession, although advanced degrees could shorten this requirement by two years; and (3) character, based on recommendations and investigations by the board.[84]

One of the most ambitious private-sector meteorological projects was the Travelers Weather Research Center, which became the Travelers Research Center (TRC) in Hartford, Connecticut, in 1960. TRC began, in 1955, as a branch of the venerable Travelers Insurance Corporation, which had suffered significant losses from claims made after hurricanes and tornadoes in 1954 and 1955. The insurance company also owned a radio and a television station and wanted to provide round-the-clock weather forecasts. Meteorologists Thomas F. Malone and Robert M. White organized TRC with a staff of more than 40, which grew to more than 120 by the end of the first year. Researchers worked on aviation weather forecasting, hurricane prediction, and photomicroscopy techniques for observing and measuring near-wall fluid-flow phenomena, and, with a grant from the Weather Bureau, prepared detailed studies of the economic impact of weather on the construction industry. When Travelers Insurance came under new management in the 1970s, TRC was broken up into several independent companies.[85]

When weather conditions play a part in crimes or accidents, meteorologists have been called to testify in court as expert witnesses since the beginning of the twentieth century.[86] Growing litigiousness, expanding automobile ownership, a rise in urban crime, and a new NWS policy prohibiting its employees from offering testimony except in cases involving the government opened a fertile new field for private forensic meteorologists. In his 1996 review of the history of private-sector meteorology, David Spiegler mentions Walter F. Zeltman's pioneering work. In the 1960s his firm, International Weather Corporation, averaged 300 cases a year. Another indication of the growing importance of private meteorology was the appearance in the AMS *Bulletin* of advice to forensic meteorologists on ethical and practical problems in 1961. "Meteorology and the law have much in common," writes Fowler Spencer Duckworth, a consulting meteorologist in Menlo Park, California, "in that neither is an exact science and both deal with disconcerting variables and their ever-altering relations to each other." One of these variables, Duckworth argues, is the constant change of temperature, cloud cover, and precipitation within a small area. Records from airport stations may not reflect the conditions in the city or its suburbs. He provides thirteen specific examples of meteorological issues in legal cases, from a driver blinded by the sun, which required establishing the thickness of the cloud cover and the position of the sun, to a case in which a controlled burn became a forest fire, which involved determining changes in wind direction and humidity. Duckworth advised his colleagues called to give expert testimony to (1) decide the pertinent facts, (2) obtain supporting documents, and (3) investigate in the field. When testifying, establish a strategy suitable to the type of court, establish qualifications, and use visual aids. He concludes with the interesting observation that presumptions of innocence coupled with the fact that meteorologists deal in probabilities make it more difficult for meteorological evidence to help convict a defendant than to acquit him.[87]

A high-profile case involving forensic meteorology occurred in the early 1960s when the owner of a horse that Ethel Kennedy had adopted sued her after the horse died in her care. Mrs. Kennedy claimed that she had found the horse abused and starving. In the trial, the horse's owner submitted a photograph showing the healthy horse grazing in a field shortly before Mrs. Kennedy's intervention. Mrs. Kennedy's attorney hired consulting meteorologist Leo Alpert, who convinced the court that the photo could only have been taken several months earlier when the angle of the sun would produce that particular shadow. Mrs. Kennedy was exonerated, and forensic meteorology received great publicity.[88]

The AMS sponsored a conference on forensic meteorology in November 1976 and has held short courses on its principles, practices, and procedures at annual meetings, but serious issues remain unaddressed. As Duckworth noted almost fifty years ago, law and meteorology operate in a world in which the variables are in constant flux. Reconstructing past meteorological conditions to prove, or even to suggest, that weather played a part in an accident, the committing or witnessing of a crime, or the collapse of a building seems iffy at best, and an epistemic fallacy at worst—the belief that if we just have enough data we will know the truth.[89]

In 1968, private meteorologists organized a new trade association, the National Council of Industrial Meteorologists (NCIM), to educate the public about the work of consulting meteorologists and to raise the standards of the profession. But not everyone saw clear skies.[90] Revised statements on public-private cooperation were issued by the NWS in 1971 and 1991, and by the AMS in 1976, but they did little to stem the tide of acrimony between some private firms and the AMS and NWS. AccuWeather's founder and president, Joel Myers, was especially unhappy with what he felt were unfair limitations on private weather forecasters imposed by the AMS certification process.

In January 1984, the AMS held a conference on hurricane and tropical meteorology with participants from public, private, and academic institutions. The questions raised at this meeting remain timely, especially in light of the conflicting forecasts issued by private forecasters and the NWS to government officials concerning Hurricane Katrina: "Can too much hurricane information be disseminated? Can or should we allow different and/or conflicting types of hurricane information to be disseminated to the public? Is the issue of conflicting, multiple sources of hurricane information a real or imagined problem? Who should be held legally accountable?"[91]

There was general agreement among the panelists that conflicting warnings might delay response, but the committee did not achieve consensus on how to avoid disagreement when a forecaster genuinely believes that a storm is going to do something different from what the NWS is predicting and feels an obligation to say so. Cooperation and discussion of differences were the only recommendations. Knowingly false weather forecasts had been made illegal by an act of Congress in 1948. Neil Frank, director of the National Hurricane Center (NHC), saw hurricane warnings as primarily a people problem resulting from the overdevelopment of beaches and barrier islands. Noting that recent evacuation studies found that it would take twenty-six hours to evacuate the Galveston area and thirty hours to evacuate the Florida Keys, he concluded: "We don't even know how long it's going to take to evacuate New Orleans, Norfolk, and the coastal areas of New Jersey, where hundreds of thousands of people are located on barrier islands during the summer."[92] Both John R. Hope, who had recently left the NHC to become a forecaster for The Weather Channel, and Robert H. Simpson, a consulting meteorologist who was former director of the NHC, were conciliatory, seeing roles for both public and private hurricane forecasts. Joel Myers was combative, claiming that AccuWeather's forecasts during Tropical Storm Agnes, June 15–25, 1972, had prepared people in Pennsylvania for severe flooding before the official warnings from the NWS and went on to list a number of NHC forecast failures.[93]

AccuWeather, The Weather Channel, WeatherBug, The Weather Underground, Inc., and similar private weather forecasting businesses are exploiting the amenity business, providing a kind of security and comfort through the flow of weather information. Joe Bastardi, an AccuWeather meteorologist since 1978, put it this way: "Americans are interested in the weather because they believe in personal responsibility. I mowed my grass yesterday because I knew it would rain today and the grass will be too long before it is dry enough to mow again." The day we spoke he was angry about a generally favorable article on the AccuWeather staff in *Newsweek,* because the author expressed doubt that anyone could make reliable forecasts more than a week in advance. The thousand-word article ran in the "Science" section of the newsmagazine, but its focus was on the weather company's more colorful personalities, meteorologists Elliott Abrams, Bastardi, and its founder, Joel Myers.[94]

Created in 1962, while Myers was beginning his Ph.D. program in meteorology at Penn State, AccuWeather has succeeded in establishing itself as the leader in private weather forecasting and other weather-related services. Its headquarters is a handsome, multistory glass box off a two-lane road about three miles from Penn State University in State College, Pennsylvania. A row of trees, a patch of lawn, a picnic table, and a parking lot give it a minicampus atmosphere, dwarfed by two large satellite dishes. Inside, a cavernous room larger than a basketball court is divided into dozens of cubicles, each with multiple computer screens. AccuWeather has more than 100 meteorologists working in three shifts, twenty-four hours a day. The ambience is subdued, far quieter than a newsroom; cubicles are decorated with posters of cloud types, historic storm track maps, and Fujita tornado scale charts. In the hallways outside the offices are posters promoting AccuWeather products such as pagers and cell phones. A 2002 brochure proclaims, "Weather Stops Traffic at Your Site . . . And Parks It There." More than 1,200 Internet sites subscribe to some version of AccuWeather's customized national and local forecasts; special services such as iSight link a site to video cameras in several cities and vacation spots, allowing subscribers to get real-time pictures of the place and zoom in on details. My guide, Jesse Ferrell, founder of Weathermatrix, now a part of AccuWeather, tells me that people like to perform in front of the cameras and that a couple in State College got married in front of the camera. When I look at these images on my computer screen, they are clear enough to indicate whether it is raining or sunny, but the amount of weather information hardly exceeds the "old Indian weather rock." Joe Bastardi caught the irony when he told me, "This country was built on confronting reality, now it's about perception."[95]

AccuWeather's real competition is not so much the NWS—although every time the NWS lengthens its long-range forecasts or offers pinpoint forecasts for areas as small as 2.5 square miles, it threatens the private sector's niche—as other private weather companies such as The Weather Channel, WeatherBug, and Weather Communication Group.

The Weather Channel, which went on the air on May 2, 1982, financed by Frank Batten, chairman and CEO of Landmark Communications, a conglomerate of newspapers, magazines, and television stations, was the brainchild of John Coleman, a Chicago television weatherman who moved in 1977 to ABC's *Good Morning America*. The Weather Channel became the dominant disseminator of weather information before the Internet began to compete in the 1990s. The past decade has seen increased and intense competition among media weather information providers. AccuWeather claimed to have 60 million different people accessing its products on 1,200 Web sites every month in 2002, while The Weather Channel asserted that it had more than 15 million daily viewers in 2000, and WeatherBug, the new kid on the block, recorded in 2006 that 80 million households use its data monthly.[96]

Founded in 1992 as AWS Convergence Technologies, WeatherBug is located in Germantown, Maryland, and now alleges that it is the "world's largest proprietary weather network with 8,000 WeatherBug Tracking Stations and more than 1,000 cameras strategically placed at schools, public safety facilities and television stations throughout the U.S." AWS began by selling specialized forecasts to television stations and developing science curricula for schools. This seeming sideline paid off as local stations realized the benefits of good public relations through educational links to their market communities. By 1998, AWS had weather stations at more than 3,500 schools nationwide. WeatherBug did not appear on the Internet until 2000 and was still losing money in 2003, but it had grown to more than 200 employees by 2004. In 2006 it joined with Send Word Now, an emergency notification service, to create Smart Notification Weather Service—a "system that generates, authorizes and delivers severe weather alerts to subscribers" when hurricanes, lightning, floods, wind gusts, and tornadoes are detected in an area as small as 3.2 square miles. The system allows user response by voice and e-mail.

One of Smart Notification Weather Service's first customers was Wal-Mart, whose director of emergency management endorsed the system as the best way to protect Wal-Mart's 3,800 stores and "mitigate impact to our bottom line."[97]

The third sector of the atmospheric sciences community, academics, also engages in the marketplace. The Weather Communications Group (WCG) in the Department of Meteorology at Penn State University is one example. In concert with the threefold mission of a land-grant university, the WCG (1) teaches graduate and undergraduate synoptic (simultaneous data from a wide geographic area) meteorology classes; (2) houses the State Climate Office that provides information to state agencies and businesses; and (3) produces a weeknight fifteen-minute weather program seen on the Pennsylvania Cable Network, daily weather broadcasts for Penn State's public television station, and weather pages for the *New York Times* and the *International Herald Tribune,* in addition to sponsoring summer weather camps for teachers and students.

Many universities focus on research and limit the hours a faculty member or researcher may devote to private consulting. However, the Bayh-Dole Act of 1980, the Stevenson-Wydler Technology Innovation Act of the same year, and their subsequent amendments allow universities to patent inventions made under federally funded research in return for the commercialization of any useful products. A good example of technology transfer from the academic to the private sector is the lightning detection system developed at the University of Arizona, which led to the establishment in 1989 of the National Lightning Detection Network (NLDN). This was merged with Global Atmospherics, Inc., into a singe private enterprise, which was then sold to a Finnish company, Vaisala, a manufacturer of precision meteorological instruments. The NWS now buys all its lightning data from Vaisala.[98]

The present state of academic- and private-sector involvement is well summarized in a 220-page National Research Council report, *Fair Weather: Effective Partnerships in Weather and Climate Services. Fair Weather* (its title less ambiguous than "Weather Is the Nation's Business") was the product of several years' work by a committee of thirteen, six of whom came from universities, four from the private sector, and one each from UCAR and the AMS. Chaired by John A. Armstrong, a retired executive from IBM, the committee was charged with examining the present state of public, private, and academic weather services and making recommendations to improve cooperation. Strikingly, there was no representative from the NWS, although many of its administrators and employees were consulted and acknowledged. Apparently, neither the National Weather Service Employees Organization (NWSEO), which represents some 4,000 NOAA employees, nor the National Council of Industrial Meteorologists, the professional society for private-sector meteorologists, nor the cooperative observers, nor the National Weather Association was consulted, although NWSEO is mentioned in an appendix. Remarkably, the NRC report contains no reference to public weather literacy, to which consulting meteorologists such as Mike Mogil and Keith Heidorn made substantial contributions.[99] It seems that some neighborhoods of the U.S. weather community fall outside the community's city limits.

Fair Weather's appendixes provide a brief overview of the major weather observation, processing, collecting, and dissemination systems currently operating in the United States; allow several members of the private sector a venue for their complaints about the NWS; and permit Roger Pielke Jr. to refine his ideas on weather policy. As Pilke summarizes:

> The question to be addressed, then, is not whether the boundary between research and commerce should blur—it has and it will. Indeed, the United States has a long history of using policy to intentionally blur this boundary, using technology policies to stimulate economic growth via public supported research, development, and technology transfer. The question facing the atmospheric sciences instead is what policies and procedures to promulgate and implant given present trends in the discipline. Since the mid-1980s, several disciplines, the medical profession being the most prominent, have been engaged in discussion and debate about conflict-of-interest policies and procedures. The atmospheric sciences have much to learn from these debates.[100]

Private meteorology in the United States exhibits, as Pielke suggests, most of the characteristics of the capitalist ethic as it developed in this country. The tensions among public, private, and academic meteorology also reveal traditional patterns of American culture. Americans have generally been skeptical of the value of knowledge if it lacks practical application. Paradoxically, weather information became more valuable as it became more theoretical. The shadow that science casts between investment and payoff often causes investors to be anxious. There may be money to be made from weather knowledge, but to get it you have to go through those damned scientists, who seem to change their minds about how the weather works every few months. Perhaps there are simpler ways to make money from the weather, as the Institute for Meteorology at the Free University of Berlin discovered in 2002 when it began selling naming rights to high (299 EUR) and low (199 EUR) pressure systems crossing central Europe. First come, first served, and first name only. Funds collected support the student-run weather observatory. A more American approach is betting on the number and severity of hurricanes and tornadoes in a season—check with the experts, but bet your hunch.[101]

Informed guesses are essentially what weather risk management, the latest fashion in weather marketing, is based on. Weather insurance for protection from damage by hail, tornadoes, wind, drought, and rain has been available in the United States since the 1880s. Sketchy as they were, Weather Bureau records provided insurance companies with a rough idea of the frequency and extent of various kinds of damaging weather. This early climate data plus the insurance companies' own records of storm losses and fatalities set the premiums. The American Meteorological Society began paying attention to weather insurance soon after its founding, noting in one instance that policies against loss from rain often did not pay off if the amount was less than 0.1 inch, or if the rain failed to fall on a specified gauge even when it was raining close by. By 1925, insurance companies were paying for hourly weather data, and in 1938 Congress created the Federal Crop Insurance Corporation, which since that time has expanded annually. In 1997, two new tools of weather risk management were invented—weather derivatives and cat (catastrophe) bonds—both essentially forms of gambling on the weather.[102]

No reader who lived through the 1990s will be surprised that weather derivatives were developed by the Enron Corporation, the Houston-based energy company that crashed spectacularly in 2001. "Derivatives," the economist Peter L. Bernstein observes, "are financial instruments that have no value of their own." Moreover, derivatives only have value in a volatile environment. Their value derives from some other asset, such as snow at a ski resort or sunny skies for an amusement park. So the ski resort buys derivatives that cover its losses if it doesn't snow. If it does snow, the profits presumably pay for the cost of the derivatives. "Enron," James Surowiecki of the *New Yorker* points out, "is not an insurance company; it prefers not to carry all the risks of the policies it sells. It's more like a bookie"—a bookie who believes, with Einstein, that God doesn't play with dice. Cat bond traders don't even bother to match up a heating company that wants to protect itself from a warm winter with a power company that wants to buy electricity in anticipation of a hot summer demand for air-conditioning, as the derivative market does. Cat bonds are a form of reinsurance that spreads huge losses among bond holders who base their prediction for losses from hurricanes, earthquakes, and other potentially devastating events on long-range forecasts and the value of the property at risk.[103]

Enter NOAA's National Weather Service. In an act of faith equal to that of the most devout Christian, the derivative traders and the cat bond investors accept weather and climate data compiled by satellites and cooperative observers, combine them with economic data, and make models of weather behavior that they bet their billions on. Convinced, like Poincaré, that "chance is only the measure of our ignorance," weather risk managers and reinsurers are working with NOAA to develop better models. Their faith in the science of risk management, as Bernstein notes, leads them to take risks they might not otherwise take. They believe that risk stimulates innovation and that every possible outcome can be predicted and insured. They are thinking like economists, not meteorologists.

A Match Made in the Heavens:
Broadcast Meteorology

Modern meteorology and wireless communication emerged at the beginning of the twentieth century. Media and weather share Earth's aura. The structure of the atmosphere makes wireless transmission, the broadcasting of radio and television, possible. Sounds and pictures are converted into electrical signals and transmitted in various frequencies into the air, where they travel, at the speed of light, until they reach the atmospheric layer 30 to 50 miles above, known as the ionosphere, where they are reflected around the world to receiving antennae, superheterodynes, oscillators, and finally individual radio and television receivers. The message is the medium. Bad weather can still mean bad reception.

The ink was barely dry on Guglielmo Marconi's patent for a machine capable of sending Morse code over land and sea without wires when Secretary of Agriculture James Wilson took note of "etheric space telegraphy" and directed the Weather Bureau to begin experiments with wireless communication. Reginald A. Fessenden, a Canadian who worked for the bureau from 1900 to 1902, was put in charge. His experiments led to improvements in the broadcasting of speech, although unstable atmospheric electrical conditions in summer continued to disrupt transmissions. The most immediate value of wireless communication for the Weather Bureau was with ships at sea, and the bureau and the navy worked closely to incorporate weather information from ships at sea and to broadcast warnings. In his annual report for 1914, Weather Bureau chief Marvin announced, "Through cooperation with the naval radio station of the Navy Department, a daily distribution of wind forecasts and storm warnings is now regularly made from two points—Radio, Va., and Key West, Fla.—for the Gulf of Mexico and the western portion of the Atlantic Ocean. During the year arrangements were completed to have a similar service for the Great Lakes." Forecasts by amateur radio operators at the University of North Dakota also began that year. Other land-grant universities joined in broadcasting the weather.[104]

By 1921, ninety-eight commercial stations in thirty-five states were broadcasting daily weather forecasts and warnings supplied by the U.S. Weather Bureau. Five years later the Boston radio station WEEI hired former Weather Bureau forecaster E. B. Rideout as its first full-time weatherman, and the Department of Agriculture was broadcasting twenty programs a week, including "Aunt Sammy, a New Radio Friend and Neighbor for the 5,000,000 Farm Women of the Nation"; "Autobiographies of Infamous Bugs and Rodents"; and "Chats by the Weather Man."[105] Weather as entertainment emerged alongside commercialization. The first man to successfully combine self-promotion, scientific forecasting, and commercialization was James C. "Jimmie" Fidler (1912–2007), who began his long career in radio and television weathercasting as a geography student at Ball State University in 1932.[106] By 1934 he was working for station WLBC in Muncie, Indiana, and in 1937 he published the first of several articles in the *Bulletin of the American Meteorological Society* describing how to sell the weather over radio.

Fidler's first article makes three salient points that remain valid. First, when people get their weather information from the radio, they begin to call the station for the temperature because they seem "to think that the radio station is the weather station." Second, when the audience became hooked on weather, "the station found itself with a fully weather-conscious public wanting more about weather and road conditions." And "the station realized that the weather could be actually featured by the station separate from news and weather reports." Third, longer pieces on how the weather works are best presented when the weather is good and unchanging; when the weather is changing and stormy, audiences want brief, up-to-the-minutes facts.[107]

Fidler, writing just a month before Orson Welles's Halloween eve broadcast of *War of the Worlds,* which caused nationwide panic, cautions that radio weathermen should be qualified to interpret meteorological data correctly, because "a radio audience is easily misled by erroneous weather reports." Promoted by his station as "radio's original weatherman," Fidler crammed his daily ten-minute broadcasts with information about temperatures and barometric pressures across the country. "Reading" the weather map over the air, Fidler had to count on his listeners' visualization of the forty-eight states. He supplemented Weather Bureau data with telephone information from truck and bus drivers and county and state police officers. In short, he developed his own community of storm spotters and neighborhood weather observers.[108]

Television did not create weather clowns, but it certainly encouraged them. In his brief but valuable history of radio and television weather broadcasts, Robert Henson asks, "Why does the very serious topic of weather continue to be treated with such frivolity?" He offers three possible answers. First, the public may not view meteorology as a science. As amply demonstrated in the first chapter, Weather Bureau employees have struggled to establish their authority over meteorology and then demonstrate that meteorology is indeed a science. Second, Henson sees a need to lighten a program filled with depressing news. Sometimes, of course, the news of violent storms is as depressing as the political, economic, and crime reports. When that happens, weather ceases to be a matter of atmospheric science and becomes "human interest." Finally, Henson wonders, will weathercasts emerge from the cloud of triviality and "grow into a broader forum on the environment as a whole?"[109] I see little evidence that this has happened in the seventeen years since Henson wrote. Although The Weather Channel, the Discovery Network, and The History Channel have produced some excellent documentaries on the larger environmental aspects of the weather, the cloud remains. The Weather Channel's hour-long magazine format show *Atmospheres,* introduced in August 2000, was replaced by the more dramatic *Storm Stories* in 2002.[110] Puppets, cartoons, clowns, and "weather girls" in bikinis no longer bring us the weather, but the rain-soaked, Gortex-clad weatherman standing beneath a wind-whipped streetlight on a beachfront in the "eye of the storm" continues their antic spirit.

In Henson's telling, the story of broadcast weather reporting reflects changes in technology and national priorities in the second half of the twentieth century. Publicly funded research and forecasts were made available to commercial broadcasters. One of the values added by radio and television is entertainment. As in the earlier era when the Weather Bureau had to sell newspapers and the public on the value of weather reports and predictions, TV stations had to capture audiences and find advertisers willing to sponsor their shows. A measure of their success is the current belief, backed by various polls, that weather is the number one reason for watching the news. Little wonder that television stations and networks, having worked hard to develop formats that make the weather important and appealing, have a proprietary attitude toward it. At one point in his discussion, Henson asks, "Whose forecast is it?" To what extent should the media acknowledge the sources of their forecasts? Television weathercasters, Henson finds, are less likely than radio weathercasters to give any credit to the Weather Service, or even the private firms they may hire to supplement their local forecasts. For Henson the possibility of conflicting and therefore confusing forecasts constituted a significant problem. A decade after his observations, a reporter for the *Washington Post* asked the same question and tried to explain why many viewers attribute the weather to the TV forecaster, confusing "the messenger and the message." Because the on-air weathercaster is the face of the invisible team that provides the temperatures, computer maps, and forecasts, he or she is recognized by the public and as such gets compared, often unfavorably, to other forecasters, including "Cindy the Weather Dog," who predicts rain by refusing to go outdoors. A local radio DJ announces Cindy's picks and claims that the dog's success rate is 94 percent. Even when they try to do the weather in a serious way, TV weathercasters end up as figures of fun. The convention, as we have seen, is old and durable. Weather Service employees may actually welcome the TV buffoon as buffer.[111]

In an effort to bring order into the emerging profession of broadcast meteorology, restore some dignity to on-air meteorology, and, not incidentally, protect the monetary value of weather forecasting, the AMS Executive Council voted to establish a committee on radio and television in 1954. Chaired by Francis Davis, an energetic young meteorologist with a master's degree from MIT, on-air experience in both radio and television in Philadelphia, and an academic appointment at Drexel University, the committee members included, among others, Jimmie Fidler, who had gone to work for the Weather Bureau in 1947 to set up a broadcasting unit; Louis Allen, a navy-trained weather observer with a master's degree in meteorology who in 1948 became Washington, D.C.'s first TV weatherman; and John Clinton Youle, a former air force weatherman who became the first nationwide weathercaster on NBC's *Camel News Caravan* in 1949. In May 1955 the AMS council accepted the report of the committee and recommended that it (1) propose criteria for the issuance of a seal of approval to radio and television weathercasters; (2) prepare a pamphlet for stations on the advantages of employing professional meteorologists; (3) compile a list of professional meteorologists interested in broadcasting; (4) call attention to the schools and universities offering instruction in media weather presentation; and (5) suggest policies to the U.S. Weather Bureau regarding the dissemination of weather information through radio and television "weather performers."[112]

The AMS, never an organization to act precipitously, took until 1957 to form a permanent Board on Radio and Television Weathercasting, which issued its first AMS seals of approval in 1960. The requirements for the seal were similar to those for Certified Consulting Meteorologists: a college degree in meteorology or five years' experience in the field. Only sixty-five individuals received the seal in the first decade of the program, leading the AMS in 1972 to lower the standards to three years' experience and the passing of a written examination. This, apparently, was found unsatisfactory, and the AMS restored the original requirements in 1981. The following year, the National Weather Association, which had been founded in 1976 by weathercasters who felt that they were not welcome in the AMS, began its own certification program. Weather Bureau policy on radio and television weathercasting was murky at best. As Robert Henson explains, "The first formal statement of the guidelines drawn up in 1954 appeared in 1970, just before the creation of the National Oceanic and Atmospheric Administration (NOAA, encompassing the National Weather Service). Section 27-13 of the first *NOAA Directives Manual* essentially restated the distinction between regular radio and regular television appearances by NWS employees, encouraging the former but forbidding the latter."[113]

Robert M. White understood this problem when he assumed his duties as chief of the National Weather Service in 1965. Addressing the annual convention of the National Association of Broadcasters on March 23, 1965, White emphasized the importance of gaining the confidence of the listening public by hiring private meteorologists to both prepare local forecasts and perform on-air. He also promised to improve access to weather hazard warnings and regular weather bulletins, and encouraged the creation of the twenty-four-hour weather broadcasts on NOAA Radio. Although they required a special radio tuned to 162.5 MHz, these broadcasts were successful, and NOAA Radio added features in the 1970s, such as "tone alerts" that warned listeners when there was a threat of severe weather in a locality. In 1997 NOAA Radio began using computer-driven voices that can be adjusted to make warnings sound more urgent than daily forecasts, an innovation that made it the object of jokes from many professional meteorologists. A new concatenated voice, introduced in 2003, is thought to be "much richer, more animated, and closer to the Holy Grail of a highly intelligible, natural-sounding voice." Psychologists have studied the impact of delivery and wording on the public's response to weather alerts, and the NWS tries to incorporate their findings.[114]

The 1960s and 1970s were, however, the heyday of what Henson calls "silly weather" and "happy weather." This was also the era of civil unrest, the Vietnam War, and the counterculture. If Willard Scott, who clowned his way through local weather in Washington, D.C., from 1967 to 1980 before becoming a star on NBC's *Today* show for more than twenty years, epitomizes everything that professional meteorologists hate, they should recall that the object of his humor was the media itself rather than weather or meteorologists. Although his delivery of the daily forecast usually followed his stand-up comedy routine and his banter with newscasters, Scott always gave a clear and timely forecast with full credit to the NWS. Some viewers may find Willard Scott, "weather girls," and weather-predicting animals ludicrous, but they function in three significant ways in our commercial society. First, they allow station managers to experiment with different formats for news broadcasts to attract viewers. Second, as suggested in the first chapter, there is a close analogy between weather and play. Both are chaotic and sometimes dangerous. A playful response is appropriate in any weather forecast; both play and weather employ semblance and deception. Finally, weather provides an element of the sublime in human life. We are awed by storms and enchanted by rainbows. "Frivolity and ecstasy are the twin poles between which play moves," wrote the great Dutch historian Johan Huizinga, and they bracket weather as well.[115]

This is not to say that all forms of media weather silliness can be explained as play. I have no doubt that there were male chauvinist station managers among those who conceived the provocative costumes and antics that characterized the performances of the so-called weather girls of the 1950s and 1960s. Henson seems vaguely embarrassed to write about these women, and David Laskin is scornful of their ignorance of meteorology, but the women hired as television weathercasters were simply seeking careers in a male-dominated, prefeminist America. There was no affirmative action and little to no support for women from the AMS at the time. It is revealing that there was not a single woman among the recipients of the first sixty-five AMS seals of approval. Currently, women receive about one-third of the television seals. In 1982, the AMS published a study of 500 women active in meteorology, about 10 percent of those receiving a B.S. or M.S. in the field and 4 to 5 percent of those receiving a Ph.D. About 20 percent were employed in the private sector, but television weathercasting is not mentioned. As late as 1993, CNN's on-air meteorologist Valerie Voss told a writer for *Weatherwise* that, although there are more women meteorologists, they "are still being considered mainly for supporting roles." Pioneer women weathercasters such as Carol Reed and Tedi Thurman in New York and Karna Small in San Francisco were smart, ambitious, and attractive. Thurman was known as "Miss Monitor" on the popular weekend talk radio show *Monitor,* begun in 1955. Thurman read the weather reports in a sexy voice accompanied by romantic music. The radio program ended in 1975, but she continued her role on NBC-TV's *Tonight Show* with Jack Paar. And there was "Bobbie the Weather Girl," who is fondly remembered by GIs who served in Vietnam from 1967 to 1969. Bobbie was a U.S. Agency for International Development secretary in Cholon who became the on-air weathercaster for U.S. Armed Forces TV. She did the weather in a swimsuit and traveled with other entertainers to remote camps on weekends. Most of the viewers in muddy barracks probably didn't care as much about the weather as they did the degree of normality that Bobbie brought them.[116]

Karna Small of KRON-TV in San Francisco earned fame by being able to write backward on a Plexiglas weather map, but that is a small part of the story. Looking for work in television news in 1968, Small had found few opportunities until she was offered the weekend weather slot on the NBC affiliate. She saw it as a way to gain entry, and so it was. She quickly learned how to get the most out of her five-minute time slot, and when the weekday weathercaster left, Small began doing the full seven days at six and eleven o'clock. It was a grueling schedule, but the skills she learned—timing, film editing, script writing—were transferable, and after four years she became a news anchor at another station. She gives credit to the meteorologists of the Weather Bureau, whom she called a few minutes before each broadcast after checking the wire services for regional weather reports. Later she took courses in meteorology at San Francisco State University. "We treated the weather seriously," she recalls, and she helped create weather maps and cloud posters for local schools. "Weather forecasting in the Bay Area isn't too difficult," she admits, "fog in the morning and high in the 60s unless there's a heat wave." Today, she scoffs at the gimmick of writing backward on a glass map, which allowed her to face the camera rather than stand to the side in profile; it was something she quickly learned, she says, but agrees that it caught viewers' attention. Popularity has its costs. Karna Small received letters from pilots offering to "show her the clouds" and from sailors eager to teach her about winds and ocean currents. Most unwelcome were letters from fans who became stalkers and had to be restrained by law.[117]

The full story of the "weather girls" and women weathercasters needs to be told, as does the story of The Weather Channel (TWC), which employs several women as on-air meteorologists. TWC's reshaping of the female weathercaster may seem unremarkable, but it is an important part of its success, as a *Washington Post* book critic noted in a twentieth birthday encomium to the cable giant: "During those 20 years the channel changed significantly. It's not so much that the on-camera 'talent' got a little prettier, though it did, as that the Weather Channel began to get into show business, just like everyone else in television."[118] TWC's story is slowly emerging from a variety of recent sources, primarily Frank Batten's insider's history and John Seabrook's piece in the *New Yorker*. Batten was CEO and president of Landmark Communications when John Coleman, a Texan who began his TV career in the Midwest and went on to become weatherman for ABC-TV's *Good Morning America* in 1977, convinced him that the country was ready for an all-weather, twenty-four-hour cable television channel. In one version of the origin story, Coleman recalls the northwesters and dust storms of his childhood and how he began to dream of a weather program that "would present on camera the great sweep of weather, as though it were theater, the drama, the history, the humor, the folklore."[119]

Coleman invested some of his own money in creating TWC and was made president, but Landmark was the principal investor and Batten was essentially his boss. Within a year of the launching of TWC in 1982, Coleman and Landmark were in court over the terms of Coleman's contract, but according to Batten, Coleman withdrew from the suit and the partnership when he failed to obtain the additional funding he had promised. Batten, who attributes his own fascination with weather to experiencing a hurricane that struck Virginia in 1933 and to his passion for sailing, assumed control and built the staff that continues to run TWC.[120]

TWC found a workable formula, I think, by taking full advantage of satellite and radar images and computer graphics, blending hard meteorological data with user-friendly airport and travel weather reports. Seabrook and D. T. Max make good points when they connect TWC success to a growing awareness of atmospheric environmental phenomena—acid rain, ozone depletion, El Niño, global warming—by Americans in the 1980s. "We watch the Weather Channel for the global read," writes D. T. Max, "the state of the planet, for reassurance that it will be O.K. . . . Where once we needed the weather, now we feel the weather needs us." Max may have picked up on a subtext in Weather Channel performance that springs from the quasi-religious character of the channel's parent company, Landmark Communications. Landmark's Web site features a statement titled "Our Culture," as well as "Landmark Core Characteristics and Behaviors," which emphasizes hard work, "creating a "shared vision," and loyalty to the company, a twenty-first-century version of late nineteenth-century "welfare capitalism." Less than ten years after launching TWC, Landmark acquired Werner Erhard's EST enterprise, created Landmark Education, LLC, and began selling its model of organizational behavior as it was selling the weather. Erhard's brother and sister manage Landmark Education.[121]

Batten's remarkably candid account of the founding of TWC asserts that Landmark's decision to enter the expanding cable television business was a major factor in the decision to gamble on an all-weather channel. It was a gamble, and TWC did not break even until 1986, by which time Landmark had invested $35 million. At this point in his account Batten summarizes:

> What did we have to show for it? We had an organization of about 160 people who had weathered the adversity together and survived. As a result of our budget woes, we knew how to stretch a buck. Meanwhile, we had

created a "mecca of meteorology," a place where bright young weather experts wanted to work. We had a home-grown technology that gave us an enormous leg up in terms of ad sales. We had a product that America had begun to accept. We had made the difficult transition from being the butt of editorial jokes and cartoons to being perceived as a vital public service, and we were beginning to emerge as a strong brand.[122]

Batten's syntax is revealing. His workers "weather" adversity—a test of their faith—and create a "Mecca" of meteorology, a place, perhaps, where weather can be understood for it spiritual lessons.

Acceptance of product (weather information) and brand name (the blue rectangle bordered in white with THE WEATHER CHANNEL in white capital letters) is indeed a remarkable achievement for a company just reaching its twenty-fifth year. Compare the image of the U.S. Weather Bureau in 1915. TWC also succeeded by continuously updating its technological ability to provide local weather information and specialized forecasts. Batten, like the successful weather salesmen before him, personalized the weather, convincing viewers that TWC's weather information was just for them. TWC's Internet site www.weather.com, begun in 1994, has enhanced this perception. "Weather for planning, weather for safety, weather for inspiration, and weather for fun," Batten's message is clear—weather is more than the sum of National Weather Service data. TWC relations with the NWS and academia seem smoother than most government-private-academic weather partnerships. TWC had a good working relationship with Richard Hallgren, NWS chief, and partnerships with NCAR to develop the Digital Intelligent Forecast System (DiCast), essentially a collection of weather and climate data analyzed for forecasts, and Weather Services International (WSI), which developed "packaging" for this information. TWC bought WSI in 2000.[123]

In 2006, TWC began a weekly program, *It Could Happen Tomorrow,* that moves toward weather science fiction, asking "what if" certain events occurred and speculating on the destruction of major American cities. Do such post-9/11 apocalyptic fantasies further blur the distinction between virtual and real, or do they encourage viewers to prepare for their own potential disasters? The answer is yes to both questions, as the public learns to live with its greater awareness of risk. In the never-ending quest for ratings, TWC is experimenting with all popular formats—*Epic Conditions,* showing extreme athletes in difficult weather conditions; *Climate Code,* an ecology-oriented news magazine; and *Abrams & Bettes: Beyond the Forecast,* a chat show on weather oddities with its own Web site to attract younger viewers.[124]

TWC is, of course, only part of the television weather scene today. The heart of weathercasting remains with local stations, with their own networks of voluntary observers, school weather stations, and audience loyalty. Each station's weather team is expected to take NWS forecasts and personalize them by making their own educated guesses. The result is that the attentive viewer may hear four different predictions of temperature highs and lows from four different stations on the same day. Weather has become "a station's brand," and, like TWC, stations spend money to promote and keep their weathercasters. Bigger budgets to promote the station's weather "team" explain, in part, why the tenures of weathercasters are longer in bigger markets.[125]

"There are 211 TV markets in the United States," George Hirschmann, weatherman for WHSV, an ABC affiliate in Harrisonburg, Virginia, since 2000, told me in January 2003. "We're 178th." Hirschmann is a large, friendly man who spent twenty-five years in the entertainment business in Los Angeles before earning a certificate in broadcast meteorology from Mississippi State University's online program. He is working toward getting a seal of approval from the National Weather Association.[126]

Hirschmann shares the problems of many small-market weathercasters. Although he gets some information from AccuWeather, his station does not have its own Doppler radar, and the NWS coverage of his service area in the Shenandoah Valley is spotty. In 2006 WHSV acquired new weather technology from WSI. Hirschmann also depends on local observers from the many towns and villages in eastern West Virginia and western Virginia that lie within the station's 100-mile coverage. He strengthens his local ties by speaking frequently before groups at schools, retirement homes, Scout meetings, and other organizations. He is also developing a meteorology course through the James Madison University Life Learning Center. Hirschmann delivers his forecast, during the 5:30 P.M. news broadcast, in a relaxed but serious manner, pointing out precipitation shown by NWS radar and reading out the day's temperatures in front of a map of the station's viewing area. He asks a "Weather Question of the Day" and gives the answer to the previous day's question, often challenging the news and sports anchors to guess first. To me, he deplored the tendency of some news teams to hype stories about storms because viewers remember the hype and not the forecast. Hirschmann is cheerfully candid about television weather being entertainment, along with sports and news. When asked why people are obsessed by weather, he shrugged and said he had no idea. "Maybe," he mused, "they've been convinced it's important by TV."[127]

Marita Sturken, a professor of communications, examined the media hype of El Niño in its West Coast context and found the commentary on "'the climate event of the century,' the 1997–98 El Niño," a complex blend of apocalyptic fantasies, regional identity, and nostalgia for a simpler past. Television's role, Sturken argues, is to create "the *experience* of control through monitoring the weather— via live broadcasts, satellite images, and endless readouts of scientific data about what the weather is doing." Following other cyberculture critics, Sturken notes the shift that has taken place, since satellite pictures have become available, of viewing clouds and the weather from far above, not from below. She thinks that "weather consumers are invited to place themselves as omniscient viewers." Weather watching, she asserts, has become a civic duty because TWC and other media carry the message that weather is crucial to everyday life. El Niño gave Southern Californians a purpose—potential shared suffering, potential mutual survival.[128] At some point the community of weather consumers becomes a community of weather victims, a topic to which I will return in the final chapter.

Designer Weather

The story of weather modification has been told in many popular articles and recently in a fragmentary way by a few historians.[129] Vincent Schaefer, a General Electric Research Laboratory machinist who, working with Nobel Prize–winning chemist Irving Langmuir on the deicing of aircraft wings, discovered in July 1946 "that small bits of dry-ice particles dusted into a supercooled cloud would quickly and completely change its nature." A short time later Schaefer seeded some clouds over Greylock Peak in the Berkshires with dry ice that resulted in a snowfall. Bernard Vonnegut, another GE scientist, improved the chances of precipitation by using silver iodide. The experiments seemed so simple that they immediately drew the attention of the press and inspired imitators. Schaefer credits several European scientists for advancing the understanding of precipitation, especially German meteorologist Walter Findeisen, who saw a connection between what was later called chaos theory and the possibility of weather modification. In 1938 Findeisen wrote: "The recognition of the fact that quite minute, quantitatively inappreciable elements are the actual cause setting into operation weather phenomena of the highest magnitude gives the certainty that, in time, human science will be enabled to effect an artificial control on the course of meteorological phenomena."[130]

Schaefer, an autodidact and enthusiastic outdoorsman, remained cautious about the potential of weather modification, acknowledging the difficulty of verifying the results of cloud seeding, but his mentor Langmuir, using his Nobel laureate status, shamelessly oversold large-scale weather control. The U.S. military was an easy sell. Environmental warfare has a long history; armies have destroyed their enemies by floods, avalanches, and fire. Military planners also knew that the Soviet Union was experimenting with weather control and, anticipating the later debate over the missile gap, sought to avoid a drizzle gap. In 1947, GE assigned Langmuir and Schaefer to Project Cirrus, a classified project supported by the U.S. Army Signal Corps, the Office of Naval Research, and the U.S. Air Force. Over the next five years, 180 field experiments involving the modification of clouds and a tropical storm were conducted in New York, California, Ohio, the Gulf states, and New Mexico, "The Land of Enchantment," which has impressive summer thunderstorms. I suspect that the high desert near Socorro was an at-

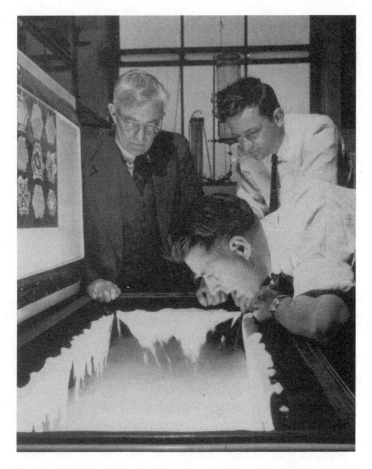

Irving Langmuir, Vincent Schaefer, and Bernard Vonnegut growing artificial snow crystals at the GE lab in the 1940s. Photo courtesy Schenectady Museum Archives.

tractive alternative to Schenectady's weather for the project's scientists. In the beginning, the experiments seemed haphazard. In one case the researchers were using regular, not dry, ice, and the shavings refroze. When the command came to "drop the ice," a 50-pound block hurled down southeast of Albuquerque, narrowly missing a plane flying below the B-17. Without verifiable results and facing increasing public opposition, Project Cirrus was terminated in 1952.[131]

Nevertheless, private cloud seeding took off. The Weather Control Research Association (now the Weather Modification Association) was organized in 1951 by the operators of commercial cloud seeding businesses. Some of these early rainmakers were colorful characters who made good news stories. Wallace E. Howell, educated at Harvard and MIT, was hired by the city of New York in a desperate attempt to end the drought of 1949–1950, which had led to the closing of swimming pools, the banning of car washing, "shaveless Fridays," and widespread prayers for rain. Howell used ground-based generators to seed clouds in the Catskills; rains came, and water restrictions were lifted in February 1951. But no good deed goes unpunished; farmers and resort owners in the Catskills filed suit, claiming the excessive rain had ruined crops and discouraged vacationers. *Slutsky v. City of New York* (97 N.Y.S.2d 238 Sup. Ct. N.Y. City, 1950), the first court case on weather modification, was a win for the cloud seeders, as the court held that the public interest of 10 million New Yorkers outweighed that of borscht belt hoteliers. Humorist E. B. White summarized the controversy in a brief essay in which he wisely observed that the remedy to the water shortage

is not the manufacture of rain but the correct use and distribution of whatever rain naturally arrives on earth. If, as the rainmakers would have it, man does invade the sky and nudge clouds, his flight will, I predict, be but the beginning of such practices, and we shall find the makers of lightning also aloft, to satisfy the desires of manufacturers of lightning rods, who may decide that lightning is in short supply and devise a way of setting more of it loose.[132]

The future of weather and climate modification and geoengineering could be clearly seen in a raindrop.

Howell went on to do cloud seeding in Canada, Cuba, and the Philippines and spent fifteen years with the Bureau of Reclamation's rain and snow augmentation project in the Rockies. He was back in court in 1965, after his cloud seeding efforts to suppress hailstorms led farmers to blame him for causing a drought. In his 1968 decision that clearly addressed the question, "Who owns the clouds?" Judge John MacPhail found

that clouds and the moisture in the clouds, like air and sunshine, are part of space and are common property belonging to everyone who will benefit from what occurs naturally in those clouds. There could be just as much injury or harm from weather modification activities as there could be from air and water pollution activities. We hold specifically that every landowner has a property right in the clouds and the water in them. No individual has the right to determine for himself what his needs are and produce those needs by artificial means to the prejudice and detriment of his neighbors. However, we feel that this cannot be an unqualified right. . . . We feel then that weather modification activities undertaken in the public interest (as opposed to private interests) and under the direction and control of government authority should and must be permitted.[133]

The Pennsylvania legislature complied and created a state weather modification board to approve projects and adjudicate disputes. Several other states have similar boards, but there is as yet no national weather modification policy.

Irving Krick was another weathermaker with a flair for self-promotion. Krick was the son of wealthy and socially prominent Californians. After graduating from the University of California with a degree in physics, he spent a few years as a piano player in Hollywood. His brother-in-law, Horace Byers, of the University of Chicago, advised him to get into meteorology, and he obtained his doctorate at Caltech in 1934, a protégé of Robert Millikan. A vigorous advocate of long-range forecasts based on past climate patterns, he regularly claimed that his forecasting accuracy was superior to that of the Weather Bureau—behavior that earned him the enmity of most of the meteorological community. After leaving his teaching position at Caltech and establishing a weather consulting firm in Denver, Krick was often in court as defendant and as expert witness. Obviously an advocate of weather modification, Krick expresses a distinctly biocentric philosophy in his book *Sun, Sea, and Sky* (1954), where he warns of the "greenhouse effect" and wonders if "Americans of only a few generations hence [will be] forced to abandon to the rising seas the present coast lines, including most of Florida, and the lower Mississippi Valley." At the same time, he notes, the nation is running out of water. Groundwater is our greatest reserve, "but in many places reckless withdrawal, 'water-mining' conjoined to soil-mining, has within a few years almost exhausted a natural supply that it took Nature thousands of years to store in the slow underground percolation from distant mountains, and will take her thousands of years more to renew." That, Krick observes, leaves two sources of future water, the sea and the atmosphere. He considers the sea as one water bank and mentions desalination projects, but the ultimate reservoir is the atmosphere, "the rivers of the sky, particularly the great river of the Westerlies above the United States" in which 70 percent of the moisture evaporates before reaching the ground. The potential rain and snowfall is enormous, Krick argues, but unless we learn to work "with Nature rather than 'conquering' her," we will destroy the planet. "[R]igid conservation rules will help," as will cloud seeding.[134]

In the historical record provided by the AMS and the Weather Bureau, Irving Krick is depicted as a disagreeable crank. Whatever the reasons for his marginalization, it is unfortunate that his vilifiers, many of them also advocates of weather modification, failed to appreciate his rhetoric. Selling cloud seeding as a conservation technique, part of a larger effort to achieve a sustainable economy, might have ultimately met with more success than the "technological intoxication" approach that characterized the promotion of weather modification in the 1950s and 1960s. Wind power promoters have had considerable success using the rhetoric of "green" energy to sell a technology as disruptive to the environment as cloud seeding. But weather modifiers, for reasons made clear by James Fleming, were caught up in the cold war rhetoric of the atomic age. The model for development was the Manhattan Project. Energy in thunderstorms and hurricanes was measured by comparison to the megatons of nuclear weapons. Bombing the clouds from old B-17s to cause a chain reaction, creating a mushroom-shaped cumulonimbus cloud with rain or snow as fallout, was seductive, especially to a young cohort of air force–trained meteorologists eager to fly into the wild blue yonder.[135]

The federal government began taking notice of weather modification as early as 1948. Senator Clinton Anderson of New Mexico introduced a weather control bill in 1951 modeled on the Atomic Energy Act, but it was not enacted. Other bills also failed, but in 1953 Congress passed legislation creating the Advisory Committee on Weather Control to review the scientific data and recommend further action. This committee, chaired by Captain Howard T. Orville, formerly of the U.S. Navy and technical consultant for the Friez Instrument Division of Bendix Aviation, included Langmuir, Vonnegut, and Schaefer. Their report, submitted at the end of 1957, told Congress that in certain conditions cloud seeding was effective in increasing precipitation and that snowfall in the western mountains had been increased by 10 to 15 percent through seeding, and urged further research. In the future, the committee assured the politicians, it will be possible to increase sunlight or cloud cover over large areas, to control water balance in the atmosphere, and to control ocean currents. The sky was not the limit.[136]

Congress responded with the Water, Cloud Modification Research Act, authorizing the National Science Foundation to initiate and support research and evaluation in weather modification. By the beginning of 1959, more than fifty grants and contracts with major universities, government agencies, and private laboratories for research on weather modification were active or pending. The 1960s brought a torrent of ventures with comic book names: Project Climax to enhance snowfall in Colorado; Project Skywater, a Bureau of Reclamation precipitation experiment in Montana; Project Whitetop, an effort to test the claims of commercial rainmakers; Project Hailswrath, an NSF-funded follow-up to the Bureau of Reclamation program; Project Stormfury, an attempt by the Weather Service and the navy to dissipate hurricanes; and, finally, from 1967 to 1972, Operation POPEYE, the military's effort to flood the Ho Chi Minh Trail through rainmaking in Laos and Vietnam. More than $20 million was spent on this project, with no verifiable results.

The history of Project Stormfury, which ran from 1961 to 1983, superficially reads like a case study of Murphy's Law applied to big science. The working hypothesis was that seeding the area around the eye wall of a hurricane would perturb its pressure and ultimately reduce its wind speeds. Only a few hurricanes were suitable, however, and the 1962 hurricane season provided no usable storms. Various technical problems and the lack of cooperation from nature delayed the project from 1963 through 1968, although advances were made during those years in numerical modeling. Finally, Hurricane Debbie, August 18–20, 1969, provided an opportunity, but the results, while encouraging, were inconclusive. Hurricanes of the early 1970s were all too weak, although an attempt was made with Ginger in 1971. A plan to move the project to the eastern Pacific or Australia met with political opposition. From what they learned about hurricanes in general—including the bitter realization that these storms cannot be controlled—the dedicated scientists have more than justified the expenses of Project Stormfury. In addition, the WP-3D aircraft developed in the course of the program helped advance many other atmospheric studies.[137]

Growing opposition to technological fixes in the 1960s brought less favorable attitudes toward weather modification, although the federal budget for various projects continued to rise through the early 1970s, reaching a peak of almost $20 million in fiscal year 1975. An NSF report in 1965 discussed possible adverse ecological consequences of weather modification, and a 1966 report by the National Academy of Sciences placed more emphasis on inadvertent climate change than on weather control. Global warming and its consequences were emerging as a concern, as were various kinds of air pollution. Symptomatic of the changing cultural climate was the establishment in 1968 of the Task Group on the Human Dimensions of the Atmosphere at NCAR with a grant from NSF. Composed of lawyers, economists, political scientists, a sociologist, and a geographer, the group considered a wide variety of problems such as the impact of weather modification on the economy, the environment, the legal framework, and the public, and recommended specific research projects, including "studies to determine how people perceive the weather and the alternative adjustments to it." Some excellent work emerged, such as Howard Taubenfeld's legal studies, Eugene Haas's opinion polling, Leo Weisbecker's work with the Winter Orographic Snowpack Augmentation Program of the Bureau of Reclamation, and Barbara Farhar's study of public attitudes toward weather modification in northern California. Yet the amount spent on these topics was never more than a fraction of that spent on laboratory and seeding experiments, and funding for all weather modification research fell to almost nothing by 1990.[138]

Perhaps popular opinion in the West was expressed by the folksinger Cinnabar Ike:

> Get your gun out, sheriff,
>   Hurry, get your plane.
> Cloud rustlers coming, coming,
>   Come to get my rain!
> My crops are dry as tinder
>   And so am I;
> Hurry, hurry, sheriff,
>   They're up there in the sky!
> Cloud rustlers, cloud rustlers,

Rustlers in the sky.
They're stealing all my rain clouds,
Goodbye, my crops, goodbye![139]

Currently, states, municipal water districts, and utility companies seed clouds from both the air and the ground to enhance precipitation. Despite the doubt expressed in the title of a 1982 article in *Science*, "Cloud Seeding: One Success in 35 Years," some meteorologists continue to beat the drum for rainmaking, and water managers in the arid states of the West hear them. The American Meteorological Society seems increasingly interested in reviving a national effort in weather modification. In 2004, Roland List of the University of Toronto summarized the current status of weather modification: "Fog and stratus dissolution are in the realm of operations. Rain enhancement has been demonstrated to work. Hail suppression has not been supported by randomized experiments. Interfering with tornadoes and hurricanes is hazardous, should it work, and is encumbered by possible legal challenges. It should await better scientific understanding. Reducing torrential rains and dealing with other disaster scenarios has not been attempted yet."[140]

A year later, a panel representing the National Research Council and the Weather Modification Association agreed that "the field of atmospheric science is now in a position to answer many of the crucial questions that have impeded or blocked progress in weather modification in the past." Obviously, the science and technology have improved since the 1970s. Whether these improvements justify federal support is a point of contention. Nowhere in the NRC/WMA discussion are the social, cultural, or environmental impacts mentioned. A bill introduced in 2005 by Republican senator Kay Bailey Hutchinson would create an eleven-member Weather Modification Advisory and Research Board to be appointed by the secretary of commerce. Members would come from the AMS, the American Society of Civil Engineers, the National Academy of Sciences, NCAR, NOAA, a university, and "at least 1 shall be a representative of a State that is currently supporting weather modification projects." The board would "develop and implement a comprehensive and coordinated national weather modification policy and a national cooperative Federal and State program of weather modification research and development."[141] This bill was approved by the Senate Committee on Commerce, Science, and Transportation in November 2005 but did not come to a vote.

Has the pendulum swung back to the 1950s, when technology ruled? The media continue to treat weather modification as they have always treated weathermen, quoting colorful characters who disagree with one another, or highlighting the uncertainties and confusion. "Can Rain Be Bought?" asks the *New York Times,* quoting a cloud physicist at the National Center for Atmospheric Research who feels that cloud seeding, "even if it's wrong, it's like buying a lottery ticket where not much investment might pay off big." A water development manager in Denver thinks it might work, remarking, "It's better than rain dancing. I'm pretty sure about that." Rain dancers are pretty sure that their magic is an improvement over weaving a bezoar into the tail of a black horse, but all these techniques derive from the same refusal to accept limits on human activity imposed by the atmosphere.[142]

Currently, ten western and Great Plains states are paying for cloud seeding programs. Utility companies and water districts seed near their reservoirs. Cloud seeding in California began in 1951 and about the same time in Utah; in Nevada it goes back forty-five years; in Texas, more than forty years. Utah and Nevada each spends more than $400,000 a year on weather control projects, and in 2006 the Wyoming Water Development Commission contracted with NCAR to begin a five-year, $8.8 million program to increase precipitation. In these states and in more than thirty countries around the world, cloud seeding has become a ritual of the changing seasons, as natural as contrails in the sky.

This naturalization of weather modification may have emboldened the new generation of technological teratologists, skillful at blending science fiction and gee-whiz experiments. An example of this is "Weather as a Force Multiplier: Owning the Weather in 2025," a "research paper" presented on June 17, 1996, to the chief of staff of the U.S. Air Force, who requested an examination of "the concepts, capabilities, and technologies the United States will require to remain the dominant air and space force in the future."[143]

"The report," reads the disclaimer, "contains fictional representations of future situations/scenario," yet it moves briskly to an executive summary that claims, "In 2025, US aerospace forces can 'own the weather' by capitalizing on emerging technologies and focusing development of those technologies to war-fighting applications. Such a capability offers the war fighter tools to shape the battlespace in ways never before possible." This is followed by an "Operational Capabilities Matrix" divided into "Degrade Enemy Forces" and "Enhance Friendly Forces." Specific tactics to achieve the first goal include "Flood Lines of Communication," "Deny Fresh Water," "Induce Drought," and "Decrease Comfort/Morale." The plot thickens, so to speak, when the reader is asked to

> imagine that in 2025 the US is fighting a rich, but now consolidated, politically powerful drug cartel in South America. The cartel has purchased hundreds of Russian- and Chinese-built fighters that have successfully thwarted our attempts to attack their production facilities. . . . In addition, the cartel is using the French *system propatoire d'observation de la terre* (SPOT) positioning and tracking imagery systems, which in 2025 are capable of transmitting near-real-time, multispectral imagery with 1 meter resolution.

By dispersing some clouds and seeding others, the air force wins the hypothetical battle, if not the putative war.[144]

This kind of gaming is a common military preparedness exercise, similar to the October 2003 paper by futurists Peter Schwartz and Doug Randall on the national security implications of an abrupt climate change (global warming), which projects armed conflicts throughout the world by 2025 over migrations from floods, scarcity of fresh water, and dwindling energy resources. By contrast, geoengineers, writing in *Climatic Change* in 2006, strain our credulity and, by extension, their credibility, with schemes for global climate control. Led by Ralph J. Cicerone, president of the National Academy of Sciences, and Nobel laureate Paul J. Crutzen, an atmospheric chemist with the Max Planck Institute in Germany, some scientists are proposing technological solutions to global warming, including launching trillions of small lenses into an orbit in space where they would bend some sunlight away from the Earth and placing large white plastic disks in the ocean to reflect solar radiation away from the seas. Other proposals include building ships with tall towers from which seawater would be sprayed into clouds to increase their reflectivity, and sending balloons into the stratosphere to release sulfur dioxide to screen sunlight. While these plans seem better suited to the pages of *Popular Mechanics* than to a journal edited by Stephen Schneider, who was warning the public about catastrophic climate change as early as 1976, they, too, may be signs of another shift in attitudes toward science and technology.[145]

To the historian, all of this sounds terribly familiar. Lucian Boia, in his survey of ideas about climate from antiquity to the present, mentions various schemes for draining the Mediterranean, diverting the Congo River to the Sahara, and warming Siberia by damming the river Ob to create an inland sea. The Soviet Union actually accomplished a fair amount of climate change by opening the Baltic–White Sea Canal in 1933 and the Volga-Don Canal in 1952, and creating the Rybinsk Sea north of Moscow.[146] Emptying the Aral Sea in Kazakhstan and wrecking the Salton Sea in California with ill-considered irrigation projects proved that the Soviets and the Americans were equally adept at destroying their environments through big technology. For sheer audacity, few men can match the now obscure Carroll Livingston Riker and his plans to change the climates of North America and Europe.

Riker was born on Staten Island, New York, in 1854, and trained as a mechanical engineer. As a boy he studied ocean waves and currents, and in 1887 he designed and built the most powerful pumping dredge yet constructed, which was used to fill Potomac Flats in southeast Washington. In 1912 he published *Power and Control of the Gulf Stream: How It Regulates the Climates, Heat and Light of the World.* The model for this Herculean task was, of course, the Panama Canal, then under construction, but Riker's goal was not commerce but climate management:

Climatic changes that would double the land value of the Northern Hemisphere can be effected by debarring the Labrador Current from crossing the Grand Bank of Newfoundland, when the Gulf Stream would flow practically unimpeded to the Pole, melting every vestige of ice thereabout, producing more uniform, warmer winters and cooler summers, and insuring a better climate to every part than is that of New York of to-day. . . . It seems the elimination of this current [the Labrador] would cause the New Jersey shore to build out, to an average of at least fifteen miles, from Sandy Hook to Cape May, and probably further south.

The southerly coast of Long Island would also undoubtedly be extended a number of miles further southward, eliminating the shoals that now exist along both coasts. A valuable piece of property to be produced by the dropping of a comparatively few stones into the ocean nearly one thousand miles away.[147]

Valuable indeed, and not so crazy that Colonel George W. Goethals, builder of the Panama Canal, would not endorse the plan and recommend it to Congress. The Maritime Association of the Port of New York also urged it. The serious journal *Current Opinion* reported the story. Maybe the plan was too audacious, or maybe Long Islanders were happy with their beachfronts, but for whatever reasons the Grand Banks jetty was never funded. Nevertheless, Riker's attention to the influence of ocean currents on climate anticipates recent concern for the threat of localized cooling stemming from changes in the Atlantic currents. Wallace S. Broecker of Columbia University's Lamont-Doherty Earth Observatory has, for more than twenty years, called attention to the possibility that global warming will affect both the temperature and the salt content of the North Atlantic, with the result that northern Europe will become significantly colder. Riker lacked the knowledge that twenty-first-century oceanographers have of the complex relationships between atmosphere and hydrosphere, but he understood that rapid climate change could take place naturally. Why not, in an age of reform and in the spirit of Mark Twain's Colonel Sellers, nudge it along?[148]

As a Progressive Party candidate, Riker ran unsuccessfully for the U.S. Senate in Virginia in 1924, then bounced back with a plan to control the Mississippi River. The 1927 flood had killed hundreds and caused billions of dollars of damage in current dollars. It changed America, according to journalist John Barry, but it did not change the way the Army Corps of Engineers managed the river. General Edgar Jadwin, head of the corps, submitted his plan for more levees, this time with "fuse plugs," which would break or be dynamited if the flood stage of the river threatened the levees. The land behind the fuse plugs would be allowed to fill as temporary reservoirs, keeping most of the land within the levees dry. The Jadwin plan also, as Barry mentions, "vastly expanded federal involvement in local affairs," as would Riker's, since it involved building a spillway, almost straight, from Cairo, Illinois, past Memphis and Vicksburg, through the Atchafalya Valley, to the Gulf many miles west of New Orleans. The spillway essentially canalized the river, with a series of dams along the spillway to control floods and generate power. Riker failed to sell his plan to Congress in 1914, but the corps' failures in the 1927 flood gave him a chance to revive it as revenge. Was the plan feasible? Probably not, but in his attack, Riker challenged Jadwin to think about what a river is. A river, Riker argued, "is simply like a gutter from the eaves to carry off the water." His analogy was crude, but it made explicit what the corps had hidden; if your gutter clogs, it is nobody's fault but your own.[149]

Weather modification, climate control, geoengineering, water management, are all names for dreams from the dawn of human civilization. They inspire innovation and derision. Remember W. C. Devereaux's Weather Distributing Administration of the year 2038, and Senator Hutchinson's Weather Modification Advisory and Research Board in 2006. To which I should add, in closing this discussion of designer weather, the Temperature Stabilization Authority (TSA), created by an unknown wit in Cambridge in 1944. Reprinted in the AMS *Bulletin*, from the *Boston Herald*, which took it from a bulletin of the Murray Printing Company, the article, "Climate Control," told the story of the TSA, a congressionally mandated agency to stabilize the temperature of New England. The TSA split, however, on both the temperature range and the method to achieve it:

> Some of the commissioners were learned in history and agreed that races living in the cooler climes were always more energetic and inventive than those living in the tropics. They accordingly proposed that the temperature of New England be stabilized between a minimum of 20 and a maximum of 50 degrees. . . .
>
> Another faction of the Authority felt, however, that while a temperature ranging between 20 and 50 would keep things moving briskly, it would demand too many sacrifices and would bear too harshly on the common man. Stabilization between 80 and 110 degrees would bring a prosperous and contented New England in our time. . . .
>
> One group proposed to bring the temperature under control by buying up the surplus heat of Louisiana and the surplus cold of northern Minnesota, storing the purchases in gigantic temperature bins somewhere in the White Mountains, and releasing just enough of the hot or cold surplus from time to time to keep the climate ever normal. But this proposition was opposed by another group, who wanted to warm things up by building a titanic system of break-waters and sluices far at sea in order to change the course of the Gulf Stream and bring it closer to old New England's shores.[150]

Satirizing ill-conceived federal projects and Riker's challenge to ocean currents, the piece is as timely now as it was then.

Today, the dystopian nightmare of everyman his own weathermaker is almost upon us. For a few thousand dollars and access to several hundred gallons of water, anyone can make snow in his backyard when it's cold enough. Companies such as Snow at Home and Snow Economics, maker of the Backyard Blizzard, sell snowmaking machines and computer software that provides a seven-day snowmaking forecast based on ZIP codes, since it is still impossible to make snow, even with the best machines, unless the temperature is below freezing. Admittedly a contributor to atmospheric warming, not a solution to it, the machines use large amounts of electricity and water. Another purpose of these expensive toys is to gratify big egos; as one backyard snowmaker remarked, "When real snow falls, my daughter thinks I made it." This is a comment chilling enough to bring back real winter.[151]

3

Seeing Weather

Ocean of Air

On June 11, 1644, the mathematician Evangelista Torricelli wrote from Florence to a colleague in Rome about his experiments with what would become the barometer, an instrument that measures the weight of air. The idea that air had weight was denied by most scientists of the time, including Galileo, who, before his death, had invited the thirty-six-year-old Torricelli to join him at the court of Grand Duke Ferdinand II of Tuscany. A few lines into his letter, Torricelli wrote: "Noi viviamo sommersi nel fondo d'un pelago d'aria elementare" (We live submerged at the bottom of an ocean of air).[1]

The discovery of atmospheric pressure is one of the great scientific achievements of human history, and the barometer, which, together with the thermometer, invented by Galileo and perfected by another Florentine in 1660, and the anemometer for measuring the speed and direction of the wind, first described in 1450 but not perfected until the middle of the nineteenth century, is one of the three legs upon which meteorology scuttled forward for the next three centuries.[2] The metaphor of the ocean of air also served the development of weather science, particularly toward the end of the nineteenth century as physicists, artists, and tourists began to reconceptualize space. As Stephen Kern convincingly argues in *The Culture of Space and Time, 1880–1912,* fundamental beliefs about form, distance, and direction were challenged by the genius of Poincaré, Einstein, Durkheim, and Cézanne[3]—to which we should add the names of A. Lawrence Rotch and Alexander McAdie, meteorologists whose studies of the upper atmosphere helped to map the ocean of air and gave Americans a new frontier. Aviation historian Tom Crouch observes, "For the newspaper-reading public of the decade 1926–1936, the stratosphere was far more than a layer of sky that began roughly ten miles above the surface of the earth at the equator. The word stratosphere conjured up images similar to those that 'darkest Africa' had evoked in the nineteenth century or the names Arctic and Antarctic in the early twentieth."[4]

Rotch (1861–1912) graduated from MIT in 1884 and established and funded a meteorological observatory at Blue Hill, ten miles south of Boston, in the same year. He was also an avid mountain climber and balloonist, pioneering in the use of kites and unmanned balloons to take temperature and barometric measurements in the upper troposphere. He worked closely with European meteorologists such as Léon Teisserenc de Bort in France and Richard Assmann in Germany.[5] Teisserenc de Bort, who made hundreds of balloon ascensions, established and named the first two layers of the atmosphere—the troposphere, or zone of turning, and the stratosphere, or level region, where the temperature begins to rise.[6]

In 1900 Rotch published his 1898 Lowell Institute lectures in a book, *Sounding the Ocean of Air,* summarizing recent work on cloud formation and classification, balloon ascents, balloon sondes (unmanned balloons equipped with instruments for measuring and recording atmospheric conditions), and kites. Unlike his British contemporary Douglas Archibald, who saw humans "like the flat-fish [that] live at the bottom of the ocean of water . . . absurdly ignorant of the condition of the atmosphere a few miles overhead," Rotch had grander visions: "The surface of our globe has been tolerably well explored, the exploration of the atmosphere by balloons and kites will continue to make great progress during the last years of the century, and at the end of the twentieth century we may confidently expect that as the seas now are a medium for transportation, so the ocean of air will have been brought likewise into man's domain."[7]

The ocean of air continues to be a viable analogy for meteorologists seeking to explain the atmosphere to nonscientists, but some of Rotch's American cotemporaries found more modern analogies.[8] For meteorologist Edwin C. Martin (1850–1915), an editor of *McClure's* magazine, "weather is simply the air's *business*—its runnings to and fro, its conflicts and avoidances, its unions and divisions and graspings and givings-up in pursuit of this one aim which it never fully achieves."[9] That aim, Martin maintains, is the "seeking of ease and never quite finding it," a goal the atmosphere shares with all nature, including man. Perhaps, in 1913, Martin was thinking of the restlessness of the Progressive reformers, the emergence of the movies as a popular art form, or artistic modernism heralded by *Poetry* and the Armory Show. Having introduced the dynamic quality of the weather, Martin proceeds to develop a link between American weather and national identity. Variety is one quality of American weather, but other countries have as much or more, writes Martin. Nor is American weather superior in either temperature or moisture. What makes American weather distinct is its respect for regional variations, while at the same time sweeping west to east and binding the country together. Moreover, like the people of the nation, the weather "is practically always in progress." Finally, Martin concludes with an observation characteristic of American political expansionism: "But while our weather is national, it is the weather of a nation of the first class and of a people highly civilized. Though national, it is not provincial. Like our political institutions and administration and our commerce, it has world connections, and it would hardly be worthy of us if it did not have them."[10] This new west-to-east perspective forces a reversal of American geographic consciousness. For 300 years human movement had been east to west, pushing west. Now, we realize, the west was always pushing back, its weather shaped the east, its fronts' daggers piercing the eastern air; "blowback" in the truest sense.

A decade after Martin, Alexander McAdie (1863–1943), former Weather Bureau meteorologist, professor at Harvard, and director of the Blue Hill Observatory, searching for an analogy to explain the location of different kinds of clouds in the layers of the air, compared the atmosphere to a three-story edifice, "but with two mezzanine floors or entresols for the accommodation of high fogs and certain clouds due to diurnal ascending currents."[11] Three years later, in a book intended for a more general audience, McAdie heightened his skyline: "Our atmosphere may be likened to a six-story building. That is, there are six concentric aerospheres or air floors. In this edifice the ground floor will be the *troposphere,* or airsphere, where there is much bustle and confusion—or, as aerographers say, much convection and turbulence."[12] McAdie's second floor is the "stratosphere," his third the unnamed "region of meteors" about which little is known. The fourth story he labels the "Kennelly-Heaviside layer" after the American and British scientists who first attempted to describe it. Today known prosaically as the "mesosphere," it occupies the stratum 30 to 50 miles above the earth. Ascending with McAdie to his fifth floor, we are in the "region of auroral displays," 50 to 80 miles above the earth. Now called the "thermosphere" or the "ionosphere," this "floor" is characterized by high ion density and influences radio waves. McAdie's sixth-floor penthouse, 400 to 800 miles above the earth, is labeled the "exosphere."

The nature of the atmosphere above the troposphere was largely unknown in 1923, so McAdie can be excused his crude attempt to construct a kind of Sears, Roebuck tower of bustling sales on the lower floors and calmer offices on the top. His purpose in invoking an architectural metaphor was chiefly to illustrate how shallow the ocean of air is compared with the size of the earth. "If we represent the distance from the earth's center to the surface by 1000 bricks laid end to end, then the thickness of the sensible atmosphere or height of the first story of our aerial edifice would be represented by a single brick; and the highest level yet reached by any human by 11 bricks; and the highest actual record that man has obtained (by sounding balloon) would be represented by 37 bricks."[13] McAdie equated a brick with 1 kilometer (.6 mile), since he later gives the record altitudes for unmanned and manned balloons, airplanes, and kites as 37,000, 10,300, 12,066, and 9,740 meters, respectively. The balloon-sonde record in 1898 when Rotch sounded the ocean of air was a little more than 20,000 meters, and the kite record only about 4,000. Progress was in the air.

The rapid development of private and commercial flying caused a major shift in perspective. Man was no longer a bottom dweller. Yet it was exactly this image that George R. Stewart revived in the popular novel *Storm* in 1941. "As a crab moves on the ocean-bottom, but is of the water," Stewart wrote, "so man rests his feet upon the earth—but lives in the air. Man thinks of the crab as a water-animal;

illogically and curiously, he calls himself a creature of the land. As water environs the crab, so air surrounds, permeates, and vivifies the body of man."[14] The half century between Archibald's flatfish and Stewart's crab witnessed a rapid mapping of the ocean of air, as Rotch had predicted. Balloons, kites, and airplanes mapped the troposphere; rockets and satellites soon mapped the stratosphere and the space beyond.

Guy Murchie (1907–1997), a writer whose life bridged the eras of flight and rocketry, was a Harvard-educated journalist, aviator, artist, and teacher whose lyrical *Song of the Sky* was a Book-of-the-Month Club selection and a John Burroughs Medal winner for best nature book. His chapter "The Ocean of Sky" begins conventionally, borrowing from Stewart but playing with the metaphor. "The sky begins at our feet. We breathe it. We are actually crawling on the sea bottom of the heavens. We are the crabs of the airy depths."[15] Within a few paragraphs, however, the sky becomes an incubated schoolhouse: "incubated because [man] is nicely balanced between the white-hot metal of subterranean layers below and the scorching vacuums of the ionosphere above, and a schoolhouse because it is divinely arranged to offer the maximum of educational experience. Most of this habitable stratum of learning is actually in the sky which is, to my thinking, the open book of knowledge, of revelation."[16] A few pages later the sky becomes a "volatile onion," with many invisible but dangerous layers. Murchie then takes his reader up through those layers to the ionosphere as seen by photos snapped from rockets—his mixed metaphors a sign, I think, of the difficulty of translating what science has learned about the atmosphere into images that make sense to nonscientists.

Murchie's canticle to the sky was both a coda to the first epic of atmospheric exploration and an overture to a new age of rockets and satellites. During World War II, V-2 rockets attained an altitude of 100 miles, well into McAdie's fifth floor and raising its ceiling. In 1949, the U.S. Army fired a V-2 with a second rocket, called a WAC Corporal, attached. This missile reached an altitude of 250 miles, establishing the upper limit of what was now called the ionosphere. The ocean of air was growing deeper and its layers more numerous and distinct. A brief article in the *Bulletin of the American Meteorological Society* in 1953 commented on "the confusion now existing in the terminology of the various atmospheric shells" and noted that the Geophysical Research Directorate at the Air Force Cambridge Research Center in Massachusetts currently divided the atmosphere into six "'spheres' or shells" and five dividing surfaces, each as much as 6 miles thick. The troposphere was the bottom 6 miles (as measured from the middle latitudes); above it were a tropopause, followed by 14 miles of stratosphere, and a stratopause. Thirty to 50 miles above the planet lay the chemosphere and the chemopause. The ionosphere rose from 50 to 250 miles before fading into the ionopause. The next 350 miles were the mesosphere and mesopause, beyond them the exosphere.[17]

Three years later, the *New York Times* confidently reported, "We know [the atmosphere] extends at least 10,000 miles upward from the surface of the earth." The accompanying diagram resembles the one published in the *Bulletin of the American Meteorological Society* in 1953 but eliminates both the chemosphere, which becomes part of the stratosphere, and the mesosphere, which becomes a layer of a greatly expanded ionosphere. The bottom of the exosphere is lowered to 400 miles. The diagram also includes a satellite to be launched during the International Geophysical Year 1957–1958 that would orbit the earth from 200 to 1,400 miles. On April 1, 1960, the successful launch of the TIROS weather satellite helped to routinize the exploration of the upper atmospheric ocean and added hundreds of stories to McAdie's little castle in the air. Currently, NOAA divides the atmosphere into the troposphere (surface to 4–12 miles), stratosphere (to 31 miles), mesosphere (to 53 miles), thermosphere (53 to 430 miles), and exosphere (430 to 6,200 miles).[18]

In an age of rockets, scientists continue to use sailplanes and balloons in their search for an understanding of the air. In 1952, meteorologist Harold Klieforth and pilot Larry Edgar soared to a record 44,500 feet (8.4 miles) while studying the mountain air waves near Bishop, California. In August 1957, as part of Air Force Project Man High I, Colonel David G. Simmons ascended 19 miles into the stratosphere in a pressurized balloon. His feat inspired Louise B. Young to write: "A fish at the bottom of the sea must have the same impression of water" as humans have of air. Young then reckons Simmons's flight to have reached the 35th floor of an 800-story (450-mile-high) skyscraper.[19]

Explorations of the ocean of air—and new ways of thinking and talking about the atmosphere—have been in progress for little more than a century. When astronauts broke through the surface of the exosphere, they began, like their earliest amphibian ancestors, a new chapter in evolution. Humans are no longer bottom dwellers, or even surface swimmers, but creatures of "space weather"—facing the effects of sunspots, solar winds, and radiation in the environment beyond the atmosphere.[20]

The wild blue yonder that separates earth from the inky blackness of the heavens, like the oceans to which it is organically tied by hydrologic and carbon cycles, is now compared more often to a membrane, the tissue of a living cell, than to an ocean or a building.[21] The latest analogy conforms to contemporary notions of biocentrism, to ideas about harmony, health, and human survival. The air is earth's aura, to be inhaled, tasted, felt, heard, and seen.[22]

Sky Awareness

What follows is the story of a quest by a loosely organized group of visionaries, teachers, artists, and scientists to know the sky, the welkin—"the blanket of air where all weather takes place"—through all of the senses.[23] It is an exemplary tale of the interplay of personality, ideas, and institutions in defining the scope of the history of weather. Sky awareness is earthbound, place-centered; it is the inverse of Earth from space, and it helps to create place identity. Place identity has become an important topic in geography, environmental psychology, and philosophy in the past few years and provides a solid theoretical basis for sky awareness.

My interest in this story was aroused by a brief paragraph in Samuel Hays's *Beauty, Health, and Permanence: Environmental Politics in the United States, 1955–1985,* which mentions a Boston television reporter, Jack Borden, who began a sky awareness program in the early 1980s. Assuming an obvious link between the sky and weather, I decided to follow up this lead as an example of popular interest in the weather. An Internet search for Jack Borden led me to his Web site, www.forspaciousskies.com, and my e-mail inquiry resulted in a meeting, a continuing correspondence, and the receipt of a large number of letters, unpublished manuscripts, and reprints from Borden's files.[24]

In 1977, Borden was working for WBZ-TV, where he had created award-winning programs on environmental topics. One day in April, as Borden tells the story, he and his wife, Jan, enthusiastic nature lovers, were visiting the Wachusett Meadow Audubon Sanctuary in Princeton, Massachusetts. He lay down on the grass, looked at the sky, and had a revelation. He saw the sky in a new way—the clouds seemed immense, out of scale—and he suddenly became aware of how unaware he had been of the sky. He soon incorporated his experience into his work, conducting sidewalk interviews and asking willing pedestrians to describe the sky while covering their eyes. When he discovered that none could describe any feature of the sky, he began his campaign for sky awareness.[25]

In 1980 Borden founded For Spacious Skies, a nonprofit organization to promote sky appreciation. He began working with elementary school teachers to develop curricula around the study of the sky, and he formed a board of directors that included photographer Ansel Adams, Smithsonian astronomer Von del Chamberlain, Friends of the Earth Foundation president Alan Gussow, National Park Service regional director Douglas Bruce McHenry, Massachusetts Audubon education director Charles Roth, and artist Eric Sloane. With the help of McHenry and grants from the EPA and several foundations, including the Edison Electric Institute and the Polaroid Foundation, Borden organized a national conference at the Grand Canyon in May 1981.

The three-day meeting, attended by about sixty artists, scientists, educators, and environmentalists including Amory Lovins, produced a number of papers, many of which were published, and which provide a useful glimpse of the ideals and aspirations of the nascent organization. Ervin Zube, professor of environmental design at the University of Arizona, presented his findings on the perception of the sky based on 625 interviews in five metropolitan areas—Seattle, Washington; Syracuse, New York; Washington, D.C.; Irvine, California; and Tucson, Arizona. This study was commissioned for the conference and was the only empirical evidence of the state of sky awareness. Zube found that those interviewed usually were not as aware of the actual conditions of the sky as they thought they were. Rather, they had stereotypes of how the sky in their region *ought* to look.[26] Conference participant Leonard J. Duhl, professor of health and of regional and city planning at the University of California, Berkeley, argued that the sky represents the wholeness of life and the place where humans can best connect with nature; healthy skies outside contribute to inner health. Duhl then cites examples of non-Western cultures that invoke the sky in their rituals of healing and spiritual wellness. He concludes with a review of work by physicists and biologists who have suggested that all life is linked at the molecular level. Duhl's efforts to find similarities between religious and scientific worldviews are characteristic of the deep ecology movement to which For Spacious Skies seems linked.[27]

While Zube provided evidence of the lack of sky awareness and suggested some reasons for it, and Duhl called attention to the historical importance of the sky in shaping the human mind, Charles Roth addressed the practical matter of educating for sky awareness. Dividing his curriculum into three major headings— "The Sky as Habitat," "The Sky as Transport," and "The Influence of the Sky on Culture"—Roth lists topics from astronomy, meteorology, physiology, chemistry, hydrology, and electromagnetism that can be taught under the headings of habitat and transport. When he turns to the influence of sky on culture, Roth clearly reveals the way in which For Spacious Skies differs from other sky-oriented education projects. Roth lists such factors as the impact of weather and climate on human evolution; the development of science and sky watching; the sky in mythology and literature; the sky as direct aesthetic experience and as art; the sky as a trigger of psychological moods; and the sky and weather in recreation, commercial, and military uses. By emphasizing the arts and humanities in sky education, Roth joins those who see environmental protection requiring changes in cultural values as well as in law, management, and technology.[28]

Meanwhile, Borden continued to recruit teachers in the Boston area to his cause. Among his first converts were a second-grade teacher and a third-grade teacher in Arlington, Massachusetts. Both teachers reported great enthusiasm from their students and improved self-confidence and creativity. By the summer of 1982, Borden and some of the teachers were participating in a program at the Boston Museum of Science called "Skyfire: A Celebration of Atmospheric Wonders," in which the children displayed their artwork, read their poems, and took part in science experiments.

Borden garnered a major venue two years later when, responding to a perceived crisis in science education, the American Meteorological Society (AMS) obtained a grant from the National Science Foundation to hold a two-day workshop titled "Meteorology as a Unifying Educational Strategy for Improving Pre-college Science, Mathematics, and Technology Education." Thirty-seven educators from several universities and some public schools, government officials, broadcast meteorologists, journalists, and representatives from scientific organizations listened to Borden and teachers from New York City and Arlington, Massachusetts, who had developed sky awareness programs. They then broke into subgroups to discuss the ways in which the study of the atmosphere could be improved by broadening traditional science education. Although no formal recommendations were made, a report published in 1986 listed twelve points on which there was general consensus. Most of the points called for greater AMS efforts in public education through publications, teacher workshops, partnerships with private and public institutions, and nontechnical programs at professional meetings.[29]

As more teachers throughout the country adopted the For Spacious Skies curriculum, Borden's program drew the attention of researchers from the Harvard University Graduate School of Education, who tested elementary students in Needham in 1985 and 1986. They concluded that students who were exposed to the sky awareness program scored 37 percent higher in music appreciation, 13 percent higher in literacy skills, and 5 percent higher in visual arts skills.[30] The Harvard study and the adoption of the program in some inner-city schools led to a story in the *New York Times* in June 1987, in which Borden mentioned that, although his program had been popular in some nursing homes, he had failed to interest prison administrators despite his feeling that prisoners were a logical audience for sky awareness, given their limited contact with nature.[31]

Within two weeks, the *Times* forwarded to Borden a letter from an inmate in the Green Haven Correctional Facility in Stormville, New York, calling attention to his short story, "A Bright Spot in the Yard," that had been published in the *Transatlantic Review* in 1976. Jerome Washington's story concerns the relationship between an older prisoner, considered strange by his prison mates because he stands by himself in the same spot in the prison yard, and a young prisoner who is led to believe that the older man can help him escape. The escape turns out to be the spot in the exercise yard where he can look at the sky and not see any of the buildings or walls. The young man learns to let the clouds in the sky take him away in his imagination. When the older prisoner is finally released, the younger man continues to free himself by imagining the clouds traveling around the world.[32]

Borden found other prisoners who had either discovered the freedom in sky watching on their own or learned about For Spacious Skies from magazine articles. One such prisoner contacted Borden from the California Institution for Women in Frontera. Sentenced to twenty-five years in prison in 1984, she acquired cloud charts and books from Borden and became a competent amateur meteorologist and devoted sky watcher who made an effort to teach other prisoners the aesthetic and spiritual values of the sky. Her success was facilitated by the cooperation of the prison administration. Massachusetts officials, on the contrary, refused to add sky-watching books to their libraries, but *Prison Life,* a magazine that reaches 100,000 inmates, now offers free cloud charts to anyone who requests them.[33]

In 1998, the Frontera prisoner responded to a request from John Day, author of several popular books on weather, to write an essay on cloud watching, part of which is now posted on Day's Web site:

> This morning each ridge was visible in the mountains to the west, and on the east there was snow on top of Mount Baldy. To the south I could see ground fog rising from the stream that meanders through a park approximately five miles from here. The fog is there on many mornings and it follows the path of the stream, zigzagging back and forth between the oaks and the sycamores. It was so clean and pristine that I could not contain myself. I had to raise my arms over my head and give thanks for the beauty that always surrounds us, regardless of our circumstances.[34]

Like the Birdman of Alcatraz, Jerome Washington and the Skywoman of Frontera bring powerful new perspectives to activities that might be considered mundane rather than celestial. Mental escape to the sky frontier as part of the rehabilitation process suggests unrecognized potential in meteorological awareness. Borden's struggle to gain recognition and legitimacy in the eyes of other environmental educators and sky watchers is evidence that the sky is too often taken for granted. After retiring to Athol in the mid-1990s, Borden continues to promote school programs by giving talks, selling and donating materials, and encouraging anyone who shows an interest in the sky. He has received numerous honors and thousands of letters from grateful discoverers of the sky. In 1996 he coauthored an activity guide for learners of all ages and worked with The Weather Channel on a resource guide, *Look Up!,* for teachers of grades three through six. In early 2003, the National Weather Service's Southern Regional Headquarters in Fort Worth placed an order for 34,000 For Spacious Skies cloud charts.

Before concluding the story, we should look briefly at two of Borden's associates in the early days of For Spacious Skies—Eric Sloane and John A. Day. Sloane is the better known of the two, a painter and author of dozens of books on colonial American crafts, flying, weather, and himself. He painted airplanes and murals at Floyd Bennett and Roosevelt Air Fields in the 1920s, then went to New Mexico, where he claimed to have invented sky painting. He enrolled briefly at MIT in the 1930s to study meteorology but left to write and paint. His first book, a layman's guide to the weather and cloud types, *Clouds, Air, and Wind,* was published in 1941. Two full-page illustrations are characteristic of his lively style. Labeled *Summer Theatre of the Clouds* and *Winter Theatre of the Clouds,* Sloane's sketches show how the meeting of a cold and a warm front produces thunderstorms and hail, and why summer air is less stable than winter and therefore less favorable for fly-

Eric Sloane, *Summer Theatre of the Clouds,* from *Clouds, Air, and Wind* (New York: Devlin-Adair, 1941). Reproduced by permission of Mrs. Mimi Sloane.

ing. Sloane illustrated flight and weather manuals for the army during the war and then focused on skyscapes and aviation art. His mural for the Smithsonian's National Air and Space Museum, *Earth's Flight Environment,* is a notable visual statement on the sky as a vital component of the ecosystem.

Sloane died in 1985, and his contribution to For Spacious Skies was minimal except as a vociferous exponent of the sky whose books and paintings continue to inspire sky watching. Sloane's flamboyance contrasts with John Day's self-effacing demeanor, which belied his importance as a promoter of sky awareness. He may be best known for coauthoring the *Field Guide to the Atmosphere* (1981) with Vincent Schaefer, but his list of publications is extensive. Day was born in Colorado Springs in 1913 and graduated from Colorado College with a degree in physics before attending the Boeing School of Aeronautics in Alameda, California, in 1936. Hired by Pan American Airways as a meteorologist, he spent the next ten years in Asia, New Zealand, and the Pacific with Pan Am and the navy during the war. Deciding that he did not want to raise his growing family in Shanghai, he returned to graduate school on the GI Bill and received his doctorate in cloud physics from Oregon State University in 1956. Two years later he joined the faculty at Linfield College in McMinnville, Oregon, retiring in 1978. Until he died on June 21, 2008, he remained active as a lecturer, newspaper columnist, and cloud photographer and maintained a Web site, www.cloudman.com. His *Book of Clouds,* published in 2002, sold more than 16,000 copies in the first three months. Day worked with Borden, and their Web sites are linked, reflecting their mutual interest in a holistic approach to clouds and the sky. Day was more conventionally spiritual and closely tied to his background as a physicist. Both men were longtime advocates for a commemorative set of cloud postage stamps; their goal was finally achieved in 2004 when the U.S. Postal Service issued a set of fifteen stamps with different cloud types—the photograph of *Cumulus humilis* was taken by Day. Borden continues to campaign for a set of sky painting stamps.[35]

The story of sky awareness is just beginning. It is much more than a footnote in environmental history; it is an episode in an ongoing tale of individuals, institutions, and the ideas they espouse to make a place for the atmosphere in environmental studies, in mainstreaming the marginal. Given the current high level of interest in weather, as evidenced by the popularity of The Weather Channel, weather handbooks for "Dummies," "Idiots," and children, Internet weather services, and weather disaster movies, it may seem that the "sky huggers" have made their point. People are looking up, but the problem of conflicting goals remains. Should we move beyond awe? If so, how? Perhaps this is what French philosopher Gaston Bachelard was getting at more than sixty years ago when he wrote: "The blue of the sky is first of all a space where there is no longer anything to imagine. But when the aerial imagination is awakened, then the background becomes active. It encourages the aerial dreamer to make changes in the terrestrial profile and to take an interest in the point where the earth communicates with the sky."[36]

In the philosophy of place developed by contemporary phenomenologist Edward Casey, the point where the earth communicates with the sky is both the *arc,* the point at which a place fades into the distance, and the *atmosphere,* the predominant mood of a place. Although Casey neglects weather as a component of arc and atmosphere, I think it is easily added to his framework for understanding the ways in which we see and know what he calls the "place-world." Casey, like those developing the field of environmental psychology, is interested in showing how we name and organize the places we encounter during our lives and how the activity of naming helps to shape our personal identity. Place is where we dwell; it is defined by memories and stories. Place is a point in space. Space is infinite; place is finite, but it is not unchanging. Weather and climate shape the topography of place, action, and reaction, and seeing the weather enables us to "see" time, even when it is passing in periods far beyond human scale. When the sky awareness advocates ask us to look up, they are challenging us to see in an active sense, to engage with the sky as a place where so much is going on that we cannot fail to learn from it. "Wisdom sits in places," an Apache man told the anthropologist Keith Basso, launching Basso on a lifelong quest to understand the ways in which place is felt and imagined. Weather is a place, perhaps the last place where we can find wisdom in a rapidly changing and increasingly disordered world.[37]

As Borden, Day, and others discovered, the sky is sometimes a hard sell. When the sky is clear, there is "nothing" to see, and when there are clouds, they often seem monotonous. Astronomy's popularity stems from there being so much to see in the night sky. The dilemma was addressed by meteorologist James Edinger when he wrote *Watching for the Wind:*

> In setting about to explore the local atmosphere for patterns of air motion we are immediately faced with a rather substantial problem: how do you see something that normally is invisible? Well, you don't see it, but ordinarily you can see something that is being affected by it—the fluttering leaves of a tree, a scrap of paper bounding along the ground, the spindrift from a cresting wave. If the wind is strong enough you will feel it. It may even snatch the hat off your head. However, if it is less than one or two miles per hour you won't be able to feel it unless you can specially sensitize yourself. Moisten a finger and hold it up. It will feel cooler on its upwind side, due to increased evaporation there. Or do as the marksman does before he takes aim. Toss a little dust in the air.[38]

After a lifetime of cloud watching, John Day condensed his passion for "those elusive, captivating denizens of the troposphere," into "10 Reasons to Look Up!" Among them: Clouds are beautiful in the combination of form, position, gradations of light and shadow, even color. They are never the same; they are an antidote to boredom. Clouds inform us about the weather to come and are the source of fresh water that sustains all life on earth. Cloud watching connects us to nature and to the rest of the globe. Clouds are a magic show, inspiring awe and wonder. Day and Borden stress the important distinction between looking and seeing—creative seeing, that fully engages the imagination, the senses, and the mind.[39]

The purpose of sky awareness, and of the metaphor of the ocean of air, is to see creatively the physical processes and the aesthetics of the atmosphere and its weather. Clouds are the most visible feature of the sky and "billboards of coming attractions," as Day calls them. Images of clouds function, to borrow the terminology of psychologist Rudolf Arnheim, in three ways: as *signs,* usually denoting rain; as *pictures,* representing different kinds of clouds and weather conditions; and as *symbols,* traditionally of emotional or political conditions. In the past 200 years, clouds have been named, analyzed, photographed, painted, filmed, and exhibited. Once "the daughter of Earth and Water, / And the nursling of the Sky," clouds have become twenty-first-century icons of the endangered planet.[40]

A Cloud No Bigger than a Man's Hand . . .

. . . "which presently overspread the heavens and watered the earth," says Elijah to Ahab in 1 Kings 18:44 of the Bible. "Never despise small things" is the usual lesson drawn from these words. When the bicentennial of the modern system of cloud classification was celebrated in 2003, we were briefly reminded of the London pharmacist, amateur painter, and astronomer Luke Howard who created it. It is a great story, superbly told in Richard Hamblyn's *Invention of Clouds.* Howard had many nephological predecessors and rivals, but his system, based on the binominal nomenclature that Carl Linnaeus had developed for plants, animals, and minerals using a Latin noun for the genus followed by an adjective describing a species, was easy to understand and widely reprinted in popular journals. Howard's earliest classification had seven genera. Currently, there are ten genera and fourteen species, all further subdivided by degrees of transparency and special forms. Howard classified his genera by appearance, from cirrus (curl or ringlet) to stratus (spread). His clouds began to loom and tower in the skyscapes of nineteenth-century poets and philosophers. Percy Bysshe Shelley and Johann Wolfgang von Goethe paid tribute to Howard's nomenclature in verse. John Ruskin challenged painters to know every cloud, and Henry David Thoreau kept almost daily records of the sky, usually inspired to play a game of free association: "The clouds, cumuli, lie in high piles along the southern horizon, glowing, downy, or dream-colored, broken into irregular summits in the form of bears erect, or demigods, or rocking stones, infant Herculeses; and still we think that from their darker bases a thunder-shower may issue."[41]

Determining the height and speed of clouds was more difficult than naming them. Before the invention of the nephoscope in France in the 1840s, measuring the speed and height of clouds was a cumbersome process requiring the observer to measure the distance to the cloud's shadow on the ground, then determine the angle between the shadow and the cloud and between the observer and the cloud. The nephoscope, basically a round black mirror and a compass mounted on a tripod with a rotatable top and a pointer or an eyepiece, went through many forms before the end of the century. Sir Francis Galton and Charles F. Marvin both contributed to its improvement. "In use," writes W. E. Knowles Middleton,

> the pointer is placed in such a position that the image of a cloud appears in line with the pointer and the center of the disk. The image is then watched, being kept in line with the pointer, and the edge of the disk. It need scarcely be said that if the instrument is set up with the 180° mark pointing north, the direction from which the cloud is moving will correspond to the graduation toward which its image moves.[42]

Other methods for determining the height and speed of clouds included balloons tracked by theodolite and balloons equipped with radiosonde (transmitter). Today, cloud tracking is done by ceilometers, satellite, radar, and LIDAR (light detection and ranging).

NASA launched two satellites specifically to track clouds and study air quality, *CloudSat* and *Calipso,* on April 28, 2006, despite the redefinition of NASA's mission by its new director, Michael D. Griffin. NASA's previous strategic plan was "to understand and protect our home planet, to explore the universe and search for life." NASA scientists like James Hansen, who studies climate change, have long seen NASA as an environmental agency concerned with both space and Earth. Griffin, in an attempt to downplay global warming and get back to the moon, abridged the statement to: "to pioneer the future in space exploration, scientific discovery, and aeronautics research." Perhaps *Calipso* (*C*loud-*A*erosol *L*idar and *I*nfrared *P*athfinder *S*atellite *O*bservatories), like its ancient namesake, will keep cloud explorers busy for another seven years, even though the mission is designed to have only a two-year life.[43]

Early cloud trackers used cameras, and by 1896, when the first *International Cloud Atlas* was published in France, there was a vast archive of photographs of clouds from all over the globe. Many of the photographers were more interested in art than in science, but for a brief time, the two cultures seemed to merge. Cloud science was simple enough for the layman to understand its principles, and cloud aesthetics was realistic enough to satisfy the meteorologists. This is apparent in a book published at the beginning of the twentieth century by the American art critic and nature writer John C. Van Dyke.

Van Dyke was born into a prominent family in New Brunswick, New Jersey, in 1856. He studied law but preferred academic life and became a professor of art history at Rutgers and an art adviser to Andrew Carnegie, to whom he dedicated his most popular book, *The Desert*. In the 1890s Van Dyke explored the American Southwest, discovering, as other artists and scientists soon would, that the high desert air produces some of the most wonderful atmospheric effects in the world. In two chapters, Van Dyke instructs his readers on the topics "Light, Air, and Color" and "Desert Sky and Clouds." His chosen analogy for the atmosphere is the ground-glass globe found on some lamps to diffuse light. Without the filter of the atmospheric globe, he cautions, life on earth could not endure. We are protected and the sky painted by the "particles of dust, soot, smoke, salt, and vapor which are found floating in larger or smaller proportions in all atmospheres." In the dry desert air where water vapor is scarce, the sky is a pale blue at lower elevations, but it darkens with elevation.

> At four thousand feet the blue is certainly more positive, more intense, than at sea-level; at six thousand feet it begins to darken and deepen, and it seems to fit in the saddles and notches of the mountains like a block of lapis lazuli; at eight thousand feet it has darkened still more and has a violet hue about it. The night sky at this altitude is almost weird in its purples. A deep violet fits up close to the rim of the moon, and the orb itself looks like a silver wafer pasted upon the sky.[44]

Blue sky is not Van Dyke's principal interest. Rather, it is the pink, lilac, rose, and fire red of the dust-strewn desert sky that appeal to him most strongly, and the intense color of the clouds, "chrome-yellows, golds, carmines, magentas, malachite-greens—a body of gorgeous hues upheld by enormous side winds of paler tints that encircle the horizon to the north and south, and send waves of color far up the sky to the cool zenith." Seeming to rely on Luke Howard's original seven cloud types rather than the international nomenclature adopted in 1896, Van Dyke begins his catalog of clouds with the low-lying rain-bearing nimbus, moving on to the cumulus that builds to 40,000 feet, then erroneously placing the stratus above them, and ending with the highest, the cirrus. Van Dyke is not concerned with scientific classification; it is the intensity of light and the color of desert clouds that capture his attention:

> They vibrate, they scintillate, they penetrate and tinge everything with their hue. And then, as though heaping splendor upon splendor, what a wonderful background they are woven upon! Great bands of orange, green, and blue that all the melted and fused gems in the world could not match for translucent beauty. Taken as a whole, as a celestial tapestry, as a curtain of flame drawn between night and day, and what land or sky can rival it![45]

Van Dyke, abetted by *Arizona Highways*, Eric Sloane, and Georgia O'Keeffe, inspired generations of artists and tourists to make sky safaris to the Southwest.

Weather Bureau meteorologist Frank Bigelow, writing for a general audience in *Popular Science Monthly* in 1902, acknowledges the beauty of clouds: "The most beautiful objects in the sky are clouds, and their daily procession from west to east in northern latitudes forms a moving tableau of living pictures for those who have eyes to see." But Bigelow then rushes into a fairly complicated explanation of "the prismatic action of the small spheres of condensed aqueous vapor that make up a cloud." Bigelow wants his readers to understand the role of dust particles and the electrons that charge each drop of condensed vapor. He simplifies the cloud classification by focusing on two types of cloud: cumulus and stratus. Cumulus begin as small, fleecy clouds about a mile above the ground, then grow vertically into cumulonimbus with tops 6 miles high. Stratus, or veil clouds, are formed when a horizontal current of warm air passes over a cold one. Charts and maps show how wind speed increases with altitude. Having established the importance of winds moving vertically and horizontally through the air, Bigelow devotes the remainder of the article to these currents in cloud formation and weather patterns.[46]

By the 1920s, many heads were in the clouds. The indefatigable Charles F. Brooks, writing as a Weather Bureau employee in 1920, reviewed the history of cloud nomenclature and made numerous suggestions for the improvement of the international classification. "Cloud names are as necessary in the work of the meteorologist as rock names in that of the geologist or plant names in that of the biologist," he begins, adding that, although it would be ideal to classify clouds based on their process of formation, lack of information forces meteorologists to rely on appearance alone. "Cloud appearance," he continues, "may be described in terms of form, coarseness, and density." Brooks then offers seven principal types of clouds, quite unlike Howard's or the International Classification:

Fibrous, characteristic of streaks of falling snow or rain as seen at a distance.
Smooth, characteristic of *sheetlike* clouds, especially when low, and when snow or rain is falling rather uniformly.
Flocculent, scaly, ice-cake like, or disk like, cloud elements in groups, which as wholes usually have smooth, curving outlines, which may be called *lenticuloid,* more or less lens- or lentil-shaped. . . .
Waved or in *rolls,* characteristic of wave movements at or near horizontal boundaries between differing winds, or near the ground, and the lines where the wind at any level shifts to a new direction.
Round-top, characteristic of the summits of locally rising air currents.
Down-bulged or *round-holed,* characteristic of localized down-currents.
Ragged, characteristic of forming or evaporating cloud in a turbulent wind.[47]

Brooks proceeds to list three degrees of coarseness (coarse, medium, and fine) and five levels of density (transparent, semitransparent, medium, dense, and very dense) to complete his classification system.

It didn't work. After experimenting with his system for fifty days, Brooks admitted failure. With about 3,000 combinations possible for each cloud, "it was hopeless to devise satisfactory names for distinguishing one combination of detailed characteristics from another." Brooks goes on to critique the International Classification of 1896 and its revisions of 1905 and 1910. Here, too, the problem is cloud complexity. Unlike snowflakes, which could be classified in less than a dozen basic types until the 1990s when scientists reclassified snow grains by recognizing both morphology and process of formation, clouds are large, three-dimensional, and distant. "In making cloud records it is seldom easy to decide on the appropriate cloud names for the appearance of the sky; and when the decision is made, one feels almost invariably that another observer might have made a different decision." The first task for the meteorologist, then, is to clearly define "the dividing line" between the ten cloud types of the International Classification. Brooks's primary example of the difficulty of finding that line is the case of *cirrostratus,* "a thin, whitish sheet of clouds sometimes covering the sky completely and giving it only a milky appearance." Sometimes, however, an *altostratus,* "a thick sheet of grey or bluish colour," when "brilliantly illuminated by the sun or moon and within an angular distance of 90° of the luminary," looks whitish.[48]

Brooks gives many more examples of the ambiguity of cloud classification, then offers a synthesis of the international system and his own, so that the ten types of the International Classification can be refined by his seven types, by density, and by coarseness. Brooks's scheme was never adopted, but the International Classification continues to be revised, the last time in 1956, and new, hybrid names continue to crop up on cloud charts and book illustrations. Some of these names are serious; some, not. It is hard not to smile over the name *Cumulonimbus mammatus,* that spectacular formation so familiar to storm watchers. One newspaper writer recently suggested the cloud name "Nebulus" and asked, "Is it a cloud or smoke or an old lady's church wig?"[49]

Manual of Cloud Forms and Codes for States of the Sky, which lists thirteen species under the ten genera, was issued by the Weather Bureau in 1949. *Cirrus* species include such novel variations as *Cirrus filosus,* "more or less straight or irregularly curved filaments (neither tufts nor little hooks and without any of the parts being fused together)," and *Cirrus nothus,* "cirrus proceeding from a cumulonimbus and composed of the debris of the upper frozen parts of these clouds." Neither of these terms appears in the 1956 International Classification, but fourteen species with names as exotic as orchids—*uncinus, spissatus, floccus,* and *mediocris*—do. In his beautifully illustrated *Book of the Clouds,* John Day names clouds arranged in parallel rows, "billow" clouds, though they look very much like what are sometimes called "herringbone" clouds. *Lenticular* clouds, well known because of their flying-saucer shape and dramatic mountaintop setting, were part of Brooks's scheme, but the even more spectacular *Kelvin-Helmholtz* cloud (named for nineteenth-century British and German physicists), which resembles the curling waves favored by surfers and occurs when a wind blows between cold and warm air masses, making the interface unstable, is a product of more recent orismology.[50] In recent years the cloudlike water vapor produced by jet airplane exhaust, called a "contrail," has been elevated to the status of *contrail* clouds.

Naming, photographing, and even altering the shape of clouds through seeding have left meteorology with only a vague notion of how clouds affect the earth's climate and weather. As one scientist working on computer modeling of clouds recently put it, "Today's stratiform cloud parameterizations are very rough caricatures of reality." The same could be said of other cloud types, because "clouds are highly variable in horizontal directions, vertical directions, and in time, so that simple interpolations may give unrealistically smooth cloud observations," and "clouds are rarely sampled directly." The complexity and elusiveness of clouds are what bring the sciences and the arts together in the pursuit of nephology.[51]

The airplane provided cloud watchers with a new tool, although they had been photographing clouds from balloons for fifty years. Now, however, they could fly up, through, and above clouds more quickly and frequently than before. One of the pioneers of American meteorology, Alexander McAdie, was one of the first to exploit the new perspective. Almost all the photographs in his treatise on clouds, published in 1923, are aerial shots. McAdie fervently believed that man, "with his practical conquest of cloudland, and the ability to explore the region where clouds form . . . is on the verge of great advances in connection with all the processes of cloudy condensation in the free air." With the Great War just concluded, the young aviators and their enthusiastic elders like McAdie found a new frontier. The terms "Cloudland" and "free air" seem to convey broader meanings.[52]

McAdie, with patriotic wit, pokes fun at the excessive use of Latin in scientific taxonomies. Shakespeare, he reminds his readers, wrote of a cloud that is dragonish, a vapor sometimes like a bear or a lion, or a towered citadel. "Instead of these descriptions," McAdie scoffs, "the modern cloud sharp would substitute for 'dragonish,' *cumulus horribilis;* for 'bear,' *stratocumulus ursus;* and for 'towered citadel,' *cumulo-nimbus-castellatus.* These rather long terms seem unnecessary. They make us think of Huxley's remark that if one had to mention a *great-beast-ium,* he could achieve a reputation for erudition by calling it a *megatherium.*" The joke, of course, is that this is exactly what one of the nineteenth-century paleontologists called the fossils of a giant sloth. McAdie writes that more than 100 Latin names have been proposed for clouds, then lists 22, some of which remain in contemporary terminology, such as *cumulus mammatus* (hanging down like breasts), and others that have not, such as *cirrus pennatus* (plumed cirrus). Kidding aside, McAdie concludes, "There is urgently needed a classification of clouds which will tell the origin of the cloud and its life history."[53]

Matthew Luckiesh was twenty years younger than McAdie and had been a pilot in World War I. After earning a DSc, he became director of the Lighting Research Laboratory at General Electric's Nela Park facility near Cleveland. He wrote several books on lighting, light and color in advertising, and vision. In 1933 he published the second edition of *The Book of the Sky: Journeys in Cloudland on the Wings of Experience and Knowledge,* which celebrated flight, sky, clouds, wind, and weather. Luckiesh's cloudland is more fanciful than McAdie's. Luckiesh even calls it a "fairyland of clouds." He is one of the first cloud chasers to notice that "the sky with its expressive clouds is the one charming aspect of Nature which follows mankind from the open places to congested cities." He comes close to suggesting that we ought to create cloud parks for the specific purpose of getting city dwellers to look up. Some artists, as I will describe later, have attempted to create clouds as part of urban outdoor sculpture.

Luckiesh expresses surprise and dismay that so few people observe the sky carefully. "How many to-night are able to recall that rugged range of vaporous mountains which floated above the southern horizon nearly the whole afternoon?" He is optimistic that air travel will make more people appreciate the sky and even hopes that everyone will be able to take an hour's sojourn by plane in evenings after work.[54]

The thought of thousands of cloud tourists buzzing around the twilight sky near a city of any size is a bit frightening today, but Luckiesh is only momentarily carried away. He provides his readers with a verbal portrait of his fairyland, explains the relation of clouds and weather, wind, lightning, and bumpiness, and devotes a chapter to what he calls "cloud capers." "Their sudden appearance and disappearance," he writes, "so appropriate to their spirit-like texture and to their invisible world, their restlessness, their endless variety, make it easy for one to become deeply absorbed in this aspect of the heavens." Like Napier Shaw and John Sloane, Luckiesh finds the sky a theater in which clouds perform:

A few days ago the cumulus clouds floating over the water were small and rather flat. The morning was hot and the cumulus clouds over the land were rearing strongly upward. As the clouds drifted from over the water and reached a position vertically above the shore, they were torn violently and rapidly upward in a variety of shapes, and all were tattered as the result of the rough treatment. This performance went on for hours and clearly demonstrated to the observer the superior strength of the upward currents over the land, which, of course, was hotter than the water.[55]

Luckiesh illustrates the battle of winds and clouds with a photograph. He recommends cumulus clouds for the novice cloud watcher because they are large and close enough to be in the same air currents with the observer.

A generation later, Luckiesh's gospel was revived by Guy Murchie in *Song of the*

Sky. Murchie, a navigator on transport planes in World War II and later for Seaboard and Western Airlines, was also a cloud watcher. Or voyeur? Murchie's clouds are distinctly feminine performers:

> Every now and then I think I sense a maidenly wile in the way some passing nimbus form waves at me as we thunder by—our four engines coolly roaring. . . .
>
> For through their sky we play the dominant, the active role. We are their hero, their goer of the go. And as the plow lusts after the gentle body of earth, so our flashing propeller blades yearn for the tender mists, whirling apart the coy veils of cloud. Our airspeed pitot tubes and loop antennae are the antlers and pistils of the sky—our phallic fuselage with its powerful driving action into the passive softness of the cloud as wonderful a symbol of ravishment as can be known.[56]

Murchie's pastoral metaphors and sexual imagery bring to mind two classics of American studies, Leo Marx's *Machine in the Garden: Technology and the Pastoral Ideal in America,* and Annette Kolodny's reevaluation of the land-as-woman trope in American literature, *The Lay of the Land.* The pastoral landscape is easily transferred to the sky with its fleecy clouds; the airplane literally becomes Hawthorne's celestial railroad, the machine in the garden of air challenging nature's beauty with the technological sublime. In Murchie's view, the virgin land becomes the virgin air, serenely waiting for the plowman. Murchie calms down after these outrageous paragraphs and runs through the usual catalog of clouds from cirrus to fog, employing somewhat less lurid metaphors, although his cumulus becomes a "fire tree," "a rising stream of bubbling hot air whose foot is a heat source: a plowed field baking in the sun, a town, or perhaps a lake whose waters are warmer than the surrounding country. . . . And remember, it is only the tiptop of the tree, above the cool, condensation level, that is visible as a cumulus cloud. The rest, like eight ninths of an iceberg, is out of sight." Tree or iceberg, fair weather to congestus, cumulus clouds are almost everybody's favorite.[57]

Closely linked to the classification of clouds and the development of aviation are the measurement and classification of visibility and of wind. The Weather Bureau tackled the former in 1919 with a proposed modification of the eleven-point British and French visibility scale. The scale ran from "0. Very bad, visibility less than 200 meters in good daylight," to "10. objects visible in good daylight beyond 30,000 meters or 18.6 miles." "6. Good, above ordinary," visibility was defined as up to 12,000 meters or 7.5 miles in good daylight. There were, of course, all kinds of limitations with this system, since it used any objects—barns, houses, or trees—that were at the desired distance. Since these objects differed in size, color, and background, they were hardly comparable, and their visibility in the same light differed greatly. Using a series of specially constructed targets with a white circle on a dark background was one solution. Visibility was a concern that warranted research.

In November 1920 the Commission on Weather Telegraphy, an international body, adopted a ten-term scale that adjusted the distances downward. Zero stood for "dense fog," with prominent objects not visible at 50 meters, while 1, "very bad," was visibility up to 200 meters; 6, or "good" on the old scale, became, 7 on the new commission scale, "visibility up to 12,000 meters"; 8 and 9, "very good" and "excellent," both meant objects were visible beyond 30,000 meters. Variability of "prominent objects" remained an issue, and vertical visibility was even more difficult to measure and define. An international standard for vertical visibility with six degrees of difference was modified in 1935 by Canadian pilots and meteorologists by adding two intermediate conditions between "good" and "fair" and between "fair" and "poor." The distinctions were based not on visibility at specified distances but on clarity from the air, from several hundred to several thousand feet. Zero, or "nil," meant that nothing on the ground was visible, while 3, or "fair," meant "Haze quite apparent. Details of objects not easy to distinguish though outlines clear; colors, except reds and yellows, showing tendency to same bluish (or brownish) tone." Little has changed. Today, vertical visibility is reported in distance to the lowest cloud layer, "ceiling," and horizontal visibility is "the greatest distance in a given direction at which it is just possible to see and identify with the unaided eye in daytime, a prominent dark object against the sky at the horizon."[58]

Wind speed measurement near the surface, while not directly related to cloud study, deserves a brief mention because wind is part of the outdoor experience of cloud watching and is a part of daily weather forecasting. Wind, like clouds, has its own devotees and a growing shelf of literature.

Middleton, the historian of meteorological instruments, identifies dozens of types of anemometers for measuring the speed of wind, but the kind that became standard in the Weather Bureau, and with which most of us are familiar, is the cup anemometer, invented by the Irish astronomer Thomas Romney Robinson about 1846. Various improvements were made over the next few years, and the whirling cups were linked to different kinds of recording devices, the most common being a rotating drum with graph paper on which a pen marked the rising and falling velocity. Weather Bureau instruments also had indicating dials in 1871. After World War II the cup anemometer was replaced in many weather stations by the Pitot tube-shaped instrument.[59]

Like visibility scales, wind scales are often based on what the human eye and body experience. When British admiral Francis Beaufort announced his system of estimating and reporting wind speeds just two years after Luke Howard codified clouds, it was a thirteen-point scale ranging from 0 to 12 and from "calm," where smoke rises straight up, to "gale" (8 on the scale), in which branches break from trees and waves form spindrift, to 12, "hurricane force" with extreme damage.

With the development of the anemometer, gale force winds were clocked at 39 to 46 miles per hour, and hurricane winds were above 73 miles per hour. In 1972, the National Weather Service developed the Saffir-Simpson Scale measuring the size and intensity of hurricanes. Sustained winds of 74 to 95 miles per hour are considered Category 1, while winds above 155 are Category 5. Complicating the interpretation of wind speed still further is the use of terms such as "sustained wind" (the average over a one-minute period), "peak wind speed" (the highest instantaneous speed during a specified period of time), and "gust" (a sudden increase in wind speed of usually less than twenty seconds).[60]

Sky, wind, and clouds draw endless comment. The torrent of books and magazine articles is recent, but the small cloud on the horizon that anticipated them was A. C. Spectorsky's wonderful anthology *The Book of the Sky,* in 1956, in which almost a hundred writers from Ovid to Faulkner express their thoughts on the atmosphere. A decade later, Edinger sought to capture the wind for the AMS Science Study Series. Then, in 1984, after the threats posed by pollution, acid rain, and climate change became clearer, or murkier depending on your ideology, the prolific South African naturalist Lyall Watson published *Heaven's Breath: A Natural History of the Wind,* a quirky, fascinating meditation on the meanings of wind in human affairs. "Of all natural forces, wind is the most enigmatic," he writes. "It is big enough and strong enough to tear the largest living things on Earth out by their roots, and yet it can seep through a hairline crack. Wind is elusive, shifty, fugitive, difficult to define—and impossible to ignore." Recently, nature writer Jan DeBlieu trailed in Watson's wake with her book *Wind: How the Flow of Air Has Shaped Life, Myth, and the Land.* Watson's countryman Marq de Villiers whistles up wilder wind yarns in *Windswept: The Story of Wind and Weather.* Most of the major contemporary nature writers—Annie Dillard, Chet Raymo, Diane Ackerman, John McPhee, Doug Peacock, Barry Lopez, Robert Michael Pyle, Gary Paul Nabhan—have written on atmospheric phenomena. Nature writing is looking up.[61]

What is written on the wind may be invisible, but to cloudmen and cloudwomen the skies are an alphabet soup of messages waiting to be deciphered. Some, like the cheerful Englishman Gavin Pretor-Pinney, author of *The Cloudspotter's Guide: The Science, History, and Culture of Clouds,* and founder of the Cloud Appreciation Society, are unembarrassed to proclaim that clouds "are for dreamers and their contemplation benefits the soul. Indeed, all who consider the shapes they see in them will save on psychoanalysis bills." Perhaps he never met Billie Alonzo, who blocked my path one day several years ago on a busy street and thrust a pamphlet into my hand. Its pages were crowded with postage stamp–size photos of clouds and sun spots purporting to be faces, with the explanation that "spirits in the sun know the symbols we use. . . . The dour faced man is probably King Saul. The 2nd king is King David. The 3rd is representative of Christ . . . tormented by a horned devil. . . . It's a work of art portrayed on a 'canvas' 864,000 miles away." Pareidolia, finding the face of Elvis in a cumulus or an Indian princess in a stalagmite, is the occupational hazard of those who look too long and hard to find something beyond the miracles that nature already provides. Japanese see animals and symbols in the unmelted snow on hillsides and call them *yukigatas*. We see them in clouds but seldom agree on their shapes, partly because the objects are too far removed to easily point out eyes, noses, and ears. More often, we lack a shared frame of reference.[62]

Has sky awareness increased? Do we appreciate clouds more? Are we watching for the wind? These questions are impossible to answer in any concrete way, but if we turn to the weather page of many newspapers, we can see the results of a promotional gimmick that shows that elementary school children are participating in what is usually called "kids' weather eye." Each day's weather page has a 2-by-3-inch drawing from a local elementary school student. The drawing is in color and identified by the name of the artist, her or his age, grade, school, and teacher. I have no idea under what conditions these drawings are produced or how they are ultimately selected for publication, but, looking at a small sample I have clipped from various newspapers over the past three years, I have a strong feeling that these children's "artful scribbles," to borrow psychologist Howard Gardner's felicitous term, tell us a lot about humans and weather. Children of six or seven, according to Gardner, are in an age of synesthesia, "a time when, more than at any other, the child effects easy transitions, across sensory systems— when colors can readily evoke sounds and sound can readily evoke colors." I notice that the raindrops in the pictures by several seven-year-olds are larger than life, big blue bubbles hovering above and around a seemingly dry child. Is this lack of drawing skill or a depiction of the full sensory panoply of rain, its sound, feel, and motion, stopped because it is too rapid to be represented as more than a blur? Most of the art depicts storms, where there is more to show. One seven-year-old boy has rain falling like bombs from a cloud that also sends lightning bolts into a tree house and nearby home. Is this autobiographical or too much Weather Channel watching? Older kids seem less bound by meteorological fact— the sun gets dark glasses in one drawing—but these young artists have all been encouraged to look up and put the weather to work in new ways.[63]

Stopping Clouds

Photographers of weather have the same problem the child artists have—how to represent rapid change in a static medium. The early twentieth-century cloud photographers were satisfied to record the forms, study variations, and illustrate types. Both photography and meteorology emerged in the early nineteenth century as part of advances in science and technology, and both remained activities limited to a few professionals and serious amateurs. By the end of the century, however, technology had liberated photography from its "professors," while science began to limit meteorology to those educated in physics and mathematics. In response, photographers began to emphasize the aesthetic dimension of their work, minimizing the literalness of photographs in favor of personal feeling, symbols, and ideas. As Sarah Greenough explains, landscape photography was ceded to the growing number of amateurs and their simpler cameras. These new photography enthusiasts were not "Kodakers" but serious hobbyists who formed clubs and wrote for journals such as *American Amateur Photographer,* which began in 1889, and *Photo-Era,* begun in 1898. Adhering to pictorialism as espoused by Alfred Stieglitz and others, these photographers sought, in their work, to go beyond "the illusory world of objectivity in favor of a deeper and more universal truth residing in the vision of the artist."[64]

Clouds and weather, heretofore either accidental background or washed out in the developing process, became a more desired subject. Though he soon rejected pictorialism as remaining too literal, many of Stieglitz's photographs contain clouds and snow as part of the composition. In "The City of Ambition," a 1910 photograph of Manhattan skyscrapers from the waterfront, steam and smoke from the buildings drift upward into a cloudy sky, an inescapable comment on the urban skyscape. Stieglitz, influenced by the German physicist Heinrich Hertz's "contention that science did not explain facts, but provided rational models of patterns of experience," frequently contrasted the geometric angles of buildings with the organic forms of water and clouds. By 1922, when he produced "Equivalence—A Sequence of 10 Cloud Photographs," Stieglitz achieved an intensely personal connection with the sky and clouds, anticipating Bachelard's observation that "clouds are numbered among the most oneiric of 'poetic things.' They are the objects of daytime oneiric experience." According to one critic, Stieglitz "described his clouds as equivalents of his most profound life experiences. He maintained that his photographs were a picture of the chaos of the world and of his relationship to it." A similar dreamlike quality is seen in Edward Weston's cloud photographs of a few years later. These photos are not the fairyland performers of Luckiesh or Murchie but atmospheric Rorschach inkblots.[65]

The cloud photographs of an undeservedly neglected photographer, Theodor Horydczak, fall between the conventions of pictorialism and art of Stieglitz and Weston. Horydczak, about whom little is known except that he served in the army signal corps in World War I, was a commercial photographer in Washington, D.C., from the 1920s until the 1950s. The Library of Congress houses about 19,000 of his photographs, 22 of which are indexed under "Clouds." One particularly striking picture, nine-tenths of which is sky, shows a layer of stratocumulus seeming to touch a row of round-topped trees, while above the low clouds puffy altocumulus appear to fly directly toward the camera. In another, identified as taken with infrared film, a robust cumulonimbus hovers above a row of light-colored trees and one dark spruce near what appears to be a tennis court. Five electrical lines stretch from unseen poles across the top of the cloud. At least half a dozen of his

Clouds, Theodor Horydczak, 1930s. Library of Congress, LC-H833-C09-014.

cloud photos are apparently marred by such lines, possibly the result of working in an urban environment, but suggesting that Horydczak deliberately used the lines, as Stieglitz did his skyscrapers, as a contrast between the geometries of culture and nature.[66]

In the decade of drought, 1930–1939, the public became aware of the dust bowl, and the sky of the Great Plains became a familiar sight thanks to the work of the photographers of the Farm Security Administration (FSA) and other government agencies, federal and local. Dramatic photographs, such as Arthur Rothstein's of a man and two children hurrying toward the shelter of an abandoned building in a dust storm near Cimmaron, Oklahoma, in 1936, helped to fix the image of rural suffering and to define the problem as nature out of control. In Rothstein's photo, the sky and the land are almost indistinguishable because the wind has mingled them so thoroughly. Dust clouds were an unwelcome addition to the lexicon of nephology. Usually the sky was an inescapable backdrop to something else that the government photographers were trying to record. A 1940 photograph by Richard Hufnagle, working for a state conservation program, captured dozens of grasshoppers clinging to a fencepost, framed by a sky filled with towering storm clouds. The photograph subtly captures the environmental threats

Dust storm, Cimarron County, Oklahoma, 1936. Arthur Rothstein. Library of Congress, LC-USZ62-11491.

facing farmers—pests, hail, and floods—while playing with the conventions of cloud photography. The twisted fencepost strains to follow the drifting cumulus, while the strand of barbed wire and the low prairie horizon help provide both a base for the clouds and a scale to judge their size. Marion Post Wolcott, one of the great FSA photographers, made some of the best skyscapes. Four-fifths of her prairie scene is Montana sky, with soft clouds drifting over a rigid row of grain elevators; the dynamics of sky and land are subtlety juxtaposed. We call it a still photograph, but it conveys motion.

For the past fifty years, photography magazines have duly taken notice of sky and cloud photography every few years, without adding much to the conventional wisdom about backlighting and avoiding cluttered foregrounds. Some advice givers prefer cumulus; others, cirrus. Nobody likes gray skies. Florida, Arizona, and Colorado are frequently mentioned as places with reliably spectacular sky effects, but Texas, the Midwest, and the far West have produced their own advocates. At least five major books of cloud photography as art have appeared in the past twenty-five years: Ralph Steiner's *In Pursuit of Clouds: Images and Metaphors* (1985), Raymond Bial's *From the Heart of the Country: Photographs*

Clouds, barbed wire fence, and grasshoppers, Tripp County, South Dakota, August 8, 1940, Richard W. Hufnagle. Nebraska State Historical Society, Condra Collection, RG3474 PH 88 4886.

of the Midwestern Sky (1991), Wyman Meinzer's *Texas Sky* (1998), and, more recently, *Between Heaven and Texas* (2006) and Richard Misrach's *Sky Book* (2000). There are surely many more, but these are representative of the genre.[67]

Bial and Meinzer are excellent photographers, and their books are well worth study. Most of Bial's color photos were taken at dawn or dusk, and his skies and clouds are bright pinks, reds, oranges, and yellows, with some vibrant blues and shimmering grays. In this they are similar to the skyscapes of nineteenth-century artists like Albert Bierstadt and Thomas Moran and to the work of some contemporary midwestern artists. Meinzer's photos capture what he calls "the fugacity of the sky," its fleeting, ephemeral nature. More interesting for the purposes of this book is Ralph Steiner, whose career and place in American photographic his-

Clouds, freight train, and grain elevators, Carter, Montana, 1941. Marian Post Wolcott. Library of Congress, LC-USF 34-058135-D.

tory give him an eminence not likely to be attainted by the others. Born in Cleveland, Ohio, in 1899, Steiner studied chemical engineering at Dartmouth College, leaving in the 1920s to study photography in New York City. Adept with a movie camera, Steiner joined the creative young cameramen who helped filmmaker Pare Lorentz with the New Deal classics *The Plow That Broke the Plains* (1936) and *The River* (1937). Both films are notable for their visual poetry in scenes of the sky over the Great Plains and storm clouds over the Mississippi. Steiner was also one of the directors and cinematographers for *The City* (1939), produced by the American Institute of Planners for the New York World's Fair. *The City* opens with an idyllic wagon ride through the New England countryside in a re-creation of eighteenth-century village life. The boy on the wagon looks up at the sky and watches cumulus drift overhead. The clouds help symbolize a pastoral life. After an active career as a photographer and filmmaker, Steiner moved to Vermont and began taking cloud photographs. As he recalled:

> I came to photograph clouds for a most un-deep reason; approaching eighty, my walking apparatus began to limit me, and I found that clouds did not need walking toward. To photograph clouds the only walking I had to do

was to avoid trees. Almost all of my clouds were caught either from our cottage on the shore of an island off the coast of Maine or from the side of a mountain near Oaxaca—both places of few trees.[68]

Steiner is putting us on. His knees may be creaky, but his mind is on the same clouds that inspired Luckiesh, Sloane, and Murchie—clouds that "play all the parts that Shakespeare dreamed of as well as menageries of nightmare monsters." When the director of the Smith College art gallery invited him to display fifty-five of his cloud photos, Steiner asked visitors to write their own captions for them. When *In Pursuit of Clouds* was published, Steiner selected five to ten of the best comments for each. The result is a unique collaboration between a photographer who photographed clouds that "spoke to him" and viewers who saw somewhat different images and ideas in each photograph. For example, a photo of what looks like cirrocumulus, a "mackerel sky," dramatically backlit, gets labeled "The big doubt," "Man's doom," "Giant pillow," "Inflated ego," "Threatened land," and "Weight." A cirrus with serrated edges is called "Loch Ness monster," "The dragon spits venom," "Feed me," "Reaching to devour," and "Out of the primeval swamp." What seems to be a lenticular cloud formed over a mountain draws this set of responses: "Reclining nude wondering where her upper torso has gone," "And she sits on the bosom of the land," "Caress of a lovely cloud," "Touch me ever so gently," "God gives Mother Earth a pat on the head," "Juno's buttocks land on Mt. Vesuvius," "Soggy celery," and "Heaven kisses the earth." Thus does a room full of Smith College art patrons see clouds and images of clouds. The clouds are not quite a Rorschach, nor does everyone experience pareidolia, but asked to respond to the stimulus of a cloud, we all cooperate, we know the game.[69]

Richard Misrach (b. 1949) is a sky and cloud photographer of a different order. For more than thirty years, Misrach has been "excavating the sky," to use a phrase of nature writer Rebecca Solnit. In numerous exhibitions and two recent books, Misrach challenges the viewer to follow his eye into the heart of a cloud, beyond its named form, to the point where visible water vapor seems so dense that it occupies the sky from within. Clouds are less transitory in his photographs than in, say, John Day's beautiful scientific illustrations. Misrach set out to "deconstruct the convention of landscape" by using his camera to create what we see. Playing with those conventions in *The Sky Book,* he labels each of the more than fifty photographs with scientific, topographic, and chronological information. For example, "Clouds (Altostratus undulates), Mono Lake 5.26.99 8:52 P.M.," in which 90 percent of an 8-by-10-inch photograph is a dark blue, late spring evening sky, with slivers of robin's egg blue and pink and a low horizon of yellow. As Solnit observes, these photographs "portray the sky too eternal, too mutable, and too ubiuitous to supply very specific information about time and place. . . . The sky is a blankness, a meditation room of sorts, and works in this tradition tend to invite contemplation of the sublime, the void, the pure visual experience." But, she continues, the titles make us think about geography, history, and biography. Misrach has called his photos a "dysfunctional journal," and there is certainly more than a hint of obsessive wandering in the Great Basin and desert Southwest.[70]

In 1997, Misrach moved to Berkeley, California, and began photographing the Golden Gate Bridge, thirteen miles away, across San Francisco Bay. Eight years and more than 2,500 photographs later (Misrach says his immediate throwaway rate is 97 percent), we can see about 70 of these photographs in his book *Golden Gate*. These photographs, mostly 8 by 11 inches and labeled only by date and time, are in some ways even more remarkable than the sky photos. When visible, the bridge gives scale and geographic coordinates to the panorama of clouds. Like Horydczak's power lines or Post Wolcott's railroad tracks, the bridge also sets the geometry of the machine against the chaos of nature. An essay in the book compares Misrach's repetitions of clouds and bridge to Cézanne's variations of Mont Sainte-Victoire, but this diminishes both artists, I think. A photographer communicates from brain to eye to film, without the vagaries of hand and brush. The photographer has only an instant to capture what he sees and feels, not the hours, days, or more, allowed the painter. The viewer of Misrach's photographs of the sky and clouds over the Pacific beyond the Golden Gate or the fog cloaking the bay and hills knows that the bay looked this way for only a few minutes of one day out of a hundred and marvels at Misrach's persistence, luck, and skill. Then, his eye is drawn through the filaments of the suspension cables into the sky beyond and the infinity of the atmosphere. Like the painters and sky artists discussed later in this chapter, Misrach maps the sky in ways that allow the viewer to move through the atmosphere and return. The Golden Gate is both a symbolic and a literal bridge between his two worlds.[71]

As amateur photographers pointed their cameras at the sky to express their feelings, scientists were experimenting with ways to make their images of the atmosphere more useful in their meteorological studies. One of goals was to photograph the whole sky. Early in the twentieth century Oliver Fassig of the Weather Bureau began experimenting with a camera that could record the whole dome of the sky. Basically a water-filled pinhole camera with film that rotates beneath the lens, the apparatus was capable of a 180° view. By the 1920s, the water-filled camera was replaced by a glass hemisphere placed behind the aperture. An alternative method is to photograph a curved mirror that reflects the sky; one researcher used a Christmas tree ornament. Improvements of lenses continue to be made. The problem of projecting the images onto flat surfaces has largely been solved by making the distance from the center of the image directly proportional to the angular distance from the zenith, according to a report by two Australian scientists. Since recording clouds in the whole sky is part of the problem of establishing "ground truth," whole-sky cameras offer a way of correcting human observer bias. Studies have shown that cloud observers disagreed with satellite data most frequently in the 3–5 okta range, seeming to prefer cloud amounts of three-eighths to five-eighths of the sky when satellite images showed somewhat higher degrees of cloudiness. Even with further technological advances in imaging systems, the whole-sky photograph will continue to fascinate and challenge photographers and meteorologists.[72]

Motion picture photography caught the attention of Secretary of Agriculture James Wilson before World War I, and he created the Department of Motion Pictures within the Extension Service of the Department of Agriculture. Within a few years the department was the largest producer and distributor of documentary films in the world. Despite this departmental "head start" and the obvious usefulness of motion pictures for the study of clouds and weather, the Weather Bureau was slow to get the message. Finally, in 1925, C. C. Clark, assistant chief of the Weather Bureau, endorsed the use of movies for education of the public. The first Weather Bureau film was *Exploring the Upper Air,* made with the assistance of LeRoy Meisinger at Scott Field near St. Louis, shortly before his death in a ballooning accident. The second film, *Watching the Weather Above,* showed the use of kites, balloons, dirigibles, and airplanes in gathering upper-air data for aviation weather forecasts. Plans were under way for films on weather forecasting, clouds, fruit frost warnings, and river and flood forecasts. Motion picture technology offered yet another tool—time-lapse photography—for cloud study. The great British meteorologist Leslie Fry Richardson, who made important contributions to numerical weather analysis, used time-lapse photography to study smoke diffusion as a cause of air pollution in 1920.[73]

Time-lapse was born with the first hand-cranked cameras. All the cinematographer had to do was crank the film more slowly than the normal twenty-four frames per second. Then, when the film is developed and played at normal speed, objects will appear to move faster than they could in human time. Early movies used the technique for comic and thrilling effects. Undercranking and its opposite, overcranking for slow-motion effects, were used by efficiency experts and other industrial filmmakers for time and motion studies. Walt Disney Studios brought time-lapse photography to new aesthetic heights in such films as *Secrets of Life,* which speeded up the process of flowers blooming. The meteorological community first took notice of time-lapse photography in 1934 when a filmmaker named Delbert M. Little screened, at the AMS meeting, his 16mm movies of fog and clouds, speeding up their movement sixty times. In *Watching for the Wind,* Edinger mentions seeing time-lapse films of fog surging up and rushing back down mountainsides, leading him to muse on the meaning of time measured in different scales. Meteorological educator Mike Mogil made a short film for children, *Our Sea of Clouds,* that he plays backward as well as forward, "just like you do in a research lab"—playing with time, playing with clouds. Today, time-lapse photography is so familiar that it seldom receives comment. We have all sat through "educational" films of seeds germinating, fruit rotting, and clouds whizzing through the sky. "I'll just put these clouds in motion," says the TV weathercaster, and we see the day's storm fronts pass over a map in less than thirty seconds, clouds exploding and blackening like popcorn in an overheated pan.[74]

Environmental historian Finis Dunaway has examined the power of such images in the environmental movement of the past century, noting the ways in which photographs of nature by Ansel Adams, Eliot Porter, and others were used by the Sierra Club and similar organizations to elevate a certain kind of nature, usually what they defined as wilderness, to the status of the sacred. Images of nature, in coffee-table books like *This American Earth* in 1955, were meant to be antidotes for post–World War II materialism and conformity, critiques of the belief in technological progress, and, perhaps, a substitute for conventional religion. Coffee-table books and posters paradoxically identified the purchaser as environmentally engaged, while converting the gloriously reproduced trees, canyons, and mountains into objects for consumption. Interestingly, the work of Bial, Meinzer, and Misrach falls into the coffee-table book category, but these photographers have not been seen as advocates for clean air. Cloud photos seem more like portraits than like landscapes. Ansel Adams took many memorable photographs of clouds, although they were usually framed by the canyons of Yosemite or served as backdrops for the Sierra Nevada. He willingly lent his support to Jack Borden's sky awareness conference in 1981, writing: "Our sky, above each human, is our window-on-our-universe, our passageway through dream and vehicle to the stars. Our air is changing from a crystalline vision to a distorting lens. How sad, and so much more so for our children and grandchildren. It does seem we have taken our living canopy of air for granted and we are killing it." Porter, on the other hand, was openly contemptuous of skyscapes, explaining that "everyone knows [the sky] is blue, so we might as well take it for granted and omit it, concentrating instead upon its indirect effect." The sky as wilderness has yet to engage the majority of American environmentalists.[75]

Screening Storms

Everybody has a favorite movie weather scene, even if the weather was totally manufactured on a studio set, which was largely true before improvements in camera lenses and film and changes in the aesthetics of cinema in the 1950s made outdoor filming feasible and expected. There were exceptions. D. W. Griffith insisted on filming all the exterior scenes of the melodrama *Way Down East* (1920) outdoors, including the climactic blizzard scene. As the film's star, Lillian Gish, remembered it:

> He wouldn't be satisfied with the fake fury of a studio storm. . . . The blizzard finally struck in March. . . . Again and again, I struggled through the storm. Once I fainted—and that wasn't in the script. I was hauled to the

studio on a sled, thawed out with hot tea, and then brought back to the blizzard, where the others were waiting. We filmed all day and all night, stopping only to eat standing near a bonfire. We never went inside, even for a short warm up. The torture of returning to the cold wasn't worth the temporary warmth. The blizzard never slackened. . . .

Above the howling storm, Mr. Griffith shouted: "Billy, move in! Get that face! That face—*get that face!*"

"I will," Billy [cinematographer Billy Bitzer] shouted. "If the oil doesn't freeze in the camera."[76]

Anyone who has seen Gish's performance can believe her harrowing account. In the conventions of the story, the innocent girl who has been made an outcast by the villain can be reintegrated into society only by surviving an ordeal like a blizzard and then being rescued by the hero. Storms—blizzards, tornadoes, hurricanes—are standard literary and cinematic devices to effect transformations in characters.

Life-threatening, destructive storms are the most spectacular weather events and give full rein to filmmakers' talents. An early and memorable example is *The Wind,* a 1928 movie by the Swedish-born director Victor Seastrom (Sjöstrom), with a screenplay by Frances Marion, and starring Lillian Gish as Letty Mason, an eighteen-year-old orphan from Virginia who journeys by train to live with her married cousin Bev and his wife, Cora, on the prairies of west Texas. Based on a novel by Dorothy Scarborough, the movie is an anthology of atmospheric effects within the limitations of black-and-white silent film. One of Letty's fellow passengers is Wirt Roddy, a smarmy fellow who flirts with her and warns her that the winds destroy women, aging them and driving them mad. He invites her to stay with him in Fort Worth, but she laughs off the proposal and questions the power of the wind. Letty is innocent and lives chiefly in her imagination, where nature does not intrude. Wirt persists by telling her about the cyclone, "a tornado of a special deadly brand we have out here," he bragged. "It's a bull buffalo of a wind that whirls in a circle like a dancing dervish, while it races ahead at the same time. It's shaped like a funnel, small at the ground, and spreading out wide at the top." In the film, sand blows against the railroad car window, etching the faces of shadowy monsters for anyone with a touch of pareidolia to see. Realizing that he has succeeded in frightening Letty, Wirt calms her by calling her attention to the beauty of the sunset.

As she looked, she saw before her a sky incredibly blue, of a clear, pure color such as she had never seen before. Far, far ahead to the west where the earth met the sky, the sun rested, a great ball of flame, its rays spreading outward and upward to the heavens it had left. In that high, clear altitude, where one can see great distances, the sun seemed at once remote and close at hand.

The wind-blown clouds above were touched to brilliant orange, rose, and gold, and all imaginable shadings of rainbow hues.[77]

Letty's meteorological education has just begun. Lige Hightower, her cousin's neighbor, tells her about a black stallion (white in the movie), which appears when the wind is howling. She learns about drought and sandstorms, and meets "Old Pedro," who predicts weather by the bellowing of the bulls, the *cantar* of the locusts, and a dead rattlesnake placed belly up across a fence. Finally, she experiences a "blue" or "cross-eyed" "norther":

As she turned toward the north, she saw a puny cloud, slight and fragile, touching the prairie's rime, a white, feathery nothing, like a ball of thistledown floating along the ground. But as she looked, it grew and darkened. Swiftly it spread over the sky until it blotted out the blue, till it hung, a black pall, over the wide heavens. It happened so quickly, with such incredible rapidity that Letty could scarcely believe it, even while her eyes watched it. . . . She heard the norther as it came roaring over the plains. . . . She saw the wind as a black stallion with mane a-stream, and hoofs of fire, speeding across the trackless plains, deathless, defiant! . . . His proud neck arching, his eyes glancing flames, he raced towards her across the sand— supernatural, satanic, the wind of the North![78]

Needless to say, Seastrom makes the most of this visually, with a montage of a galloping white horse, clouds, blowing sand. In the opening scenes Gish wears a floppy, wide-brimmed hat that lashes her face. Movie palace pianists and orchestras rattled the chandeliers with accompanying wind and tornado music. Scarborough's descriptions of the wind and their effect on Letty are vivid, but Gish, hand clutched to her ears, eyes rolling, hair literally standing on end, conveys the force of the wind and the mental collapse of the heroine in forms unavailable to the novelist. We begin to understand that the wind she hears is as much internal as it is external. Scorned by her cousin's jealous wife and desperate for protection from her fears, Letty marries Lige even though she does not love him.[79]

This deceit and Letty's repressed sexual desires lead her to believe that she is stirring up the wind and sand because she thinks about the wind constantly. Watching the "whirling curtains of dust," Letty is horrified to discover that "the wind was no longer naked and invisible. It had clothed itself with those swirling veils that revealed its obscene antics, its horrific gestures. It was a thing unbearable to *see* the wind!" Letty's madness convinces her that she can control the wind by whispering spells and making signs. She clinches her fists to capture the wind. It is in this mental state that the lustful Wirt visits her and puts more fears in her head. Guilt-ridden, she asks her husband to let her go away for the winter and then tells him that she does not love him. Lige leaves hurt and confused. Wirt returns, and Letty realizes that he is part of her problem, but fear of the storm and her own wayward passion prevent her from sending him away. In the formulas of both novels and movies of the 1920s, a night spent together is tantamount to having sex, and Letty wakes up filled with remorse. When Wirt continues to taunt her, she shoots him and drags his body out of the house and tries to bury him, but the wind keeps uncovering him. "The wind was angry that she read its thoughts. . . . It defied her, challenged her, mocked her. . . . The wind was even now whispering the truth in [Lige's] ears—shouting it at him!" At this point in the novel, the wind rises "almost to cyclonic fury," and the "curtains of sand" are "writhing with hideous suggestion." Letty runs out toward the tornado, "the wind that was at last to have its way with her." No yellow brick road for Letty. The MGM studio, convinced that movie audiences would not like this bleak ending, ordered Marion and Seastrom to reshoot. In the new version, Lige returns and reassures the hysterical Letty that the wind has not exposed the corpse. "Wind's mighty odd," Lige says on the final title card, "if you kill a man in justice—it allers covers him up!" He takes her in his arms, and they face the wind and their future together.[80]

Melodramatic and a bit corny even in 1928, Seastrom's movie is still one of the best visualizations of weather to come out of Hollywood. Filming not in west Texas but in the Mojave Desert of Southern California using eight airplane engines as wind machines, Gish again endured some painful temperatures and considerable discomfort. Audiences in newly air-conditioned theaters could empathize with Gish's suffering and still enjoy the benefits of artificial weather.

If sensual cyclones were not enough for audiences of the 1920s and 1930s, Hollywood was ready to produce hurricanes. John Ford's movie of that name, appearing in 1937, is still considered a remarkable technical achievement. *The Hurricane* stars Dorothy Lamour as Marama and Jon Hall as Terangi, Polynesian lovers who are separated for eight years by an authoritarian French colonial governor (Raymond Massey) who sentences Terangi to prison for fighting with a white man and keeps adding to his sentence each time he tries to escape. The climax to the story, based on a novel by the authors of *Mutiny on the Bounty*, Charles Nordhoff and James Norman Hall, is a search for Terangi during a hurricane. Because Terangi helps the wife of the governor (Mary Astor) to survive—he lashes her, his wife and daughter, and himself to a palm tree—he is finally allowed to escape with his family.

Although Ford shot 140,000 feet of film, enough for several movies, in American Samoa, the hurricane was created by special effects technicians James Basevi and Robert Layton, who built a replica of the native village and a 200-yard lagoon on the United Artists studio backlot. Using several twelve-cylinder Liberty motor wind machines, wave machines, fire hoses, and holding tanks, they poured thousands of gallons of water on the hapless actors. Lamour and Astor suffered almost as much as Gish, and audiences judged the movie a great success. Thomas Moulton won an Academy Award for Best Sound, Thomas Mitchell received a nomination for Best Supporting Actor, and Alfred Newman a nomination for Best Score. The sounds and sights of the final minutes of the movie are indeed horrific. A decade later, John Huston directed a credible hurricane in *Key Largo,* but Humphrey Bogart, Lauren Bacall, and Edward G. Robinson got to ride it out in a hotel. Like other movie storms, the tropical cyclones in *The Hurricane* and *Key Largo* are symbolic of nature's power over human will, with the ability to thwart the best-laid plans. The storms deliver justice to oppressors, free their victims from injustice, and allow the survivors a fresh start in life—a Hollywood dream shattered by Hurricane Katrina.[81]

A few American movies of the cold war period and even later—*Our Man Flint* (1966), *The Spy Who Loved Me* (1977), *The Avengers* (1998), and *Storm* (1999)—feature weather as a weapon. Others—*White Squall* (1996), *The Ice Storm* (1997), *The Perfect Storm* (2000), and *Cast Away* (2000)—used storms in traditional ways, as tests of character or reflections of emotional states. Looming behind these re-creations were, of course, the nightly television dramas provided by tornado season (spring and summer), hurricane season (summer and fall), and blizzard season (fall and winter) in which nature provides the special effects. It fell to television to provide, for those still not storm-sated, a series of weather thrillers in which tornadoes and superstorms wreak devastation on various parts of the nation. *Tornado* (1996), *Storm Chasers: Revenge of the Twister* (1998), *Devil Winds* (2003), *Lightning: Bolts of Destruction* (2003), *Category 6: Day of Destruction* (2004), and *Category 7: The End of the World* (2005) are examples. The latter two, produced by CBS TV, linked their apocalyptic visions to global warming.

In *Category 6,* actors Brian Dennehy and Randy Quaid play a meteorologist and a storm chaser, both fed up with bureaucratic inertia and power-company greed.[82] In *Storm Chasers* and *Lightning,* female meteorologists battle gender barriers as well as government corruption and public indifference. In *Tornado,* a drama produced for Hallmark Entertainment, an auditor sent to pull the plug on a tornado research project gets her comeuppance. The ancient formula of women in peril works for tornado melodramas just as it did for vampire movies. Scientists are depicted as well-meaning but often unhinged by their obsession with the weather. An early exception to this stereotype, and the only movie I know featuring the importance of meteorology in World War II, is *Destination Gobi* (1953), starring Richard Widmark as a navy weather observer stationed, with his crew, in the desert. They radio their forecasts, despite windstorms and Japanese attacks, with the help of friendly Mongolians to whom the officer promises saddles for their assistance. The request forces the navy to cooperate with the army cavalry, adding a layer of bureaucratic comedy to the movie. Based on a purportedly true story by Edmund G. Love, *Destination Gobi* is unfortunately no longer available for screening.

Made-for-TV storm operas owe a great deal to two popular movies, *Twister*

(1996) and *The Day after Tomorrow* (2004), which continue to generate heated discussions among meteorologists and climatologists. They usually express contempt for these movies, but they cannot stop talking about them. *Twister* has received more attention, beginning with science reporter Keay Davidson's book *Twister: The Science of Tornadoes and the Making of an Adventure Movie.* Davidson tells the story of tornado research and gives the movie just a few pages, mostly focused on its special effects. We learn that the director, Jan De Bont, hired a meteorologist, Vince Miller, to warn the 250- to 300-member cast and crew of wind or lightning hazards while they filmed in Wakita, Oklahoma, and Eldora, Iowa. Davidson describes how Wakita (pop. 300) was partially destroyed and rebuilt by Warner Brothers–Universal Studios, how the 3,500-pound replica of a tank truck was raised 75 feet above the ground and exploded, and how the tornadoes were produced digitally. That the script was written by the prolific novelist Michael Crichton and his wife, Anne-Marie Martin, and that Crichton was one of the producers, explains both its tornado-as-monster plot and the greedy-scientists-must-die subplot.[83]

Recently, Crichton has gained some notoriety for his novel *State of Fear* (2004), in which evil environmentalists fake evidence for global warming to advance their own agendas. Scientists in Crichton's earlier work are devoured by alien viruses, gobbled by dinosaurs, and clobbered by tornadoes—nature is scary. *Twister* was deemed so scary that it received a PG-13 rating "for intense depiction of very bad weather." "In *Twister* there are good scientists—Bill Harding, played by Bill Paxton, and his soon to be ex-wife, Dr. Jo Harding, played by Helen Hunt—and bad, Dr. Jonas Miller (Cary Elwes), who has copied Bill's invention for measuring tornadic winds. Bill, having failed to raise money to complete his research, has become a TV weatherman and fallen in love with a reproductive therapist, Melissa (Jami Gertz). When he returns to Oklahoma to get Jo to sign the divorce papers, he discovers that she is in a race with Jonas to deploy the storm sensors. Bill joins her team, and they begin to chase tornadoes. Melissa tries to keep up with Bill during one chase but realizes that she will never understand or approve of his tornado obsession and leaves him. Jonas is killed in his attempt to penetrate a tornado, but Jo and Bill make one last and successful attempt to scatter their sensors (tornado reproductive therapy?), and they survive the tornado by strapping themselves to a water pipe in an outhouse! Tornadoes are a visceral and sexual experience, and this movie nails it.[84]

Hostile critics of the movie miss the point. Meteorologists Daphne and Richard Thompson complain that tornado chasers do not whoop, "Whooeee! We've got ourselves an F3!" But Keay Davidson quotes one chaser who coined the word "stormgasm" for the "animal-like sex sounds that chasers typically make while watching a tornado. Anyone who has seen the more popular chase tapes can recall the 'oh baby . . . oh baby' grunting and howling sounds." The Thompsons also complain that the sensor technology ("Dorothy") shown in the movie was out-of-date when the film was made, but Crichton wrote his script in 1986, after learning about an invention by meteorologists Al Bedard, of the Wave Propagation Laboratory in Boulder, and Howard Bluestein, of Oklahoma University (OU). Bedard and Bluestein's TOTO (Totable Tornado Observatory) was a 396-pound instrument package designed to be deployed from a pickup truck in less than thirty seconds into the path of a tornado. Bluestein and his students experimented with TOTO from 1980 to 1985 but abandoned this approach when it became apparent that it was too difficult to place the instrument directly in the path of a tornado. Bluestein claims TOTO was the inspiration for "Dorothy," but, during the 1986 tornado season, Fred Brock and his associates, also from OU, deployed smaller instrument packages (called "turtles") that were closer to the "Dorothy" concept.[85]

Whatever the inspiration for Crichton's script, the National Severe Storms Laboratory (NSSL) in Norman is happy to take credit for making "several positive contributions to the production of the film." NSSL also consulted with Universal Studios on the studio's *Twister* ride in Orlando and was allowed to add severe weather safety tips to the tornado-experience program. Applications to the meteorology department at OU soared, and Universal Studios gave the university a grant to develop a mobile radar unit. The Twister Museum in Wakita, Oklahoma, and the Twister Bed and Breakfast in Eldora, Iowa, are also trying to market their association with the movie, proving again, if further proof were needed, that Americans can find ways to profit from disasters, even simulated ones.[86]

Film scholar Scott MacDonald, viewing *Twister* in the context of American landscape art, sees it as the quintessential American movie, exhibiting the historical paradox, familiar in most Westerns, of idealizing nature's wildness while seeking to control it. Bill and Jo are aroused by tornadoes, but the knowledge they acquire will ultimately make these storms more predictable and familiar. It is even more American (despite director De Bont's Dutch origins), he argues, in its use of helicopter shots of moving vehicles. "In the European films [of the 1960s], travel in an automobile was filmed either from inside the vehicle or from directly in front of the windshield, creating (at least for Americans) a sense of claustrophobia. In American films, automobile travel was filmed from outside the vehicle, often from above the vehicle and moving along with it, creating a sense of exhilaration and freedom." In their use of special effects to create a believable tornado outbreak, MacDonald argues, the filmmakers reveal another paradox. Though the movie praises scientists who risk their lives to discover the nature of a storm for pure science, the real heroes of the movie are the special effects technicians, who, with the goal of making money for the studio, have equaled or surpassed such storm chasers' videos as seen on The Weather Channel. A further irony, he points out, is that the credits for the film, including the numerous digital artists at George Lucas's Industrial Light and Magic studio, roll over the only shots of natural skyscapes used in the movie. Perhaps historian David Nye is correct in arguing that the sublime in nature has been largely supplanted by what he aptly calls "the technological sublime," which includes the movies. Weather, I think, combines the natural and the technological, since it requires both the personal experience of storms in the landscape and their translation through the electronics of computer modeling and computer graphics to fully appreciate its terror and beauty.[87]

The relatively apolitical *Twister* encouraged the National Severe Storm Laboratory to identify with the movie. The more overtly political movie *The Day after Tomorrow,* written and directed by German-born Roland Emmerich and produced by 20th Century Fox in 2004, moved NASA to send a memo to its employees forbidding them to do "interviews or otherwise comment on anything having to do with the film." The memo was sent on April Fools' Day, but that irony was lost in ensuing battles between NASA scientists, who typically have spoken out on climate change, and NASA administrators, who have not. *The Day after Tomorrow* was loosely inspired by a book, *The Coming Global Superstorm,* by Whitley Strieber, a writer of horror novels and a self-described alien abductee, and Art Bell, a radio talk-show host specializing in the paranormal. The movie is a simple disaster story about a series of hurricanes, tornadoes, blizzards, and tidal waves that destroy most of the major cities of the United States. A TV weatherman is clobbered by flying debris, and wolves escape from a zoo and roam Central Park. The cause of these storms is attributed to anthropogenic global warming.[88]

Dennis Quaid plays paleoclimatologist Jack Hall, whose warnings about a possible snow blitz are heeded by NOAA but dismissed with a sneer by a Cheneyesque vice president, played to perfection by Kenneth Welsh. When New York City is struck by a tidal wave—the movie's opening-day ad shows water covering all but the upraised torch of the Statue of Liberty (that enduring Hollywood icon of doomed civilization)—Jack fights his way north to rescue his son, Sam, played by Jake Gyllenhaal, who has taken refuge with three of his high school classmates, a librarian, and a homeless man in the New York Public Library. There, they survive wrapped in coats from the lost and found and subsisting on vending machine snacks. The serious consequences for civilization of climate change are revealed in a scene in which Sam's classmates begin collecting books to burn in order to keep warm. They debate whether to burn the works of Friedrich Nietzsche—opponent of religious dogmatism and advocate of the übermensch—but avoid that decision by finding shelves of tax codes, which, having been made useless by the new ice age, are more appropriate fuel.

Predictably, many climatologists dismissed the events in the movie as impossible, while some conceded that disastrous climate changes will take place over the next several centuries unless current trends are halted. Fox News commentator Stephen J. Milloy, ignoring the fact that the film was produced by the movie division of the company for whom he works, dismissed the plot as "nauseating elitism." Patrick Michaels of the Cato Institute denounced the movie and the science that supports global warming. Greenpeace, seizing the opportunity to point out that the Cato Institute, a right-wing think tank, is supported by ExxonMobile, created a parody Web site, "The Day Is Today," "a disaster produced and directed by ExxonMobile and George W. Bush."[89]

The appeal of natural-disaster movies and novels may be what writer Mark Svenvold calls "catastrophilia," a peculiarly American mix, according to Svenvold, of millenarianism, the sublime of the wilderness, and the fear of banality. He contends: "The secret goal of all catastrophe writing is this . . . to bring us as close as possible, by proxy, to the edge of an abyss, and thereby render the world changed, an aftermath, as if to say, by way of conclusion, Yes, the world is tragic, but at least it is not banal." Further, "Catastrophilia is decoupled from any sort of vision. It is an absolute end, a phenomenological cul-de-sac, a sideshow tent, presenting a dead-end tautology, like a dog with five legs, or like a celebrity—famous because it is famous. It offers a distraction from history, rather than an engagement with it."[90]

This seems too clever by half. Some fundamentalist ranters may welcome Hurricane Katrina as a sign of the apocalypse, and some environmentalists, like abolitionists before them, want you to feel the sting of nature's lash so that your vicarious suffering will get you to change your driving habits and vote for green candidates. Ultimately movies such as *Twister* and *The Day after Tomorrow* appeal to audiences for different reasons. If there is one common element in their allure, it is fantasy play, the challenge of imagining dire situations and finding ways out of them. We are satisfied or disappointed with movie disasters to the degree that they resolve their threats in ways that meet our expectations. We imagine how we would succeed in getting "Dorothy" into the tornado and how we would save a loved one from a blizzard. "Once upon a time," begins the commonest of childhood stories, someone is in peril, and someone is rescued, or not, the end. The natural disaster movie does engage us in history, a history as old as humankind.[91]

Another category of weather movie also needs to be considered—the weatherman movie. In *The Weather Man* (2005), Nicholas Cage's character, the evocatively named David Spritz, is a successful weathercaster at a Chicago TV station. Though Spritz has received a tentative offer from *Hello America,* a network morning show in New York City, he is profoundly depressed because his father, a prize-winning novelist (Michael Caine), is dying of cancer. David is separated from his wife, his fifteen-year-old son is in drug counseling, and his twelve-year-old daughter is overweight and wears sexually provocative clothes. Each attempt to reconnect with his family goes awry. He is good on camera but does not understand the mechanisms of the weather he forecasts. "It's all wind," is all he understands. This is beautifully visualized in a sequence in which he takes up archery in an effort to find something he and his daughter can do together. The wind's effect on the arrows is a metaphor for the chaotic forces of illness and marital discord in his life. He is also frequently pelted with junk food by people who recognize him as a celebrity, leading him to the realization that TV weathercasting is itself a kind of mental junk. The fact that he is successful in the virtual world of television, but a mediocre writer and a failed husband and father in the real world, plays with the idea that the former has replaced the latter as a measure of success. In an inconclusive ending, David reconnects briefly with his son and daughter before leaving for a new start in New York.[92]

The ultimate weatherman movie, *Groundhog Day,* has been called a deeply religious experience by Buddhists, Jews, Christians, and the Falun Gong, as well as a quasi-Jungian crisis of masculinity, and one of the greatest comedies of all time. NOAA's National Climate Data Center (NCDC), not to be outdone by the National Severe Storms Laboratory in cashing in on the movies, created a "Special Climate Report" for Groundhog Day 2006, complete with a history of the ceremony and a comparison of the marmot's forecast accuracy compared with NCDC data since 1988. While this level of attention to what appears to be another Hollywood screwball romance can be explained by the ever-escalating hype needed to sell popular culture products, the lively script by Danny Rubin and Harold Ramis, coupled with Ramis's assured direction, gives the movie the ineffable feel of a classic. Phil Connors, played by Bill Murray, is an egotistical Pittsburgh television weatherman sent to cover the annual February 2 celebration sponsored by the town of Punxsutawney, a small coal town in western Pennsylvania, in which a groundhog named Phil is removed from his burrow each year to predict whether there will be six more weeks of winter or an early spring. The celebration, which has been going on since 1887, thus predating the U.S. Weather Bureau, is one of the country's most enduring examples of "fakelore," invented traditions designed to enhance community identity and attract tourists and business. Conceived by Punxsutawney's newspaper editor in the spirit of the numerous public celebrations and historical pageants—Memorial Day, Arbor Day, Labor Day, Mother's Day—begun in the years between the Civil War and World War I, Groundhog Day was inherently a parody of these and similar events.[93]

To parody a parody requires a deft touch, which the filmmakers have. Phil forecasts that an approaching snowstorm will miss western Pennsylvania, then travels to Punxsutawney with Rita Hanson, his producer, played by Andie MacDowell, and a cameraman, Larry, played by Chris Elliott. Phil is scornful of the entire event and makes snide remarks during the taping. After completing their work, they depart for Pittsburgh but are stopped by the highway patrol because snow has blocked all the roads. Forced to turn back to Punxsutawney, Phil retreats to the bed-and-breakfast he stayed in the night before, awakening on what he thinks is February 3, only to discover that for everyone else it is February 2. He is even surlier for this broadcast, much to the consternation of his colleagues. Snowbound, they spend yet another night in town. Forced to repeat Groundhog Day a third time, Phil simply refuses to do the broadcast and tries to tell Rita that he is experiencing the day again and again. She sensibly suggests he see a doctor, the doctor sends him to a psychiatrist. Phil considers the implication of endlessly repeated days, of life without future consequences, and begins to misbehave. Many

Groundhog Days pass; we glimpse Phil repeating his encounters with the citizens of Punxsutawney with endless variations. Increasingly desperate, he tries to kill himself in various ways and even kidnaps the groundhog and drives off a cliff. Finally he seems to accept his fate, begins to do good deeds, and falls in love with Rita. The meteorological signs of the seriousness of their romance are the snowman they build and the ice sculptures that Phil learns to make. Snow has long been a metaphor of beautiful and unobtainable women; a snowball thrown by such a woman is a taunt and a promise. Chaplin uses it this way in *The Gold Rush,* and Frank Capra follows suit in *It's a Wonderful Life* when George Bailey throws a snowball at his future wife. Snow also signals that George has returned to life. When Phil kisses Rita for the first time, it begins to snow. The snow that Phil failed to predict trapped him in time. His acceptance of snow and his love for a woman who represents nature restore the harmony he had disrupted by mocking nature and claiming that he "makes the weather."

For the religiously and philosophically minded, *Groundhog Day* seems to be "a stunning allegory of moral, intellectual, and even religious excellence in the face of postmodern decay, a sort of Christian-Aristotelian *Pilgrim's Progress* for those lost in the contemporary cosmos." So thinks Michael P. Foley, a professor of theology quoted by Jonah Goldberg in the conservative political journal *National Review.* Goldberg finds other enthusiasts, who see the groundhog as the resurrected Christ and the movie as an exponent of "the doctrine of God's 'middle knowledge,' first put forward by the 16th-century Jesuit theologian Luis de Molina, who argues that human free will is possible because God's omniscience includes His knowledge of every possible outcome of every possible decision." *Groundhog Day* was chosen by the Museum of Modern Art in New York City for the series "The Hidden God: Film and Faith," but Goldberg prefers to view it as a commentary on Nietzsche's doctrine of the eternal return. How would knowing that life is nothing but endless repetition shape your behavior? My own response to the movie was to be reminded of the Spanish philosopher Miguel de Unamuno's dictum: "If it is nothingness that awaits us, let us make an injustice of it . . . let us fight against it quixotically."[94]

This is the decision Phil eventually makes, after enjoying a period of nihilism. February 3 finally arrives; Phil wakes up with Rita beside him. By embracing the natural and rejecting the inauthentic, Phil is allowed to rejoin the human race. If *Wall Street* (1987) was, as many have claimed, the perfect movie expression of the Reagan era, *Groundhog Day* is the epitome of the Clinton years, the sardonic comeuppance of those who disdain the ethics of their office. Moreover, like Phil the weatherman, Americans had to endure, ad infinitum, television news commentary on thongs and the semen-stained dress.

Hollywood movies are, of course, not the only cinematic way of seeing the weather. Educational filmmakers have made numerous contributions. In May 2006, the *Bulletin of the American Meteorological Society* published lists of weather films compiled by meteorologists, climatologists, and librarians. They mention dozens of films about tornadoes, hurricanes, and climate change. *Cloud Catcher: A Sky Symphony in 10 Movements* (2004), a thirty-minute documentary using time-lapse photography and classical music to dramatize the movement of different types of clouds, is meant to be both educational and artistic.

An earlier educational film, now of more value as history than as science, is *Unchained Goddess*, written, produced, and directed by Frank Capra for the Bell Telephone Laboratory Science Series in 1958. Obviously inspired by the Disney Studios, the film mixes animated characters representing a mythological weather goddess, Meteora, and her court of storm, wind, rain, and snow deities, with two human characters, a science teacher, Dr. Frank C. Baxter, playing himself, and a befuddled television producer played by Richard Carlson, fresh from his Hollywood triumph in *Creature from the Black Lagoon*. The animation was under the direction of Shamus Culhane, who worked for Disney for many years. The meteorological consultants were Bernhard Haurwitz, a major figure in theoretical meteorology who was at this time on the faculty of New York University, and Morris Neiburger, an atmospheric scientist at UCLA.[95]

The film opens with time-lapse clouds roiling in the sky as the title and main credits roll past. Next we see Carlson in a studio rushing to complete an educational movie on weather. He and Baxter, seeking help, pull back the curtains over a screen to reveal Meteora, who offers to assist them. The cartoon Meteora, with flaming red hair and a long, slinky black dress, manages to be both impossibly thin and curvaceous, strikingly like a Barbie doll, who made her appearance the following year. When Meteora learns that Dr. Baxter is trying to understand weather, she expresses her love for him because, so far, no man has really understood her. This seemingly trivial subplot, a romance between nature and science, sets up the old dichotomy: nature as woman who willingly surrenders her secrets to man the scientist. Late in the hour-long movie, Capra and his advisers seem suddenly to realize that some of the children forced to watch *Unchained Goddess* may hold more fundamentalist religious beliefs. The book of Job is quoted, and the viewer is reassured that God wants humans to find the answers to the mysteries of his creation. Nature's and God's laws have nothing to lose but their chains, a liberation that can only be accomplished by the Weather Bureau and Bell Labs.

Dr. Baxter and Carlson call on a variety of cartoon characters to explain why and how wind blows, how clouds are formed from dust and water, and how positive and negative electrical charges cause lightning. "Professor Coriolis" (Gaspard Gustave de Coriolis, 1792–1843, was a French engineer who identified the force that acts on moving objects because of the rotation of the earth), speaking in a comic French accent, explains why winds curve, using a carousel and two boys throwing a ball. The blending of the animated Coriolis and the filmed boys on the carousel is especially well done. The basic facts out of the way, the movie shifts to explaining the work of the Weather Bureau, data collection, the drawing of isobaric maps, and the meaning of "fronts." We glimpse a picture of Vilhelm Bjerknes and see the inside of the National Weather Analysis Center in Suitland, Maryland, where, we are told, a staff of "weather psychiatrists" is busy trying to understand the weather. This is followed by films of tornadoes and hurricanes, which, the narrators tell us, are the least predictable weather, but with the aid of new computers and brave pilots who fly into the storms, their mysteries will soon be solved. A photograph of John von Neumann and Jule Charney flashes on the screen. Weather control is mentioned, and the question is raised about what would happen if we changed the Gulf Stream. As the movie speeds to its conclusion, one of the narrators mentions unintentional climate change and global warming but returns immediately to the happy cartoon characters who break into laughter when it is revealed that Dr. Baxter's forecast for sunny skies is washed out by heavy rain. The weatherman remains an object of fun. *Unchained Goddess* is a very watchable visualization of weather, its science still basically sound despite fifty years of new knowledge.

Experimental and art filmmakers have also been attracted by weather. Scott MacDonald examines the work of several of these filmmakers in his book *The Garden in the Machine: A Field Guide to Independent Films about Place,* but only two films seem relevant here. One is Larry Gottheim's *Fog Line* (1970), a ten-minute, 16mm, silent movie shot near Gottheim's home in Binghamton, New York. The opening minutes are purely abstract, a milky green blob on the screen. Slowly the fog clears enough that trees emerge, then four power lines that cross the screen above the trees. Only after repeated screenings, or when they are pointed out, do viewers see the horses moving slowly in the distance in front of the trees and a bird flying through the upper third of the scene. The film ends abruptly; and there are no credits. MacDonald praises this film for its references to early photography and for the questions it raises about landscape art. The wires are a deliberate device to force the viewer to look above and below them as the fog thins, a reference point for measuring time. Moreover, they suggest the intersection of the technological and the natural. The knowledge that Gottheim filmed the scene with a telephoto lens, flattening the landscape and distorting the size of the horses and trees, leads MacDonald to compare Gottheim's work with Thomas Cole's famous painting *View from Mount Holyoke, Northampton, Massachusetts, after a Thunderstorm, or The Oxbow* (1836), which contrasts wild nature under the towering dark clouds with the sunlit valley filled with cultivated fields and farms. Humans were just beginning to make inroads into the American Garden of Eden in Cole's day; now the "vestiges of that garden, or really metaphors for it, are safely contained within the grids of roads, fields, and power lines."[96]

This is both over- and underreading Gottheim's images. MacDonald has forgotten the weather. The fog never completely lifts during the film; it remains, an atmospheric fairyland, like the sky in Luckiesh and Murchie. As in Horydczak's still photos of the sky, Gottheim's electrical lines recall the parallel rails disappearing into the distance in popular nineteenth-century paintings of the way west. The lines are a kind of open road inviting the viewer to travel though the wilderness. Only the weather can provide the experience of unsullied nature. The power lines represent the machine in the garden of ether, their visibility controlled by fog, the lowest of the clouds, but part of their semitamed frontier nonetheless.

Unlike Gottheim's studied minimalism, the other films MacDonald considers, George Kuchar's *Weather Diaries,* provide a lesson in wretched excess. Kuchar has devoted more than twenty years, 1986 to the present, to filming what he calls *The Weather Diaries.* There are currently fifteen diaries, ranging in length from twenty-four to eighty-one minutes, all shot in color video. MacDonald compares these videos to Henry David Thoreau's *Walden,* though they are closer to Thoreau's journals in their apparent random observations on weather and life. Kuchar is a Bronx native who relocated to San Francisco in the 1970s. Crossing the country, he watched his first severe storm from a motel in El Reno, Oklahoma, and subsequently made a six-minute film, *Wild Night in El Reno* (1977), a montage of postcard scenes of Oklahoma, revealed—like the weather in *Unchained Goddess*—by pulling back a curtain from a screen. We see grass blowing gently in the wind, followed by a violent thunderstorm, and then ending with the morning calm. Kuchar returned to El Reno for several of his "diaries." *Weather Diary 1* (1986), the longest of the episodes, combines sequences of storms and skyscapes with shots of his daily routine in his motel room, including his toilet habits. This mixture of the sublime and the unwatchable, MacDonald thinks, follows Thoreau's combination of intensely personal reflections on living with only the barest essentials and his poetic evocations of nature. Kuchar's flushing toilet, MacDonald writes, is "a metaphor for the funneling air of the tornadoes he's come to be near, and of course, these spiraling motions are two versions of the same gravitational pull." "Professor Coriolis" would be surprised to see how his "effect" gets twisted in popular lore.[97]

The succeeding *Weather Diaries* are ever more self-referential and bizarre, though weather remains a theme. MacDonald contrasts the storm chasers in *Twister,* who seek to understand tornadoes, and the storm squatter (Kuchar's own sobriquet) in *Weather Diary 1,* who wants to reinvigorate "the beautiful and the sublime within his everyday experience." Kuchar wants to understand himself and to show his audience that they can reclaim some of the lost pleasure of sky watching for very little investment of money or time. Insofar as most seekers of the sublime want knowledge of both themselves and nature, Kuchar is probably as good a guide as anyone. Interestingly, Kuchar told MacDonald that he became interested in the weather at an early age through Eric Sloane's books. Kuchar's first job was doing graphics for New York City television weatherman Frank Field.[98]

Viewing *Weather Diary 1* (1986) and *Weather Diary 4* (1988), I found a good joke gone stale. The eighty-one minutes of the first weather diary are relatively amusing and occasionally insightful regarding the interplay of television weather and real weather and the attitudes of Oklahomans toward tornadoes. Visually, *Weather Diary 1* is a montage of shots of storms approaching and clearing, messy motel room interiors, clips from television programs (Gary England, KWTV weatherman in Oklahoma City, appears frequently), and Kuchar eating and speaking into the camera. Kuchar adopts the persona of a city boy bemused by both the weather and the culture of small-town Oklahoma. He pronounces it an alien world of fat people and storms. Anticipating Nicholas Cage's weatherman, he thinks he may be in junk food alley, not tornado alley. Alternately frightened and aroused by a night of thunder and lightning, he declares the night's storm "a wet dream." While storms rotate outside, he watches the water swirl in the toilet. Kuchar's view of the weather is a kind of shaggy-dog story (there are a number of shaggy dogs in the film), the point of which, if there is a point, is that most people, represented by the citizens of El Reno, are totally indifferent to the weather, the beauty of the sky, and aliens from the Bronx. There is no tornado shelter for the residents of a nearby trailer park, but a sign on a telephone pole advertises, "Storm Shelters $1450." The irony may be lost on the El Renoites, but Kuchar and his hip audience enjoy it. In *Weather Diary 4,* the weather context is a summer heat wave and drought in the upper Midwest, and global warming seems to be partly responsible for the grotesque antics of his colleagues at Lawrence University in Appleton, Wisconsin, where Kuchar is conducting a film workshop. Kuchar pissing into the toilet bowl and commenting on "golden showers" dispelling the drought, seems less funny than his tornado-inspired jokes. Nevertheless, the scope and persistence of *The Weather Diaries* confirm the often contradictory nature of weather talk in American life.[99]

It is instructive to compare experimental and documentary weather-themed films with the large body of nature films that began in the silent era and became powerful instruments of wildlife conservation in the post–World War II era. Television shows such as *Zoo Parade* and *Wild Kingdom* promoted, according to historian Gregg Mitman, a domestic ideal that incorporated pets into the nuclear family, but the narrow and often distorted emphasis on domestication left a legacy of misunderstanding about animals in the wild. Nature films, like *National Geographic* magazine, promote the preservation of pristine nature, without facing the hard choices of wildlife management. These films are, by turns, sentimental and violent, but they have one thing in common: every effort is made, according to Mitman, "to eliminate the urban and human setting from the scene. If this vicarious experience of nature—lived through the camera lens—is to feel intimate and pure, the activities of civilization must be hidden, for any sign of artificiality would destroy the illusion of this recreated nature as God's place of grace."[100]

Although sky and weather films share some of this false ideal, weather films more often acknowledge the interconnections of humans and the atmosphere. Unlike the charismatic megafauna (whales, baby seals, polar bears, elephants, and rhinoceroses) featured in nature films, the chaotic megameteors (winds, tornadoes, rain, and snow) that play their part in weather movies seem important only if they affect humans. Except for the sky symphony films, weather in film is not pristine nature but a frontier we resettle every day.

Painting the Sky

Writing about weather and sky in painting and other fine arts presents some challenges. First and most obviously, these objects are intended for visual consumption before any written analysis can interfere with the artist's intent. While this is true of photographs, movies, and exhibits as well, it presents a special problem in a book with a limited number of reproductions. Even if I were able to reproduce in color as many paintings as I would like, there would still be serious distortions of size and tone. My purpose in including the topic of sky art, however, is not so much aesthetic as historical and heuristic. If I succeed in getting readers interested in art as a way of understanding American perceptions of sky and attitudes toward weather, then the next step is a visit to a museum or gallery to look at skyscapes. My second challenge was selection. Almost every landscape includes some sky, but not all landscapes are skyscapes. Weather is implied in paintings—by light, vegetation, clouds, and representations of rain or snow—but most landscapes seem to exist in a moment before or after something has taken place in the atmosphere. The problem raises the question, What are we looking for in art when we look for weather? Hans Neuberger, professor of meteorology at Pennsylvania State University, took a broad view when he personally surveyed 12,000 paintings in forty-one museums in the United States and eight museums in Europe in 1967, declaring that 53 percent of all the paintings contained meteorological information. My approach is to begin broadly and then focus on a few works that seem especially relevant to the perception, marketing, and management of weather.[101]

Meteorologist Stanley David Gedzelman, surveying cloud painting before Luke Howard, concluded that the earliest known representation of a cloud, found in a Neolithic site in Anatolia, dates from about 8,000 years ago, but that the conventions of art remained more important than scientific accuracy in cloud painting until the nineteenth century. Neuberger noted a correlation between clear skies and paintings from Mediterranean countries. In his sample, skies painted by British and American artists tend to be overcast. Luke Howard's classification of clouds and the sciences of optics and meteorology changed artists' approaches to clouds and the sky in several ways. In England, as art historian Kenneth Clark claims, J. M. W. Turner stripped his paintings of clouds and storms of recognizable objects but made his audiences feel the weather through shades of color. No American artist, not Frederick Church, nor Albert Bierstadt, nor Martin Johnson Heade, attempted Turner's intensity and focus on weather. Weather in nineteenth-century American painting is usually a statement about place, a record of one of the marvels of American scenery. Regional artists excelled at advertising their local climates—for example, the Monterey, California, artists Julian Rix, *Foggy Morning Near San Rafael* (1881); William Keith, *After California Rain* (1890s); Granville Redmond, *Fog Burning Off* (1905); Francis McComas, *Sky* (1906); and Mary Neale Morgan, *Flying Clouds* (1920s).[102] Every region of the United States has produced its gallery of sky painters, and, in some regions, as noted earlier in connection with photography, the sky is iconic.

In her brilliant study of American landscape painting in the nineteenth century, Barbara Novak neatly sums up the contribution of American artists to sky painting:

> Sky in American art has a clearly identifiable iconography. The painters of a culture deeply imbued with transcendental feeling found in it the purity of renewal, the colors of hope and desire, heavenly reflections of earthly nostalgias. They also cast upward to the sky the shrewd empirical glance that constantly corrected their powerful system of ideal beliefs. The sky is a finely tuned paradigm of the alliance between art and science. In that mutable void, the landscape artist's concerns—poetic, ideal and symbolic, empirical and scientific—were sharpened rather than blurred. As the source of light, spiritual as well as secular, the sky relieved absolutism with infinite moods, unchanging ideals with endless process. No wonder the artists fixed their particular attention on those moist cargoes that described the void in brief but repeated combinations: clouds.

American artists such as Jasper Cropsey, Thomas Cole, and Frederick Church, influenced more by John Constable than by his countryman Turner, made daily sketches of clouds and kept detailed weather diaries, for reference for their paintings. Americans, Novak writes, were attracted to clouds of the middle region of the sky, cumulus and cumulonimbus, clouds that were often in rapid motion, although Heade often depicted cirrus and John Kensett's frequently cloudless skies conveyed Goethe's identification of "the pure cloudless sky with the calm of eternal bliss."[103]

The twin legacies of these painters—empiricism and mysticism—can be seen in many forms in the art of our time. In the past century, however, our understanding of clouds, sky, and nature itself has changed. Photography—time-lapse, fish-eye lens, satellite—has made it possible to examine clouds in ways an artist cannot hope to imitate, yet the belief that only the genius of an artist can add meaning to the facts of science remains strong. Physics has largely explained how water molecules in the air constantly pass through phases from vapor to liquid to solid, creating as they do visible clouds and precipitation. So far, so good, but philosophy has raised the question of the relationship between what we call "cloud" and the gob of water molecules floating above. A picture of a cloud, whether pho-tographed or painted, is, like the word "cloud," a representation or "signifier" of the presumed reality. A picture of a cloud is both less and more than the visible molecules. Less, because, even if it is three-dimensional (a sculpture of a cloud), it does not have the same complex atomic structure or display the physical processes. More, because clouds symbolize heaven, dreams, ecstasy, danger, and hope and inspire awe, joy, fear, and wonder in the human mind. Thus, when some art historians write about clouds, they resort to a typographical solution to indi-cate what they mean, for example: /cloud/ is the word or picture, the signifier, while "cloud" is the real thing, the signified, and *cloud* is all the symbolic stuff that gets stuffed into the word or picture. Like the vapor, liquid, and solid phases of water molecules, the relations among the three terms are dynamic. The degree to which a twenty-first-century artist is aware of this philosophical problem and the degree to which she chooses to deal with it determine how seriously the artist is considered by her peers.[104]

There is a group of sky and cloud painters whose work resembles wildlife art, maritime art, and aviation art, specialized paintings of recognizable objects of ven-eration, idealized seals, sails, Cessnas, and sunsets. Their work is frankly commer-cial and often produced in different mediums—oils, lithographs, prints—and sizes to offer a wide price range. This does not mean that their work is not attractive, nor their love of the sky impure. An artist such as Peter Allen Nisbet (b. 1949) has a well-developed philosophy of sky painting; it just doesn't concern itself with ab-stract theories of the relationship of representation and reality. Nisbet was taught to paint in watercolors by his mother, switching to oils when he left the navy, moved to the Southwest, and began to paint professionally. Working from a studio near Santa Fe, New Mexico, in the 1980s, Nisbet produces a variety of cloud forms in canvases ranging from 7 by 10 inches to 40 by 72 inches with titles such as *Head-ing South,* depicting a contrail arching over the desert; *Aspects of Ascension,* a mag-nificent cumulus congestus; and *Magic Kingdom,* wisps of cirrus over distant mountains. In a 1987 interview Nisbet acknowledges the influence of nineteenth-century Romanticism and believes that his paintings of the sky and vast horizons "represent the ultimate resting place, a final destination, an end to the yearning for place." Nisbet's conflation of space and place allows him to idealize the sky as an untrammeled frontier without examining the ironies of contrails.[105]

Among Nisbet's sky painting predecessors, two stand out, the reclusive Helmut Siber (1903–1984) and the very public Eric Sloane (1905–1985). Their stories are instructive in understanding the motivations of those who become enthralled by the sky and its clouds. Siber was born in Bremen, Germany. He studied art and was apprenticed as a lithographer but wanted to travel and became a merchant seaman, visiting Africa, Southeast Asia, and China in the 1920s. He sketched and worked in watercolors, specializing in storms. He immigrated to the United States in the early 1930s, married, and lived in the Poconos in Pennsylvania, where he supported himself as a laborer. Later, he worked as a fire lookout in California, southern Appalachia, and finally Vermont, where he settled in the late 1950s and became associated with the Fairbanks Museum in Saint Johnsbury, which has about a hundred of his watercolors. When his wife died in 1970, he decided to return to Germany and destroyed many paintings and his autobiography. Siber achieved some recognition during his lifetime, with two one-man shows at the American Museum of Natural History in New York City, in 1951 and 1954, and several exhibits at colleges, libraries, and art galleries from California to Maine.[106]

Siber apparently made numerous weather observations daily, beginning at sunrise in order to plan his day. He kept a weather journal of which the following entry from 1955 is a sample:

Monday, May 5—The fresh green of wild maple, birch & poplar has reached the 1800′ elevation, the higher slopes still retaining the purplish hues during daytime of bare woods.

10:30 am—Rows of altocumulus castellanus and cumulus. Wild spring flowers in the woods in full blossom. Ferns partly up.

Noon to 4 pm—Towering cumulus rising out of blue haze but shearing off just before state of cumulonimbus.

5:00 pm—From cumulus to stratocumulus and fracto cu. Whitish blue sky. A painting of this state of nature should show birches with brilliant white trunks, delicate whitish green foliage, a branch of hickory with opening buds and a songbird—all against purplish higher ridges and whitish blue sky with white cumulus on westerly wind.[107]

And so on, with entries every fifteen minutes to half hour until 6:30. Comparing Siber's soft, almost pastel colors with his journal entries suggests that he found beauty and personal fulfillment in the muted skies of the Northeast, that he would not have been as happy as Nisbet and Sloane in the dazzling light of the Southwest. "Whenever possible," Siber wrote in an essay titled "On the Work Itself," "I try to observe a storm throughout its life cycle, from the early signs of its birth to its eventual exhaustion and death. Only thus can one gain a total impression of the life of storms." He also believed in experiencing the atmosphere with all the senses, feeling changes of temperature and humidity, hearing sounds in the forest, seeing the clouds. The paintings, usually 5 by 7 inches or 16 by 22 inches, have an oriental tone and softness. For Siber, weather was neither national symbol nor exuberant expression of life's possibilities, but the provider of clouds that were familiar and trustworthy signs of renewal.[108]

As if to prove that no two people see the same sky, Eric Sloane (born Everard Jean Hinrichs) looked at the same raw material as Siber and saw clouds as performers in the theater of the sky. Siber, the mariner, began looking at storms from the bridge of a ship, feeling the force of winds that shaped both the waves and the clouds. Sloane, aviation enthusiast, history buff, and gadfly meteorologist, may have had as intense a feeling for clouds as Siber, but he developed as an artist in very different ways. After prep school and a stint at Yale, young Hinrichs studied at the Art Students League and was influenced by George Luks and John Sloan, whose name he modified and took as his own when he decided that his given name was too German-sounding. In 1925 Sloane went west and discovered the sky around Santa Fe, where he determined to become "America's first cloud painter," later claiming that he coined the word "cloudscape." He went to MIT to study meteorology with Sverre Petterssen, who told him, echoing Edwin Martin, that "the United States is a melting pot of weather; like your people, each weather pattern is in that typical all-American hurry." Sloane also began to study the weather diaries kept by early American farmers, which led to his books on colonial weather lore, barns, and tools.[109]

Sloane, unlike Siber and Nisbet, did not paint out-of-doors, preferring to work from memory because the passing of time, he claimed, intensified his emotions. Sloane's cloudscapes, as we have seen, range from the cartoons of his 1941 effort to explain meteorology, *Clouds, Air, and Wind,* to his 75-foot mural, *Earth Flight Environment* in the National Air and Space Museum in Washington, D.C. Sloane was a gifted writer, with a knack for explaining the weather by analogies and with anecdotes. In his book *Look at the Sky . . . and Tell the Weather* (1961), he follows what he calls the "continental polar air-mass broom" across the continent, describing the ways in which fourteen people in different occupations—bush pilot, weatherman, airline pilot, birdwatcher, farmer, sailor, doorman, and author—experience the weather in thirteen states and Canada.

His doorman at the Hotel Biltmore in New York City is, I think, his cleverest creation. "You know," the doorman tells Sloane, "that barometer you gave me works like a charm. When the pressure goes up, people tip like mad. When the pressure goes down, they keep their hands in their pockets." The doorman had become adept at predicting the weather even though he could only see a small patch of sky above Forty-third Street. Sloane illustrates with a drawing of whirling air lifting thermal cells of warm air above the skyscrapers. Sloane's oils and acrylics, usually 2 by 3 feet or larger, are solid compared with Siber's, masses of sharply etched forms but subtly shaded. Often, he included a small airplane to emphasize the massiveness of the clouds. Unlike Nisbet's and Siber's, Sloane's clouds float in a timeless and placeless sky, though occasionally he includes the recognizable landscape of the Southwest or Midwest, or a landmark like a New England church steeple. Clouds, after all, are everywhere, and, Sloane tells us, they serve a dual purpose. "In nature," he writes, "there are all sorts of warning signals designed to be seen (or in the case of camouflage, designed not to be seen). But I think that nature didn't stop just at warning signals: I am convinced that the sky and its clouds are things designed for man's information and spiritual contentment. I am certain that the sky was designed to be seen by man."[110]

Seen but, like Columbus setting foot on an island in the Caribbean, hopelessly lost until the letters home, the maps and sketches, begin to tell a story about the place. The cloud cartographers, the skyscape photographers and painters, challenge their viewers to find order in the chaos of the atmosphere. The philosopher Edward Casey, pursuing his theory of the importance of place in human existence, has examined the work of a number of contemporary landscape artists for what he feels is their "renewed respect for the earth, its hidden potential and unsuspected power." Drawing an analogy between cartography and landscape art, Casey tries to explain the significance of the abstract paintings of Jasper Johns and Richard Diebenkorn and the massive land art of Robert Smithson and Michael Heizer in terms of four methods of mapping. His approach is relevant, I think, for analyzing contemporary sky art. In brief, Casey's four styles of mapping are as follows:

> *mapping of.* To make a map *of* something is to make a map of a particular place or territory in the effort to capture its exact geography, its precise structure, its measurable extent.
> *mapping for.* Mapping is *for* something, and not strictly *of* something, when it is designed expressly for some particular purpose.
> *mapping with/in.* Where the first two kinds of map are *indicative* signs by their very nature and function—each being a subsistent particular that points to the presumed existence of another such particular (i.e., object, place, or region)—a map *with/in* proceeds by adumbration rather than by indication: by indefinite indirection rather than by definite direction. . . . Such mapping concerns *the way one experiences* certain parts of the known world: the issue is no longer how to get there or just where "there" is in the world-space, but *how it feels to be there, with/in that very place or region,* whether the feeling itself is one of amazement or boredom, duress or ease. I divide up the word *within* in order to signal the internal complexity of this experience.
> *mapping out.* To the degree that I find myself with/in the living landscape, I am part of that landscape, just as it is part of me. . . . But just because of this mutual incorporation of self and earth, the human subject must find a way *out* if he or she is to re-present the experience of deep immersion in such a manner as to make this experience accessible to others.[111]

Casey's proposed model for mapping the earth is not as difficult as it may first seem. His first "way" is essentially what Luke Howard attempted and what subsequent cloud atlases (the use of the term "atlas" confirms this) and picture books of clouds continue. Casey's second kind of mapping is nicely illustrated by The Weather Channel's "Travel Delay" map. In his words, such maps "do not pretend to cartographic accuracy; rather, their point is to provide a schema of how to move most efficiently across a portion of (implicitly or explicitly) gridded space." While the third and fourth ways of mapping are more complex, several of the artists discussed here provide examples. "Mapping with/in" and "mapping out" are simply the degree to which an artist enables a viewer to follow the process of discovery and integrate the insights gained into his own worldview.

It seems obvious to me that all serious artists try to achieve all four levels of "mapping," what they might call representation of sky, clouds, storms, and weather. While Casey's first two types are more literal and less symbolic depictions and the second two are more symbolic and less literal, all four employ conventions of cartography and/or art to achieve their goals. The absence of any recognizable signs and symbols in a map or a work of art renders it indecipherable. For example, a blank piece of paper may represent a cloud or a polar bear in a snowstorm, but it does not allow the viewer to find his or her way in or out of the experience. Sky and cloud artists such as Siber, Sloane, and Nisbet try to balance their mapping of and for with with/in, though not, I think, with mapping out. Their commitment to representing recognizable forms in the sky means that they are not challenging the viewer to move much beyond the passive experience of walking through a gallery or turning the pages of a book. In some of his painting of clouds from above, Sloane forces his viewers to reorient themselves, to imagine how it feels to be within the region. Scientists who paint, like Graeme Stephens, professor of atmospheric science at Colorado State University and one of the developers of NASA's CloudSat orbiting satellite, may be inhibited by their knowledge of what they don't know. Stephens's paintings incorporate the flashy "look at me" quality of Nisbet with the respectful harmonies of sky and earth in Siber, but they do not ask the viewer to do more than share a momentary experience.[112]

Stephens explicates, in a valuable way, the connections he sees between weather art and science in "The Useful Pursuit of Shadows" (2003), an essay in *American Scientist*. Appreciating the potential, and the limits, of both painting and meteorology to explain the visual beauty and physical laws of clouds, Stephens seeks to unite the inspiration of the artist and practical modeling of water systems. He notes with approval that Luke Howard's address "On the Modification of Clouds" (1802) sought to elevate the study of water suspended in the atmosphere from "a useless pursuit of shadows" to an understanding of the universal laws of the atmosphere. Two centuries later, computer models are producing "numerical prediction of the movements of *invisible* air." Mapping the invisible—mapping with/in, to use Casey's terms—brings science and art closer together. Stephens concludes by calling our attention to what a contributor to the *Quarterly Journal of the Royal Meteorological Society* in 1939 called "landscape meteorology," which is the record in photography, painting, and writing of "the scenic influences of sky, atmosphere, weather and climate." The writer insisted "that the scientific and artistic methods are only different modes of approach to the phenomena of the self-same world, and that each, if cautiously but resolutely coordinated with the other, is bound to open up wider vistas of truth." This is the challenge addressed, I think, by many of the landscape meteorologists working at the beginning of the twenty-first century.[113]

Talented regional artists who have fully absorbed the weather of the places they paint are able both to offer a recognizable map of the sky and to suggest the experience of journeying through it, anticipating and reflecting on the panorama of clouds and storms. This is apparent in the work of Kansas artist John Steuart Curry (1897–1946) and contemporary prairie artists such as Keith Jacobshagen (b. 1941) and April Gornik (b. 1953). Jacobshagen's glorious sunset in salmon, apricot, and gray reflected off scattered clouds, titled *Crow Call [Near the River]* (1990–1991), is almost 4 by 8 feet. The painting maps place; the title suggests directions and invites the viewer to wander mentally into its landscape. Gornik's more abstract *Gyre* (1989), a 6-by-9-foot oil, of a squall line beneath a roiling cumulonimbus that covers three-quarters of the canvas but shows only a fraction of the storm, takes the viewer on a journey into a place where the double meaning of the title—gyrating and malignant spirit—may be revealed. In her almost schematic representation of a classic supercell, the land beneath the clouds is featureless, little more than a canvas to absorb the gray and tawny colors of the thunderstorm.[114]

Gornik's drawing of a rainstorm at sea, *Storm along the Coast,* orders nature into six clearly visible horizontal bands of alternating shades of dark and light. The sky and water are joined by vertical columns of falling rain, but within the apparent balance there is movement and chaos in the frothy, shallow waves striking and receding from the muddy beach, new squalls forming on either side of the falling rain, and the roiling clouds. Her sky occupies more than half the image, drawing our attention to the downpour and making us feel the veering winds and plunging temperatures of the storm.

Still greater abstraction is found in the skies of the Canadian artist Emily Carr (1871–1945) and the American Georgia O'Keeffe (1887–1986), where the meteorological is magical. Carr's clouds are in a perpetual dance with her towering pine trees. In *Above the Trees* (1935–1939), the viewer feels as if he were leaning into a pool of buttermilk being stirred by the tops of the trees. In the painting *Logger's Culls* (1935?), clouds writhe in agony over a stump-filled clearing. One cloud tilts toward the upper right side of the painting, resembling an unfinished Haida or Tlingit totem pole. Carr lived among the Indians of the Northwest coast and incorporated some of their sensibilities into her work. O'Keeffe seldom provides such clear keys to her maps of the sky. In *Light Coming on the Plains, II,* a small, 7-by-9-inch watercolor of 1917, a deep blue domed sky fades to tan above a blue earth. A thin tan horizon separates earth and sky, cracking the eggshell-like arc of the sky. The dome is stationary, architectural; the light is in motion as if the sun is rising behind a dust storm. In a much later and much larger (8-by-23-foot)

April Gornik, *Storm along the Coast,* 2000. Charcoal on paper. Permission of the artist.

painting, *Sky above Clouds IV* (1965), rows of brick-shaped white clouds diminish in size as they recede into an opaque horizon. Installed by itself in a stairwell of the Art Institute of Chicago, O'Keeffe's painting disorients the viewer who ascends or descends under her sky, participating momentarily in her mapping from within.[115]

Artists of Casey's fourth stage of mapping, "mapping out," take us into their private visions of sky and weather but also provide us with ways to continue that experience in our own work. Three contemporary artists seem to me to aspire to Casey's fourth level with some success. The oldest and probably most important of the three is Walter De Maria (b. 1935), who studied history and art at the University of California, Berkeley, before moving to New York City, where he began to make sculptures in the minimalist style. He was also briefly a drummer with the rock group the Velvet Underground. In 1977 he completed construction of his best-known work, *The Lightning Field,* in western New Mexico. Consisting of 400 stainless-steel poles averaging about 20 feet in height and arranged in a grid measuring 1 mile by 1 kilometer, the poles are spaced 220 feet apart. Commissioned and maintained by the Dia Foundation in New York City, *The Lightning Field* can be considered land art similar to Smithson's and Heizer's, but focused less on the land and more on the weather, which he views as an "intense, physical and psychic experience." Visitors to *The Lightning Field* are limited to half a dozen at a time, and they are required to stay overnight in accommodations provided by the foundation. The site is open from May through October, but the artist and the foundation prohibit photographs, leading some critics to complain that it is impossible to make an independent assessment of the work because of these and other restrictions.[116]

De Maria's installation is especially spectacular during a thunderstorm when lightning strikes the poles, but it is more a subtle mapping of the earth-atmosphere interaction twenty-four hours a day, with continual changes in light, sound, and mood. De Maria's purpose seems to be to force on visitors a heightened awareness of the complexity of nature in an apparently minimalist environment. Visitors who come with some knowledge of the federal government's Thunderstorm Project and Project Cirrus in the desert Southwest will place De Maria's work in one context; those who view *The Lightning Field* through the lens of Richard Misrach or the canvases of P. A. Nisbet will place it in another. De Maria's use of the grid, his careful topographic measurements, and even his authoritarian control of the visitor's experience will remind some viewers of the history of settlement in the area and the meaning of power, both electrical and political. De Maria summarizes his work in a number of brief statements: "The land is not the setting for the work but a part of the work. The sum of the facts does not constitute the work or determine its esthetics. Because the sky-ground relationship is central to the work, viewing *The Lightning Field* from the air is of no value. . . . The light is as important as the lightning. . . . The invisible is real."[117]

James Turrell (b. 1943), also supported in part by the Dia Foundation, began an even more ambitious and costly project, which remains unfinished, in the Arizona desert north of Flagstaff in the late 1970s. Known as the *Roden Crater,* it will eventually be a series of observation points for viewing the sky, sun, and stars. Currently a 150-foot tunnel under one side of an extinct volcano crater culminates in a towering keyhole-shaped door that leads to an elliptical room from which the viewer exits either into the crater itself or into a circular room with a circular opening in its roof for viewing the sky.

These "skyspaces," as Turrell calls them, are about light and its absence. He has done other skyspaces, notably the Live Oak Friends Meeting House in Houston, which has a 12-by-12-foot aperture in its ceiling. Visitors to *Roden Crater* are encouraged but not required to stay overnight at Turrell's nearby lodge. In addition to rooms, Turrell says he is selling colored air and blue skies, emphasizing the physical and psychological importance of light for humans. Where De Maria was influenced by Duchamp and the Dadaists, Turrell's inspiration comes from Constable, Turner, and the Impressionists. In contrast with the work of De Maria, *Roden Crater* is meant to be seen from the air. (Like Sloane, Turrell loves flying and claims to have flown monks out of Tibet during the Vietnam War period.)

Turrell seeks to emphasize the sky as an enclosure and chose a crater as the site of his construction rather than imposing a plan upon the landscape. Where De Maria's *The Lightning Field* recalls the grid of the Northwest Ordinance, Turrell's *Roden Crater* appears to follow in the tradition of enhanced natural wonders, like a tightrope walk across Niagara Falls or a via ferrata in the mountains of Utah or West Virginia. Despite the differences, *Roden Crater,* like *The Lightning Field,* is meant to be a spiritual journal in which the artist provides a partial map to be completed by the traveler.[118]

Throughout his artistic career, Turrell has been primarily interested in the effects of light. "Light is a powerful substance," he writes. "We have a primal connection to it. But, for something so powerful, situations for its felt presence are fragile. . . . My desire is to set up a situation to which I take you and let you see. It becomes your experience." His principal metaphor for his own work is meteorological—warm front, cold front, occluded front, stationary front—but some critics have stuck with the trope of the topographical survey. Mario Diacono, for example, sees *Roden Crater* as ritual space, "a secret 'temple,' invented to the ends of a higher, extra-historical experience of being on the part of an eventual observer. That observer will be closer to the shaman 'skywatcher' than to the visitor of contemporary art museums." This may be true, but it misses the science that Turrell emphasizes throughout his work. He selected a high-altitude site (more than 5,000 feet) so that the sky would appear a deeper blue. Unlike most meteorological landscape artists, Turrell seeks clear, not cloudy skies. Using the crater of a volcano is meant to call attention to volcanic ash as a shaper of atmospheric effects and climate; one of the special effects he hopes to provide is a view of the "green flash," a rare phenomenon that occurs as the sun sets if dust and moisture in the air scatter and bend the orange, red, and blue light waves, leaving the green. As Turrell's most astute critic, Craig Adcock, points out, "When all of the levels of the crater are taken into consideration, the permutations of order and sequence become almost infinite." Turrell combines the natural fumaroles of the volcano with his man-made tunnels to position the observer according to seasonal atmospheric effects. Place and space are mapped and remapped by artist and visitor.[119]

"Mapping out" does not always require earth sculptures or the compulsiveness of middle-aged men to succeed. The cool, apparently uncomplicated acrylics of Dozier Bell bring the viewer to a mutual incorporation of earth and sky in a more traditional setting. Bell, a native of Maine and a 1981 Smith College graduate with an MFA from the University of Pennsylvania, specializes in atmospheric phenomena, clouds, fields of stars, and the haze of the sky, incorporating the technologies of cartography, LANDSAT satellite images, radar, and sonar. As a brochure accompanying her spring 2004 exhibit, Aether, at the National Academy of Sciences in Washington, D.C., pointed out, "Bell draws a correlation between these modern methods and timeless mental processes of detection, the placing of oneself in the cosmos, and impulses toward destruction and deliverance in paintings that encompass both our present-day awareness of both the potential for destruction on an unprecedented scale and the corresponding vastness of the ethereal." In a painting titled *Seraphim 2,* for example, part of a cumulus is rendered in the right side of the canvas, while the left is empty space with a few stars. Showing clearly against the white of the cloud are plus signs, coordinates that indicate the grid of a satellite image. *Trace,* a 4-by-5-foot acrylic on linen painting, is marked on all four sides (at the 12, 3, 6, and 9 o'clock positions) with straight black lines that extend a few inches into the painting, providing both a spatial and a temporal orientation, although the painting could be hung upside down or sideways without losing its effect. The cool blues and pinks form cloudlike shapes, but they appear insubstantial, ephemeral products of the dynamics of the atmosphere. Bell makes us feel the physical and the metaphysical lapse rate as we ascend in her painting.[120] Bell's syncretism of the celestial orders of ancient religions and tools of contemporary science may be jarring to some, but her efforts to "map out" the territories of air, clouds, and space unexplored by artists is part of a larger movement in the arts.

The early twenty-first century has experienced increased activity in sky art that challenges the viewer in interesting ways. In the winter of 1999–2000, the Hirshhorn Museum and Sculpture Garden exhibited some photographic images of cloudy skies by the Cuban American artist Felix Gonzalez-Torres (1957–1996), one of which, *Untitled (Aparición),* was duplicated on 45-by-30-inch paper and left in a stack on the floor. Visitors were encouraged to take one copy, a gesture that can be interpreted in many ways, but for me, still in the beginning stages of research for this book, it was an opportunity to add a piece of the sky to my growing collection of weather-related art. Placed on the ceiling of my office, the photograph was intended as an inspirational conversation piece, which, alas, few visitors noticed or commented on. Another lesson might be that the apparition of the clouds is lost on most people, their skies "untitled." Or else, my visitors simply lacked the habit of looking up.

The Columbian American artist Iñigo Manglano-Ovalle (b. 1961) has, in recent years, collaborated with geneticists, architects, climatologists, and others to produce weather- and climate-related works, such as polished titanium clouds, and video installations. Among the latter, *Climate* (2000) is a set of three video screens showing a meteorological/financial analyst reporting on weather derivatives; an unseen individual unloading, cleaning, and loading a gun; and a woman waiting in a lobby. All the scenes are identified as being shot in Chicago's Lake Shore Drive Apartments, designed by modernist architect Ludwig Mies van der Rohe. *Climate* attempts to break down the boundaries of atmospheric and political climates. Surveillance of nature and humans, Manglano-Ovalle suggests, has created a third place where we represent both worlds to ourselves.[121]

Representing—"mapping"—the skies in ways that mimic the evanescence of nature has been attempted by multimedia artists Vik Muniz (b. Brazil, 1961), Cai Guo-Qiang (b. China, 1957), and Anish Kapoor (b. India, 1954). Muniz, noted for his parodies of great artworks and the humor of his "junk" sculptures, creates what he calls "Pictures of Clouds" by hiring a skywriting pilot to draw generic clouds above a city. In February 2001, he had clouds drawn for four days above Manhattan, the only permanent work being the photographs taken on those days. "Art for me," he told Deborah Solomon of the *New York Times*, "is not about saying things. It's about discovering ways to say them. There's no way to discover without being involved in the making of it, and through the process, you start to realize the mechanics of representation and you start to use them better." Cai's *Clear Sky Black Cloud,* displayed on the roof of the Metropolitan Museum of Art at noon on Tuesdays through Sundays from April through October 2006, is similar in its performance aspect to Muniz's *Clouds.* Cai uses gunpowder and fireworks in many of his artworks, and *Clear Sky Black Cloud* is intended to provide an ersatz nephological experience for viewers. While Cai can provide the dark cloud, nature needs to furnish the clear sky. Its failure to do so on a weekly schedule may, in fact, be Cai's point.[122]

Kapoor, considered a major British sculptor, recently installed *Cloud Gate,* a 125-ton stainless steel convex mirror, in Chicago's new Millennium Park. The piece reflects the Chicago skyline and, when conditions permit, clouds. *Sky Mirror,* a new piece, was part of a temporary installation in Rockefeller Center in September and October 2006. As Randy Kennedy writes, Kapoor presents these pieces as "a conceptual variation on landscape painting: a kind of living, self-creating landscape painting that literally holds up a mirror to nature, as Stendhal said of the landscapes of Constable, whose cloud studies inspired Mr. Kapoor." "Upside-down clouds really do look upside-down," says Kapoor, who, like Muniz and Cai, wants his audience to think about their environments in new ways. Apparently he partially succeeded. As one critic observed, "For a few moments it [*Sky Mirror*] adds a mood of wonder and delight to this otherwise cold, oppressive setting." That these artists come from three corners of the world to practice their art in the United States is one other link in the chain of art and science that this book emphasizes. Like the atmospheric sciences, sky art is part of the globalization of knowledge. Mapping the topography of the ocean of air cannot be a matter of artificial boundaries and spheres of influence; it is about seeing the weather.[123]

A notable collaboration between artists and atmospheric and environmental scientists occurred in 2006 when Iowa State University's Brunnier Art Museum opened its exhibit "Obsessed: Images of Weather." Artists and meteorologists collaborated to produce new artworks and public programs emphasizing the ubiquity of weather in human life. This show was followed soon after by one in Boulder, Colorado, in the fall of 2007, under the sponsorship of the Boulder Museum of Contemporary Art in cooperation with EcoArts, the National Center for Atmospheric Research, and others. The exhibit, Weather Report: Art and Climate

Change, matched several dozen artists and scientists who shared a curiosity about the representation of nature. In one installation, hydrologist Sheila F. Murphy, geologist Peter W. Birkland, and artist Mary Miss used flood data on Boulder Creek to place bright blue 6-inch discs on trees, telephone poles, and the facade of the museum to show the high-water mark of a 500-year flood. Other exhibits envisioned the retreat of Alpine plants up mountainsides as the planet warms and effects of climate change on pikas and butterflies. The exhibit is a modest beginning, perhaps, but one that can be replicated in many locations where scientists and artists are trying to raise public awareness of environmental issues.[124]

Exhibiting Weather

At the beginning of the twenty-first century, the National Oceanic and Atmospheric Administration created the Science on a Sphere (SOS) program, in which computer graphics and real-time satellite data are projected onto a 30-foot globe to illustrate in three dimensions the geologic evolution of the earth and its weather patterns. Spectators get a satellite-eye view of our planet, floating, as it were, on the surface of the ocean of air. The reversal of perspective may have both positive and negative consequences as it comes to be fully appreciated. Certainly, as NOAA claims, SOS can present research data in compelling new ways. On the other hand, miniaturization, as critic and philosopher Susan Stewart explains in *On Longing,* her meditation on description and metaphor, always tells us both more and less about a subject: "The miniature offers a world clearly limited in space but frozen and thereby both particularized and generalized in time—particularized in that the miniature concentrates upon the single instance and not upon the abstract rule, but generalized in that that instance comes to transcend, to stand for a spectrum of other instances."[125]

Even though SOS can show the movement of clouds in the atmosphere in almost real time, the viewer, standing inside the sphere, sees the earth from space as a detached observer and has no temporal reference point. Clouds swirl, the earth rotates, but a viewer (this viewer) has difficulty seeing the connection between the apparent clarity of the sky seen from space and the hazy view he gets from the ground. The "Blue Marble," as the *Apollo 17* astronauts termed the Earth they photographed from space, is, like a dollhouse or a toy train layout, an object of nostalgia, not a place where we live, or a planet under extreme environmental stress. Viewed from where we live, sky and clouds are gigantic; we are connected to past and future as part of nature. The miniature, Stewart argues, is a cultural creation; the gigantic is natural. The atmosphere viewed from Earth holds us in place, gives our lives spatial and temporal meaning.[126]

The Smithsonian Institution collected Weather Bureau instruments, but they were seldom displayed, and as the emphasis shifted from exhibiting technological history to cultural history in the 1970s, the history of meteorology and the atmosphere disappeared completely from view in the Smithsonian. Even the opening of the National Air and Space Museum in 1976 did little to restore meteorology as an exhibit topic; the emphasis was on aircraft, not the flight environment. A weather exhibit may be too controversial for the Smithsonian in the current political debates over global warming, but the Marian Koshland Science Museum of the National Academy of Sciences in the capital nevertheless produced an informative exhibit, Global Warming Facts and Our Future, where a visitor can press a button and see how much of tidewater Maryland and Virginia will be flooded when the sea level rises a meter.[127]

If weather and climate have been neglected by the Smithsonian museums, weather disaster has not. Just a month after Hurricane Katrina, a curator and a photographer from the National Museum of American History, in partnership with George Mason University's Center for History and New Media and the University of New Orleans, went to Houma, thirty miles southwest of New Orleans, to begin collecting objects that they felt represented Katrina. They hoped to acquire a "dropwindsonde" used to measure the wind speed and air pressure of the hurricane, handmade "Help!" signs, part of a levee, and other objects from the storm. A second collecting trip to New Orleans yielded 58 artifacts and 1,731 photographs for the museum and the Hurricane Digital Memory Bank (http://www .hurricanearchive.org/index.php), sponsored by George Mason University. Among the now numinous objects are pieces of a floodwall, a Fisher-Price toy castle that had been buried in the mud of someone's home, a pair of lace curtains from a house showing the water level, a Coast Guard rescue basket, and purses made by local artists from the plastic bags of MREs (meals ready to eat). The National Weather Service donated two dropwindsondes. When and how the objects are to be exhibited will be decided in the future, but a similar collection of 9/11 artifacts has drawn favorable attention at the museum. We seem to live in a time when the public likes to relive tragic events, and when museums are valued more for their memorial function than their educational role. Weather-disaster exhibits could, of course, fulfill both tasks, and it will be interesting to see what emerges from efforts such as the Smithsonian's.[128]

Some regional museums celebrate, or try to explain, the distinctive weather of their localities. The Minnesota Historical Society assembled an amusing exhibition, How Cold Was It? in the early 1980s on Minnesota's climate and winter carnivals. In 2002 the society opened a sequel, Weather Permitting, focusing on how Minnesotans have experienced their weather in clothing, in sport, and through television. A prominent part of the exhibit is a replica of the Weatherball that was fixed to a tower on the Northwestern Bank Building from 1949 to 1982, when the building was destroyed by fire. Promoted as the tallest illuminated sign between Chicago and the West Coast, the twelve-story ball changed colors to forecast the weather. In 2006 the Johnson County Museum in Kansas advertised for stories, photos, and other items to be used in the exhibition Wild & Wacky Weather, apparently unconcerned with its potential impact on real estate and tourism. National Science Foundation grants encouraged other museums to consider "wild weather exhibits." The Virginia Air and Space Center in Hampton, Virginia, also boasts a Wild Wild Weather exhibit created in partnership with NASA Langely Research Center. The National Climatic Data Center in Ashville, North Carolina, an archive of historical climate data, maintains an exhibit of early meteorological instruments.[129]

Weather seems to fare better in children's museums, such as the Weather Discovery Center of the Mount Washington Observatory in Conway, New Hampshire, funded by NOAA, Subaru, and memberships, where children can experience various wind forces, learn about weather maps, or pretend to be a TV weathercaster. A much more elaborate exhibit has been created by the Children's Museum of Houston. Funded by the NSF and developed in cooperation with Scholastic Entertainment, NOAA National Weather Service, the Oklahoma Climatological Survey, and the AMS, the museum has two identical traveling exhibits called Kicks Up a Storm. With more than two dozen weather-related activities for children aged five to twelve, the exhibit requires a minimum of 2,500 square feet and costs $60,000 to rent for three months. Opening in Houston in February 2003, it has also been on display at the Liberty Science Center in Jersey City, the Calgary Science Center, the Minnesota Children's Museum, and the Children's Museum of Atlanta. The activities are deliberately low-tech, using balloons, posters, and homemade weather instruments more than computer screens and video, though these are included to provide local weather data. As with many children's museums, a TV studio for producing weather broadcasts is a prominent feature. Kicks Up a Storm comes with teacher and family guides and "marketing tools to maximize community participation."[130]

Houston may be the capital of the weather museum business. The Weather Research Center, founded in 1987 by Jill F. Hasling and her father, Dr. John C. Freeman, as a resource center for public education about weather and weather safety, opened a new facility, the John C. Freeman Weather Museum, in April 2006. Focused more on weather camps, teacher workshops, seminars, and conferences than on collections of artifacts, the museum does have all the bells and whistles an amusement park–savvy public expects—the experience of walking through a tornado, "Tame the Tornado"; a surround-sound movie, *Hurricane in the Round;* a flash-flood table, "Rain, Rain—Go Away"; and, inevitably, a mock TV studio. The institution appears to be well funded, with levels of support identified by cloud types, although it is not clear whether the "Cirrus Club" members (Halliburton, for example) donate more than the "Towering Cumulus Club." The center's Web site is impressive, if a bit wonky. A link to a page titled "Dvorak Cyclone Classification" shows the eight-point scale with corresponding wind speeds, but nothing about its history or application. Since the Dvorak scale is not mentioned in the AMS *Glossary* of *Weather and Climate,* in *Weather for Dummies,* or in the *Encyclopedia of Hurricanes, Typhoons, and Cyclones,* all but the initiated are left in the dark.[131]

Another kind of weather exhibit is the re-created environment found in zoos and botanical gardens. The Montréal Biodôme, with four ecosystems, each with a distinctive climate, is a spectacular example. Visitors enter a tropical rain forest with temperatures maintained between 21°C and 28°C, depending on the season, and humidity of 70 to 80 percent. The dense forest is populated with capybaras, caimans, frogs, and more than 350 bats. The next environment is the Laurentian forest, where the temperature drops to 4°C in December and rises to 24°C in June. River otters, lynx, and beavers prowl the birch and maple forest. Continuing the Canadian climate, the third room is the St. Lawrence marine ecosystem, with temperatures similar to those of the Laurentian, but featuring aquariums filled with cod, mackerel, halibut, salmon, and other Atlantic species. The final room includes the Arctic, where the temperature is kept between 12°C and 15°C, and the Antarctic, ten degrees colder. There are no polar bears, but auks and penguins abound. No other museum experience that I have had has had the impact of the Montréal Biodôme. The visceral experience of heat, cold, and humidity in the context of the flora and fauna adapted to the climate reminds the visitor that weather is experienced in many ways.[132]

If there is such a thing as a folk museum of weather, it is Roger Brickner's Museum of American Weather in Haverhill, New Hampshire. A retired history teacher, Brickner traces his fascination with weather to his experience of the 1938 hurricane that devastated much of Long Island, not far from where he was living. The museum, which opened in 1992, is housed in a renovated shed next to his home in a quiet New England village, off the beaten path. When I visited in the summer of 2002, Mr. Brickner led me through the "galleries," the first containing a three-panel mural by his friend Mark Carroll, depicting the destruction caused by the 1938 hurricane on Long Island and in Connecticut and Rhode Island. The painting has the quality of ex-voto drawings left by petitioners in Mexican churches. Across from the mural is a partial reconstruction of his aunt's Fire Island summer house, which was destroyed in the storm, decorated with some of the salvaged furnishings. The next display is of the northeastern blizzards of 1888, 1947, and 1978, with vignettes by Carroll. The 1888 painting shows a tunnel of snow dug in Peterborough, New Hampshire. A loan from Yale's Peabody Museum illustrates, in photographs and U.S. Geological Survey maps, the path and damage caused by the tornado of July 10, 1989. Transcriptions of eyewitness accounts and drawings are attached to some of the sites. Two other tornadoes are chronicled: the three-day storm that began in Acadia, Nebraska, on June 7, 1953, and ended in Worcester, Massachusetts; and the famous "Tri-State" tornado of March 18, 1925, that killed more than 700 people in Missouri, Indiana, and Illinois. Privately funded and maintained, without video or teachers' guides, the Museum of American Weather comes closer to capturing the meaning of weather for most

Americans than do glitzier displays. Weather's personal meanings are archived here, as in Kuchar's home videos.[133]

Not all weather exhibits succeed. The Charles and Henriette Fleischmann Atmospherium and Planetarium of the Desert Research Institute (DRI) in Reno opened in November 1963, in partnership with the University of Nevada, with ambitious plans to unite the day and night skies. The dream of Wendell Mordy, an enterprising meteorologist and science administrator, the Atmospherium is a futuristic building, designed by Reno architect Raymond Hellman, with a hyperbolic paraboloid roof covering a 9,900-square-foot building. The centerpiece of this concrete flying saucer is a small, fifty-five-seat theater with a 30-foot domed ceiling (the same diameter as NOAA's Sphere) onto which films photographed with a fish-eye lens are projected. Mordy, who held a doctorate from the University of Stockholm, had been a weather forecaster for the air corps in World War II and then worked as a meteorologist for pineapple and sugar growers in Hawaii before becoming the founding director of the DRI in 1960. He left in 1970, spending the next five years working on science policy at the Center for the Study of Democratic Institutions in Santa Barbara. He was appointed president of the Science Museum of Minnesota, in St. Paul, in 1977, a position he held until his retirement in 1984. He died in 2002. In his remarks at the opening ceremony of the Atmospherium, Mordy could not restrain his excitement:

> The Atmospherium is really a science theater in which pictures of the sky surround you, just as the sky surrounds you in nature. The motion pictures, sound effects, music, colorful lights, and poetry all can be woven together.
>
> I hope tonight's program will give you a hint of the Atmospherium's potential. In the month since our equipment has been operating, we have captured some very lovely cloud sequences near Bishop, California, which I think you will enjoy. But what you see tonight will not be the whole story.
>
> As you watch the whole-sky, time-lapse film this evening, you will realize that this technique can be used to photograph anything that happens in the sky, anywhere in the world. Carribean [sic] hurricanes, Midwest tornadoes, dust storms, water spouts, ice storms, the movements of the earth's shadow on the atmosphere, the aurora, lightning, Bishop's Rings, and rainbows all can be documented and shown to you here. We even hope to be able to take the camera under water to show sea life.[134]

The AMS covered the event, predicting that the Atmospherium will do for meteorology what the planetarium has done for astronomy. In his remarks, Mordy dismissed the planetarium half of the project because "most of you know what a Planetarium can do." These statements allude to some deep divisions between the day and night sky aficionados, which would plague the Atmospherium through its troubled history. The inaugural brochure gave equal notice to atmospherics and astronomy, but "Atmospherium" appeared in type twice the size of "Planetarium," and the first page highlighted only *weather* phenomena, asking, "What causes thunderstorms? How do they form? What makes sunsets red? How does a Sierra Wave Cloud develop?" and claiming, "The world's first 'atmospherium' brings this knowledge within the reach of everyone." Twelve years later the Atmospherium/Planetarium was bankrupt and closed its doors. Mordy was gone, the DRI was uninterested in continuing its administrative relationship, and the Fleischmann Foundation was unwilling to subsidize what was supposed to be a self-supporting institution. A committee of community leaders was formed to save the building. The University of Nevada–Reno agreed to assume the administrative oversight. The Fleischmanns pledged a matching grant of $450,000, if the committee could raise $350,000 by December 15, 1976. In its press releases, the committee made clear that the Atmospherium would change programming to emphasize community needs, specifically elementary and high school science education, and focus more on astronomy. It was successful in saving the building, but the films and school programs in the following years were mostly about space exploration, dinosaurs, and constellations.[135]

In the next thirty years, the renamed Fleischmann Planetarium and Science Center was threatened many times by the expansion of the university campus and population growth, most recently when the university proposed tearing the building down to construct additional parking for the nearby Lawler Event Center, a sports-entertainment complex. Historic preservationists intervened successfully, and the striking building clings precariously to a hillside on the east side of North Virginia Street, dwarfed by a parking garage and facing a strip of gas stations and motels. On my last visit, I was one of two persons in the audience for a screening of *Adrenaline Rush,* a SkyDome 8/70 movie about skydiving and base jumping. It was a stomach-churning experience to watch skydivers soar upward in the dome as a lot of blue sky flew by the swiftly gliding adventurers. When I left, I noticed that the xeriscape around the building was well maintained by the university and that the vault of the afternoon sky, patched with scattered cumulus and stitched by contrails, arched gloriously from the Sierra Nevada to the Great Basin above the glow of the neon city.

The failure of the Atmospherium to draw sufficient paying audiences and wealthy patrons can be explained by many factors. The sky is not an easy attraction to market. First of all, as John Day points out, it's free. So is the night sky, of course, but there you want to see those little twinkling stars in more detail. It's harder to find those bears, and archers, and scales in the stars than it is to see castles, and dragons, and Elvis in the clouds, so we photograph clouds, speed them up with time-lapse photography, and amplify their storms with surround sound. We move the sky indoors, and it loses its appeal. Sometimes less is less.

4

Transcribing Weather

E Pluvius Unum

Is there a poet so contrary to tradition that she or he has not written at least one line incorporating some element of weather? Possibly. Is there a reader or reviewer who can recall a favorite rain, snow, or wind poem that I have omitted here? Absolutely. My excuse for failing to mention anyone's best-loved weather poem is simply my effort to select both verse and prose, from more than a century of weather awareness, that reflect my purpose in this chapter, to discuss creative works that use weather in interesting and entertaining ways. These poems and novels also illuminate the basic themes of this book, that weather is part of a national discourse on American identity—spiritual, regional, psychological, and political—and that weather links the meteorologist and the artist in their efforts to understand order and chaos in nature.[1]

A chapter on weather in American literature is more than justified by the complex relationship between poetry, fiction, and science, especially in the recent past. "Science," the Dutch literary scholar John Neubauer notes, "may appear in poetry as theme, as underlying worldview, and as a model for poetics."[2] Weather in American poetry, whether verse written in traditional forms and accessible to the public, or avant-garde poetry intended for readers with considerable knowledge of literary history and theory, falls on a continuum from realistic to symbolic. As symbol, weather offers virtually limitless possibilities, from the analogies with human emotions to the "language" of nature. As literary historian Eduardo Cadava observes in his exemplary essay, "Literature and Weather," in the *Encyclopedia of Climate and Weather:*

> The quickness with which the weather becomes a subject of conversation suggests its linguistic character. A nebulous space of communication, literary weather constantly produces language, conveys messages, and serves as the condition for narrative. From messages carried by the wind and blown away in storms to the rainbow that signals the covenant between God and his chosen people, from the tempests that indicate the turbulent emotions of a hero to the "dark and stormy night" without which there could be no gothic novel, the weather has been a medium of communication. . . .

The link between language and the weather is also evident in our understanding of the weather as a kind of text, a system of signs to be read—even if these signs are the nebulous texts of wind velocity, cloud formations, and vapors.[3]

Cadava's wordplay with the adjective "nebulous" illustrates his point. Both scientists and poets are fascinated by the paradox of weather, its seasonal regularities and daily uncertainties, its physics and metaphysics. Meteorology is a good place to begin to mend the gap between the sciences and the humanities. Weather, elusive and allusive, challenges both ways of knowing. The evolutionary biologist Stephen Jay Gould believes that the sciences and the humanities are mutually dependent. In support of this he quotes novelist and lepidopterist Vladimir Nabokov, who wrote, "There is no science without fancy, and no art without fact." Somewhat earlier Nabokov declared, "I cannot separate the aesthetic pleasure of seeing a butterfly and the scientific pleasure of knowing what it is."[4]

Among American poets, Walt Whitman (1819–1892) stands out as a pioneer observer and interpreter of weather. "The Voice of the Rain" (1885) is as succinct a statement in verse of the hydrologic cycle as we are likely to find:

And who art thou? Said I to the soft-falling shower,
Which, strange to tell, gave me an answer, as here translated:
I am the Poem of Earth, said the voice of the rain,
Eternal I rise impalpable out of the land and the bottomless sea,
Upward to heaven, whence, vaguely form'd, altogether changed, and yet the
  same,
I descend to lave the droughts, atomies, dust-layers of the globe,
And all that in them without me were seeds only, latent, unborn;
And forever, by day and night, I give back life to my own origin, and make
  pure and beautify it:
(For song, issuing from its birth-place, after fulfillment, wandering,
Reck'd or unreck'd, duly with love returns.)[5]

The falling rain, after gathering in the atmosphere through evaporation and transpiration, is like the poet who gives life to latent and unborn ideas through his love of humanity, which compels him as surely as the laws of condensation and precipitation. Whitman's use of the archaic word "recked," meaning "heeded," in the sense of recognized, like his choice of "atomies" to refer to the smallest particles of dust rendered visible by light (e.g., "a mote in a sunbeam"), underscores his careful experiments with language, finding the precise word, as a meteorologist would find the appropriate mathematical formula. Whitman "translates" the physical laws of nature into words, as Edmund Halley, Daniel Bernoulli, and others translated them into numbers.[6]

At the beginning of the twentieth century, poetry, like meteorology, was entering a period of rapid change. In "Winds of Texas," Harriet Monroe (1860–1936), founding editor of *Poetry* magazine, depicts the winds as wolflike, blowing fiercely from the Gulf through Galveston, San Antonio, Austin, and the Panhandle: "The wolf-howl of the winds of Texas sweeps through the wide and aching night." Closer to Whitman in its panorama of Texas geography than to the imagistic poetry she encouraged, "Winds of Texas" is, nonetheless, built of striking lines such as "Searching the sky, pricking eager ears for the footsteps of the sun on the plains, / Shouting for joy of shaking out the stillness from wide and lofty spaces." Monroe's wind has shape and sound within the laws of solar heat and acoustics.[7]

Considering the city's reputation for wind, it is not surprising that Carl Sandburg (1878–1967), Monroe's fellow Chicagoan, also found inspiration in the wind. "Wind Song" is a brief narrative poem in which the speaker tells the reader that he learned to sleep in an apple orchard, where "the wind swept by counting its money and throwing it away." The branches of the trees "trapped the wind into whistling, 'Who, who are you?'" The speaker concludes:

> Long ago I learned how to listen to the singing wind and how to forget and
> how to hear the deep whine,
> Slapping and lapsing under the day blue and the night stars:
> Who, who are you?
>
> Who can ever forget
> listening to the wind go by
> counting its money
> and throwing it away?[8]

The lessons of this wind, then, are first how to listen and learn, and second how to forget and let go, but who can forget the profligate wind that gets and spends without thought of the future? Sandberg's choices of "slapping and lapsing" for wind sounds are packed with meanings—a blow, a fall, "lapse rate," a passing from higher to lower status, a loss of religious faith, a lapse of memory—all these connotations and more are suggested. The speaker identifies with the wind, its wealth counted in leaves from a tree, his in moments of illumination.

Wind holds manifold interests for poets. May Swenson (1913–1989), known for incorporating science in her poems, follows the tradition of seeing weather and wind as a wild horse, but she plays with the idea, enlarging it into a metaphor for an apocalyptic struggle between nature and technology. The poem begins simply, "I hope they never get a rope on you, weather," and builds to the final two stanzas:

Reteach us terror, weather,
with your teeth on our ships,
your hoofs on our houses,
your tail swatting our planes down like flies.

Before they make a grenade of our planet
I hope you will come like a comet,
oh, mustang—fire-eyes, upreared belly—
bust the corral and stomp us to death.[9]

Swenson, writing at the beginning of the era when weather satellites and cloud seeding seemed to promise control of storms, laments the loss of beauty and mystery in weather.

A younger poet, writing about the same time as Swenson, recovers the mystery, if not the beauty, of wind and weather by rediscovering what Whitman knew—that the messages in wind are cryptic and confidential. Charles Wright (b. 1935), a prize-winning poet whose poems often address religious questions, puts his "Weather Report" in the form of a letter to a friend, William Brown. He asks his friend to think of a wind blowing from the Arctic across the prairies, many miles, until it reaches his house, where

> . . . rising upon
> the windows, as if, somehow,
>
> it wished inside and I must
> now, open the door, let it
> enter . . . Brown, what is this wind?

Wright's question achieves eschatological importance with its emphatic, "*this* wind," which explains both why earlier in the poem the poet imagines that the wind comes "from so removed / a place to arrive, as it does, / with such emotion (as though / distance were ever a guide)" and why he must open the door. Wright's wind spills "downward across the tundra, / the bleached plains, the cold mosses, / working its way still deeper / into the mainland," terminating its journey at Wright's desk. The message may be undecipherable, but it is urgent.[10]

A simple synthesis of wind and language occurs in David McCord's (1897–1997) lilting quatrain "Weather Words":

> I know four winds with names like some strange tune:
> Chinook, sirocco, khamsin, and monsoon.
> Like water over pebbles in Lost Brook:
> Sirocco, monsoon, khamsin, and chinook.[11]

By taking words for winds from four parts of the world, McCord highlights the global nature of weather. That three of the words are from Arabic—the hot, dry Moroccan sirocco, the equally hot and dusty Egyptian khamsin, and the wet monsoon of the Indian subcontinent—says much about the importance of these winds in arid regions and the way in which words travel like winds, scrambled to other cultures. Chinook, a Native American word, is a warm, dry wind that blows into the Rockies and Great Plains in the late fall, often melting snow.

In "A Theory of Wind," Albert Goldbarth (b. 1948), a Chicagoan transplanted to Texas, begins with the poet/speaker comparing his confused feelings about the wind to a stack of galley proofs, which, despite the words printed on them, cannot understand God any more than the authors who wrote them. As he feels the wind's "rough tongues" exploring his body—"Everywhere, objective and efficient: / its assessment"—he says, "No wonder tribal life thought / God behind such touch." The next two stanzas reveal that he has been correcting the proofs in a room in which a woman, his lover, is gravely ill. The wind against the window cries, why? The power of the wind to destroy makes the speaker in the poem think of how errors occur in typesetting and in life:

> And then
>
> the long hours, correcting. If it's true for this
> level it's true for that level, but anyway no shaman
>
> or oncologist explains it. *Why?* The wind
> is come chill through her room tonight. *Why?* And what
> was my page supposed to say? Our lively lust,
> list?, last?, our lovely dust . . . it's so hard
> to remember. It's always night, somewhere.
> Some tree is always unsafe. And all we can do
> is pray for ours, in our backyard, that we thought
> so pure of form, its bark moire like Persian lamb . . .
> but what do we know? The wind is going over
> everything tonight, proofing for error.[12]

Wind as a representation of an indifferent god or nature remains a powerful symbol in an age in which natural hazards seem to be increasing.

If wind is often depicted as a wild animal or a representation of the divine, rain is frequently personified as a woman. In Robinson Jeffers's (1887–1962) "Distant Rainfall," the squall line reminds the poet of "mourning women veiled to the feet / Tall slender rainstorms walk slowly against gray cloud along the far verge."[13] Though born and educated in Pittsburgh, Jeffers spent much of his life and wrote most of his poetry at Big Sur on the California coast.

Hildegarde Flanner (1899–1987), his near contemporary, grew up in Indianapolis but settled in Berkeley, California, in 1922. Her poem "The Rain" (1927) casts the precipitation as a woman seeking a lover. The opening lines establish this storm as a product of northern California topography and vegetation:

> The rain upon the earth falls down
> Long and naked, straight and wan;
> Still and white along the hill,
> Bending to her miracle,
> Slow and silver in the grass—
> Lover, maid and sorceress.
> She lays a yellow jasmine blossom
> In the centre of her bosom,
> Leaps and flutters like a moth
> When the wind is from the south,
> Bows and murmurs like a fountain
> When the rain is from the mountain.
> Have me now—oh, take me, love!
> She sighs along the orange grove.
> Here my arms, here my lips!
> She cries among the eucalypts.[14]

By falling, of course, the rain dies in the dust of the dry hills. Rain gives life through her death.

The Canadian poet Dorothy Livesay (1909–1996) domesticates rain, comparing it with her grandmother: "Alive with herself and her voice, rising and falling— / Rain and wind intermingled." The rain triggers a brief recollection of her grandmother and transports the poet to her childhood, another kind of re-generation:

> I remember long veils of green rain
> Feathered like the shawl of my grandmother—
> Green from the half-green of the spring trees
> Waving in the valley.[15]

The rains of Jeffers, Flanner, and Livesay, seen as a line of veils, long hair, "the feathery fringe of a shawl," are weather as observed in open spaces, where the poet can see the relationship between sky and land. Rain for the urban poet is often experienced as sound, not sight, but it retains its gendered allusions.

Robert Creeley (1926–2005) is an experimental poet who mixes colloquial language with abstract meditations on the sense of loss in modern life. His poem "The Rain" suggests that the life-giving quality of rain can be psychological as well as hydrological:

All night the sound had
come back again,
and again falls
this quiet, persistent rain.

What am I to myself
that must be remembered,
insisted upon
so often? Is it

that never the ease,
even the hardness,
of rain falling
will have for me

something other than this,
something not so insistent—
am I to be locked in this
final uneasiness.

Love, if you love me,
lie next to me.
Be for me, like rain,
the getting out

of the tiredness, the fatuousness, the semi-
lust of intentional indifference.
Be wet
with a decent happiness.[16]

The internal half-rhymes "persistent," "insisted," "decent," and the suffix "-ness"
repeated five times bind the poet's mental state to the rain, internalizing what
begins as external. The rain triggers thoughts of his misery and his need for a
lover to bring, like rain, "decent happiness."

Because rain is so elemental, fundamental to life, it also evokes thoughts of
death and the end of natural cycles. Anthropologist and nature writer Loren
Eiseley (1907–1977) imagines his death when the rain "tells you the years are
done":

Rain is the world's intent, it lashes every furrow,
stifles all cries of parting or farewell
beneath the sound of eavespouts and of gutters.

Now windcocks spin, bolts split, skies open;
this rain is driving toward the end of time.
No sudden hush, no light toward morning ever
will break this steady pouring. I depart
just as I came, at midnight, with rain falling.
It is the rain that speaks last to the heart.[17] *

If the world's purpose, desire, is rain, then humans must accept that their voices, their lives, will be stifled, extinguished, as easily as their feeble efforts to understand or control weather—"windcocks spin."

For the poet and literary historian Jay Parini (b. 1948), rain as part of the process of growth and decay does not need to be a "steady pouring," merely "mizzle," a Middle English word combining "mist" and "drizzle," but also connoting disappearance. In his poem "Mizzle," "the sodden weather of an early spring" brings the sights and sounds of an awakening natural world, but also awareness of perpetual decay:

A thousand stumps that stud the swampland
once were something you might want to climb;
I watch them sink to silt and crumble,

decomposing in the same sure way
that everything we love at last undoes
its laces, sighs at ease, then lays

its head down, nuzzles into loam,
relinquishing the lovely stand of life
to meld, the slow atomic mince and slaughter.[18]

Parini's choice of words, alliteration, and internal rhymes creates an intricate pattern that mirrors the complex process of humification, "the slow atomic mince and slaughter" of fallen leaves, dead plants and insects, and saprovores in the soil. At death, humans, as part of the organic world, decompose and "nuzzle" into the loam, aided by "mizzle." The half-rhyme links the two natural processes. This is a brilliant synthesis of meteorology, ecology, and metaphysics, with weather not merely reflecting mood but contributing to our understanding of the deeper relationship of weather and human life.

---

*Loren Eiseley, "It Is the Rain That Tells You," reprinted with the permission of Scribner, an imprint of Simon & Schuster Adult Publishing Group, from THE INNOCENT ASSASSINS by Loren Eiseley. Copyright © 1973 by Loren Eiseley. All rights reserved.

Salvation by rain is literal for California rancher and poet Linda Hussa, whose poem "Give Us Rain!" celebrates the end of a drought. The poem begins by comparing rain to "a worthless uncle, a tawdry aunt" who are still greeted warmly when they show up, then "They whisk off too soon, / just a scent of store perfume on the pillow." When the rain comes, the poet addresses it as if it were a shiftless relative:

> Oh! We've missed you, rain!   Give us more!
> Fill the tanks!
>
> Lap the banks!
> Wake the seeds!
> Even weeds!
> Drown the fish!
>
> More hay!   Meadows jumping!
> Turn those skinny old cows into butterballs!
> Don't you know how happy we are to see you   beautiful old rain?
> God!   We are happy![19]

The deliberate simplicity of the concluding lines underscores the elemental importance of rain, even in an age of irrigation and cloud seeding. Rain is as much a kinsman of the rancher and farmer as her blood relatives. The exclamation points splash across the page like raindrops.

Elizabeth Madox Roberts (1885–1941) was once a much-admired experimental novelist and poet. Her novel *Jingling in the Wind,* partially concerned with rainmaking, will be discussed later in this chapter. Her nursery rhyme–like poem "The Sky" makes the reader notice the atmosphere in a new and unsettling way:

> I saw a shadow on the ground
> And heard a bluejay going by;
> A shadow went across the ground,
> And I looked up and saw the sky.
>
> It hung up on the poplar tree,
> But while I looked it did not stay;
> It gave a tiny sort of jerk
> And moved a little bit away.
>
> And farther on and farther on
> It moved and never seemed to stop.

I think it must be tied with chains
And something pulls it from the top.

It never has come down again,
And every time I look to see,
The sky is always slipping back
And getting far away from me.[20]

The speaker in the poem is deliberately naive, perhaps a child, but the reader easily imagines the experience of being disoriented by the vault of the sky after looking up suddenly. The motion of birds or clouds, seen as shadows, in contrast to the apparent stillness of the sky, requires the speaker to refocus, beyond the poplar tree, and creates the illusion of an endless receding sky that must be pursued.

The skies of certain seasons and places are notable. Leonard Nathan (b. 1924), a respected poet and translator living in California, catches a winter sky in "The Sky Hung Idiot Blue":

The sky hung idiot blue all windy day
Plucking the leaves off huffy sycamores;
Women whirled as if their skirts danced them
Through leaf-gold alleys for a dreamed ballet.
The fruitless season uttered its apothegm
In keening flues or shot-quick slam of doors.

Why "idiot" blue? I think the answer may lie in the etymology of the word, which is related to idiom, as in "peculiar" or "special"; the poet wants to shake up his readers and make them pay attention to the meanings of this sky. The connotation of "idiot" as opposed to near synonyms like imbecile, fool, and dolt is usually that of someone who knows better but has acted stupidly. The sky, clear, blue, and seemingly innocent, unleashes a wind that forces everyone to prepare for winter. The second stanza describes preparations taken by the weather-wise. The final stanza returns to the meteorological conditions, which have now occluded, turned violent with a passion beyond sanity:

Winter and even the West is hunched and slow;
Lightning charts the white and smoting kiss
Of shuddering fronts that thunder as they mate.
The wiseman's face has thinned in the woodfire's glow,
And tugging thick shawls, the ballerinas wait
As if there will be dancing after this.[21]

In "El Paso Sky," Naomi Shihab Nye (b. 1952), a translator of Arabic and a poet who lives in San Antonio, captures the special quality of the sky of a place:

> When it's no good on earth I look up. When the cups on my
> table all have chips around the edges and I can't get that
> feeling of what to do next, I press my eyes into the skinny
> pink stripe melting under the blue rumple that rolls and rolls
> and the dark corner growing over the mountains. I say to
> myself, "It's happening without you." If I had the biggest
> arms in the world, I couldn't hug that. When I think of the
> people who are dead now, who weren't dead just a little
> while ago, and how easy it would have been to pick up the
> phone and talk to them by dialing a number—I look at the
> sky. It's all one piece now.[22]

On the page opposite this poem is a photograph by Wyman Meinzer of a sunset over a range of mountains with five bands of color, from inky black hills in the foreground to an apricot sky above the last visible crest of the mountains. While the photograph does not match my mental image of "pink stripe melting under the blue rumple," it is an appropriate image for the feeling of the poem, in which the speaker finds solace in the sky. Shihab Nye engages in an interesting experiment in this poem. The opening line makes us want to know why "it's no good on earth" and what looking up means. The first answer is elliptical but recognizable as the speaker stares deeply into a sunset, the thin pink light of the sun fading as the dark blue sky wrinkles and rolls over the mountains. The enigmatic "It's happening without you" links the sunset with the recently deceased, whose deaths also "happened" without the speaker's participation. The sky darkens to blue; a cycle ends. This is the kind of meditation on the sky that For Spacious Skies has long encouraged, but we should also remember the words of the greatest of weather poets, Wallace Stevens (1879–1955), who in the poem "A Clear Sky and No Memories" questions the frequent association of sky and sunshine with memory in a line that demolishes the easy metaphor: "Today the mind is not part of the weather."[23]

Clouds are as popular a topic for weather poetry as wind, sky, and rain. Amy Lowell (1874–1925) was as active exponent of Imagism, a form of poetry that emphasizes compression and immediacy, which is highly effective for meteorological effects such as storms and clouds. Although it retains some of the quaint anthropomorphism of nineteenth-century verse, "Night Clouds," published about the time the Weather Bureau was beginning to explore the upper air, is a good example of an emerging cloud consciousness:

> The white mares of the moon rush along the sky
> Beating their golden hoofs upon the glass Heavens;

The white mares of the moon are all standing on their hind legs
Pawing at the green porcelain doors of the remote Heavens.
Fly, mares!
Strain your utmost,
Scatter the milky dust of stars,
Or the tiger sun will leap upon you and destroy you
With one lick of his vermillion tongue.[24]

In addition to being a relatively rare view of clouds in the night sky, Lowell's poem captures the drama of a sunrise—"golden hoofs" of the horselike clouds, "green porcelain" sky, "vermillion tongue" of the sun—in a half dozen exquisitely chosen words.

Robinson Jeffers, writing a decade later in the mid-1920s, also sees in clouds a source of dreams and visions. His "Clouds at Evening" begins:

Enormous cloud-mountains that form over Point Lobos and into the sunset,
Figures of fire on the walls of to-night's storm,
Foam of gold in gorges of fire, and the great file of warrior angels:
Dreams gathering in the curded brain of the earth,
The sky the brain-vault, on the threshold of sleep: poor earth, you like your
   children
By inordinate desires tortured make dreams?

After a few more lines of examples of such dreams, Jeffers unexpectedly concludes: "I have grown to believe / A stone is a better pillow than many visions." In other words, nature provides diverse spectacles, some of which may be alluring but misleading. Weather in many of Jeffers's poems is the enemy of permanence, wearing away stone and shaking trees to their roots, yet weather remains an essential part of nature's instructions to humans. The sky is the "brain-vault" for the earth's "curded brain," suggesting a physiological as well as a symbolic relationship.[25]

Later poets have treated clouds more matter-of-factly. Shortly before his death, the Montana poet Richard Hugo (1924–1982) lived in Scotland and wrote a series of poems about the island of Skye. In one, "The Clouds of Uig," dedicated to a former teacher, Hugo begins by noting the way clouds change the colors and shape of the land, but notes:

They have no form. No one would mistake
the shade of one on water for a boat.
They never slow down and they never run out.
When one sky leaves, taking with it the rain
that couldn't make anyone wet or leave grass

dry very long, another sky follows close behind,
the loud blue interval between. . . .
recess in a crowded school.

The final stanza compares the clouds with students moving through the teacher's life, the meaning of their perpetual motion not quite understood, but accepted:

We can live under them.
They move certain as blood. Under their shade
the bay locks complete and, deep in that cloudy water,
many lives go on.[26]

Hugo's reverie on clouds and the passage of time befits an era of satellite imagery and nephanalysis. In "A Big Clown-Face-Shaped Cloud," Kenneth Koch (1925–2002), a poet of very different sensibilities, shows that clouds are still fun to watch and provide simple pleasures:

You just went by
With no one to see you, practically.
You were in good shape, for a cloud,
With perhaps several minutes more to exist
You were speaking, or seemed to be,
Mouth open wide, talking, to a
Belted angel-shaped cloud that was riding ahead.[27]

Clouds may be ephemeral and mundane, but their oneiric power, as Bachelard observed, remains strong, and they continue to carry messages even if only to one another.[28]

Familiarity with the science and history of clouds leads contemporary poets to treat them as specimens of a larger natural world. Billy Collins (b. 1941), poet laureate of the United States in 2001–2002, expresses this sensibility in "Student of Clouds," which begins by describing John Constable's method of sketching and painting clouds, after which Collins declares:

In photographs we can stop all this movement now
long enough to tag them with their Latin names,
Cirrus, nimbus, stratocumulus—
dizzying, romantic, authoritarian—
they bear their titles over the schoolhouses below
where their shapes and meanings are memorized.

In the final stanza he returns to Constable and the feeling Collins has that the painted clouds are moving out of the painting and the nineteenth century and over the meadow where he is walking, "bareheaded beneath this cupola of motion, / my thoughts arranged like paint on a high blue ceiling." Collins cleverly turns the dome of the sky into a cupola, an architectural element often found on churches and one-room schoolhouses, and then paints its ceiling with cloud-shaped thoughts. He has imaginatively linked cloud science and art.[29]

These examples must suffice to make my point about American poets and their engagement with the weather as part of personal, regional, and national identity. The cumulative effect of a century of poems about weather phenomena is a recognizable meteorological identity in which the elements, as the weather is sometimes referred to, reveal significant American characteristics. These are seldom regionally specific, as Harriet Monroe's "Wind of Texas," or Charles Wright's wind in the prairie heartland. Rather, they are the "nebulous," psychological and spiritual characteristics exemplified in Whitman's desire to "translate" messages from nature and Creeley's pursuit of a "decent happiness." Weather, like region or ethnicity, is a building block for a larger edifice. Yet some scientists think that the variegated weather of North America prevents a sense of shared meteorological experience, that too many weather patterns are going on simultaneously for Americans to feel united. This assumption misses the point; variety is what unites us.[30]

Even before the Weather Bureau made them aware of the unusual extremes of American weather, a mobile population knew that "normal" American weather was always "abnormal." As meteorologists call for "variable cloudiness," indicating frequent changes during a forecast period, poets anticipate "variable identities" within the national character. For meteorologists, "variables of state" are "the minimum numbers of descriptors of the physical state of a thermodynamic system," such as the temperature, density, and mass of air. For poets, variables of state are the descriptors of a nation, such as weather, which bring symmetry if not order out of the chaos of individual diversity. In a double sense, the poets suggest, we live in "variable states."

"Pray Hard to Weather"

Snow, as I have argued elsewhere, has provided poets with copious material. Snow is a symbol of nothingness, isolation, silence, timelessness, death, and the unknown, but also of life, love, purity, and fragility. Snowstorms evoke memories of past winters and stimulate the imagination. Snow represents order in its hexagonal flakes, and chaos in blizzards. Because, according to tradition, no two snowflakes are alike, snow requires constant renaming, or as William Matthews (1942–1997) puts it: "And here comes snow, a language / in which no word is ever repeated."[31]

Jun Fujita (1888–1963) is an interesting weather poet who was born in Japan and came to Chicago in his twenties to study mathematics. To pay for his education he worked as a photographer for the *Chicago Evening Post*. Abandoning math, he became a professional photographer and watercolorist and wrote poems in English, often using the Japanese tanka form, similar to haiku. "Snow" is part of

> Fainter than hushed feet,
> Stealing through my hazy dream,
> Snow—
> In the midnight woods.[32]

This evanescent observation captures a moment of waking as snow falls outside. The quatrain seems nothing more than a simple description, but upon reflection, the hushed feet and the midnight woods inspire a narrative, one a reader can easily share, of snowfalls dreamed and lived.

If Fujita offers a moment of awakening to the mysteries of snow, May Swenson welcomes its somnolence and its power to transform the familiar landscape. The title of her poem "The Fluffy Stuff" connotes both the cliché of TV weathermen and something inconsequential, which, nevertheless, may literally terminate life. After fourteen lines describing her backyard filling with snow, the poet concludes:

> What began gauzy, lazy, scarce, falls willingly now.
> I want it to race straight down, big, heavy, thick,
> blind-white flakes rushing down so plentiful, so
> opaque and dense that I can't see through the curtains.[33]

Snow's transformative power, to reveal and to hide, is part of its appealing paradox, recognized by snow scientists and poets.

Peter Viereck (1916–2006), a professor of Russian history and a poet best known for his World War II poem "Kilroy," explores the connections between snow and the wished-for order of nature in "A Walk on Snow," which begins:

> Pinetrail; and all the hours are white, are long.
> But after miles—a clearing: snow and roundness.
> Such circle seemed a rite, an atavism,
> A ripple of the deep-plunged stone of Myth.
> I crossed that ring to loiter, not to conjure.
> Stood in the center as in melodrama.
> Wondered: if this center were a gate?
> A gate from earth to non-earth? Gate where fingers,
> Where rays perhaps, are fumbling signals through?
> Or are stars cold for all their brightness,

Deaf to our urgencies as snowflakes are?
Then magic blazed: a star spoke through the gate:
"I am not cold; I am all warm inside."

A walk in the snow brings the speaker to a point, physically and mentally, where the weather evokes a series of increasingly difficult questions about humanity's place in nature and the meaning of life. The next two stanzas begin to formulate an explanation based on an understanding of science and a rejection of myth and magic. The poet listens to the storm, "Like chord-joined notes of one sky-spanning octave," and plays on the word "tellurian," meaning both "an earth dweller" and the apparatus that illustrates the tilting axis of the planet, the cause of seasons.

Hoping to find answers to his questions about appearance and reality in both art and science, the poet ends the third stanza expecting that his discovery of a magic gate by which he can communicate with the universe will end his search. But in the fourth stanza those hopes are dashed, a single snowflake mocks his longing:

Art, being bartender, is never drunk;
And magic that believes itself, must die.
My star was rocket of my unbelief,
Launched heavenward as all doubt's longings are;
     It burst when, drunk with self-belief,
I tried to be its priest and shouted upward:
"Answers at last! If you'll but hint the answers
For which earth aches, that famous Whence and Whither
Assuage our howling Why with final fact."

At once the gate slammed shut, the circle snapped,
The sky was usual and broad and silent.
A snowflake of impenetrable cold
Fell out of sight incalculably far.
Ring all you like, the lines are disconnected.
Knock all you like, no one is ever home.
(Unfrocked magicians freeze the whole night long;
Holy iambic cannot thaw the snow
They walk on when obsessive crystals bloom.)
Shivering I stood there, straining for some frail
Or thunderous message that the heights glow down.
     I waited long; the answer was
The only one earth ever got from sky.

Written in the 1940s at the dawn of the nuclear age, Viereck's poem reflects the hopes and fears of that era. Weather, especially a snowstorm, provides the perfect metaphor for the failed quest for meaning. Snow and the sky beyond taunt truth-seeking poets, the "unfrocked magicians" with their "holy iambic." With echoes of both Robert Frost's "Afterflakes" and Wallace Stevens's "The Snowman," Viereck's poem uses the snowpack, where "obsessive crystals bloom," as an eternal enigma, unthawable and unknowable.[34]

No snow poem discussion is complete without the poetic acknowledgment of the ways in which most twenty-first-century Americans experience inclement weather, not as precipitation but as anticipation. Billy Collins captures this perfectly in "Snow Day," a poem that describes a man waking up to a radio report of a snowstorm that has closed all the city's offices and its many nursery schools; he is "as glad as anyone to hear the news":

> that the Kiddie Corner School is closed,
> the Ding-Dong School is closed.
> the All Aboard Children's School, closed,
> the High-Ho Nursery School, closed,
> along with—some will be delighted to hear—
> the Toadstool School, the Little School,
> Little Sparrows Nursery School,
> Little Stars Pre-School, Peas-and-Carrots Day School
> the Tom Thumb Child Center, all closed,
> and—clap your hands—the Peanuts Play School.[35]

The sheer lunacy of the names matches the "revolution of snow, / its white flag waving over everything," and when the poet sees three girls playing outside, "plotting . . . which small queen is about to be brought down," he extends the metaphor of the snow as liberation. His juxtapositions of the antic snow, the fantasylands of the schools, and the real world of sometimes cruel play establish a place where weather and behavior are perfectly matched.[36]

The literary critic and poet Radcliffe Squires (1917–1993) grew up in Salt Lake City, Utah, and taught at the University of Michigan most of his career. In "Storm in the Desert," he uses the distinctive weather and topography of the Great Basin to contemplate the relation of humans to the vastness of geologic time:

> No one has ever said he goes "out" into the city,
> We always say we go out into the desert.
> It is out; out of self, out of presences.
> Absence is what we have here.
> In deserts we stand under something
> That is not there, a sea, a mountain,
> And we feel it, but, whether it is prophecy

Or memory, we cannot say. We only know
The absence is waiting for something stranger
Than the future, more familiar than the past.

Look, the storm rises as a lavender
Light around these stopes whose grief
For something that is not there
Is the more bitter for the pale ungrieving
Springtime of their color.[37]

These stanzas are, like the geology of the Wasatch, layered. From the suggestion of exile ("going out into the desert") and the multiple connotations of "presences" (act of being present, the presence of other people, the space in our vicinity, a divine or other invisible being, the consecrated elements of the Eucharist), the poet prepares the reader for a spiritual experience, a revelation through the storm. The colors of the storm and the abandoned mining excavations ("stopes") are incongruously bright before the storm gathers strength and darkens the landscape. Typical of desert storms, the rain ceases after a few minutes and:

The water runs away in the orderly whorls
Of stone-fingerprints, and small brown birds,
Jerking in the gravel, emit thin songs. Below us
In the valley those pale tall stopes
Stretch vernal-green, vernal-rose all the way
To the horizon. They are quite unchanged.
The storm's rhetoric, the civil grammar of the birds
Were only interruptions in their love of a vast absence.[38]

Squires's description is meteorologically correct, and the meanings of the storm linger in the reader's mind. As Cadava points out, weather can be a language with various rhetorics. To me, the storm speaks briefly to the desert wanderer of the hope of order and "civil grammar" in life, but this modern anchorite realizes that what he witnessed was merely an "interruption," leaving only the absence of belief.

Another poet who sees a language in weather is Samuel Hazo (b. 1928), a Lebanese American, who, in "Rains," plays with the names of rain and the hydrologic cycle:

Shower, drizzle
or storm, it's all a matter
of seas sun-siphoned to the clouds
and then returned aslant or straight
as plummets to the bullseye world.

Then, after descriptions of different kinds of rain, concludes:

> In all the languages
> of rain they say there's still
> a place for order, even
> in bluster, even in passion.[39]

The poet's choice of interesting words such as "plummets" and "bluster" works well here because it gives the rain more complex aural and visual dimensions. A seemingly simple observation on variety and pattern in life is confirmed by the analogy to nature.

In an earlier poem Hazo describes driving in a lightning storm and listening to a news report from Washington of the civil war in Lebanon in the 1980s. The lightning is "like rocketry, a scar of fire / slashes down the sky":

> The lashing rainfall wails
>   In Arabic for this Guernica
> In Beirut . . .
>   I think of Lorca
> who believed the lightning-worlds
> of love and poetry could have
> no enemies.
>   He never dreamed
> of lightning-chevrons on black
> shirts, lightning-wars
> and lightning-zigzags crayoned
> on a map that named a war
> that scarred a generation . . .
>   This
> generation's condors storm another
> Spain.
>   The rain's a litany
> of Lorcas bulldozed into pits.
> The voice from Washington is no one's
> And the world's.
>   *Viva la Muerte!*[40]

As we have recently seen, it is difficult for many Americans to empathize with the victims of war in the Middle East. By comparing the civil war in Lebanon with the Spanish civil war and invoking the spirit of the great Spanish poet and dramatist Garcia Lorca, Hazo hopes to convince his readers of the gravity of this conflict. He effectively uses the image of lightning to remind us of the Nazi SS and blitzkrieg. "Condors" evoke both the vultures drawn to war's slaughter and the "Condor Legion," the German bomber squadron that participated in the massacre of civilians in Guernica and other Spanish cities in 1937. War and weather have been linked for centuries by the sound of thunder and the flash of lightning.

Lightning as an analogy for sudden insight is probably as old as language, but two poets, A. R. Ammons (1926–2001) and Charles Martin (b. 1942), play with the idea in astutely different ways. Ammons's "Standing Light Up" comments on the way in which what is secreted in the mind may reveal more than what is obvious in nature by contrasting the way thunder reverberates long after lightning strikes. The sky at the beginning of his poem is dark and without shape, "none of the / lightning's veins shows." The thunder could be avalanches in obscured mountains. It does not matter, says the speaker in the poem, because

> with a little ink and type produced: take the
> truth that in a drizzle drops tickle leaves so
>
> it's a pause whether it's a breeze: who cares
> about a truth like that: nearly all, maybe
>
> all, most truth doesn't matter a tittle of rubble
> or rain: what matters is that sometimes the
>
> spirit halts and listens for what outleaps
> the insides of summits thunder's rumble has
>
> never jarred: what is to be seen within
> scares the eye brighter than any lightning.[41]

Ammons's use of unconventional punctuation paces the reader's eye and forces a reading strategy that makes the conclusion inevitable and acceptable. Lightning and thunder are real, but "truth doesn't matter a tittle." This wonderful word, which means both a small punctuation mark, such as the dot over an *i*, and by extension, any small particle, and a whisper or gossip, captures the double meaning of the speaker's experience of hearing an unidentified rumble and then realizing that he can make the sound stand for other things he imagines. It is art, human creativity, which gives shape and meaning to nature. The elliptical title can now be read not as an adjective and a noun but as a compound noun suggesting a light that is either permanent, as in a "standing committee," or temporary, as in "standing rules." The ambiguity of reality, discovered or invented, is the subject of the poem. Ammons's poetry is often arcane, but so is much science.

Martin, on the other hand, is quite clear and direct. "Reflections after a Dry

Spell" does not try to hide either the literal or the metaphoric meaning of "dry spell." He begins his poem with a quotation from the poet Randall Jarrell:

> *A good poet is someone who manages, in a*
> *Lifetime of standing out in thunderstorms, to be*
> *Struck by lightning five or six times.*

Martin then contrasts a poet who took Jarrell's advice literally and spent his time chasing storms hoping to be struck, with one who dismissed the advice.

> But the one who knew it was nothing more
> [That flash of lightning] than a metaphor,
> And said as much, as he went out the door—
>
> Of that one, if you're lucky, you just may find
> The unzapped verse or two he left behind
> Of the confusion between World and Mind.[42]

The mocker gets shocked. Martin seems to be in conversation with Ammons on the question of the relation of literal and metaphorical weather. A dry spell for a writer means something different from a drought for the earth, but the consequences may be equally severe for the individual. Jarrell's advice is clearly ironic; getting struck by metaphoric lightning may be a good thing, but few survive more than one stroke of either inspiration or an electrical discharge from the atmosphere.

As for other kinds of storms, the distinguished African American poet Robert Hayden (1913–1980) turns an ice storm into a prayer for himself and, perhaps, all people who have been oppressed:

> Unable to sleep, or pray, I stand
> by the window looking out
> at moonstruck trees a December storm
> has bowed with ice.
>
> Maple and mountain ash bend
> under its glassy weight,
> their cracked branches falling upon
> the frozen snow.
>
> The trees themselves, as in winters past,
> will survive their burdening,
> broken thrive. And am I less to You,
> my God, than they?[43]

A version of Woody Guthrie's (1912–1967) "Dust Storm Disaster" is included in the second volume of the Library of America's *American Poetry: The Twentieth Century,* while appearing in an abbreviated form on NOAA's "Art and Poetry" Web site. Whether it is considered a song or a poem, it is a vivid portrayal of the storm and its effects.

> On the fourteenth day of April
> Of nineteen thirty-five
> There struck the worst of dust storms
> That ever filled the sky.
>
> You could see that dust storm coming
> The cloud looked death-like black
> And through our mighty nation
> It left a dreadful track.
>
> From Oklahoma City
> To the Arizona line
> Dakota and Nebraska
> To the lazy Rio Grande.
>
> It fell across our city
> Like a curtain of black rolled down
> We thought it was our judgment
> We thought it was our doom.
>
> The radio reported
> We listened with alarm
> The wild and windy actions
> Of this great mysterious storm.[44]

Guthrie continues, listing more affected towns and states and describing the reactions of those driven from their land. Guthrie's verses confirm historian Donald Worster's thesis that dust bowl farmers were woefully ignorant of their environment. They interpret the storm in eschatological terms, they get their news from the radio rather than their own community, and they are clueless as to the causes of the storm. The farmers' misuse of their land and misunderstanding of their climate exacerbated the devastation of the drought and dust storms, according to Worster. Guthrie, of course, was an entertainer, not a climatologist, and he is interested in the high drama of disaster. Most disaster poems focus on human suffering, not on responsibility.[45]

Turning to tornadoes, another plague of the Great Plains, we find some notable poetry. In "Tornado Warning," Karl Shapiro (1913–2000), who spent many years at the University of Nebraska, provides an exacting portrait of a tornado about the time the Weather Bureau established its Natural Disaster Warning System:

> It is a beauteous morning but the air turns sick,
> The April freshness seems to rot, a curious smell.
> Above the wool-pack clouds a rumor stains the sky,
> A fallow color deadening atmosphere and mind.
> The air gasps horribly for breath, sucking itself
> In spasms of sharp pain, light drifts away.
> Women walk on grass, a few husbands come home,
> Bushes and trees stop dead, children gesticulate,
> Radios warn to open windows, tell where to hide.
> The pocky cloud mammato-cumulus comes on,
> Downward-projecting bosses of brown cloud grow
> Lumps of lymphatic sky, blains, tumors, and dugs,
> Heavy cloud-boils that writhe in general disease of sky,
> While bits of hail clip at the crocuses and clunk
> At cars and windowglass.
>   We cannot see the mouth,
> We cannot see the mammoth's neck hanging from cloud,
> Snout open, lumbering down ancient Nebraska
> Where dinosaur lay down in deeps of clay and died,
> And towering elephant fell and billion buffalo.
> We cannot see the horror-movie of the funnel-cloud
> Snuffing up cows, crazing the cringing villages,
> Exploding homes and barns, bursting the level lakes.[46]

Shapiro's generally plain language and matter-of-fact description of the formation of a tornado stand in contrast to his comparison of the tornado to a disease such as cancer and his evocation of the prehistoric past, which comes to his mind when the funnel of the tornado resembles a mammoth's trunk, but the progression prepares the reader for the almost comic ending, the Hollywood tornado that carries cows and houses up into the air.

A generation later, Carol Muske (b. 1945), who grew up in Minnesota, turns a childhood memory of a tornado into a meditation on the meaning of art in a moment of crisis. The contrast between Shapiro's detachment and intellectualizing of the storm and Muske's intensely personal treatment provides a glimpse of the range of possibilities we can all find in weather. Muske prefaces her poem, "An Octave above Thunder," with a quotation from T. S. Eliot's "What the Thunder Said," from *The Wasteland,* then commences:

She began as we huddled, six of us,
in the cellar, raising her voice above
those towering syllables . . .

Never mind she cried when storm candles
flickered, glass shattered upstairs.
Reciting as if on horseback,
        she whipped the meter,

trampling rhyme, reining in the reins
of the air with her left hand as she
stood, the washing machine behind her
        stunned on its haunches, not spinning.

She spun the lines around each other,
her gaze fixed. I knew she'd silenced
a cacophony of distractions in her head,
        to summon what she owned, rote-bright:

      Of man's first disobedience,
        and the fruit . . .
      of the flower in a crannied wall
      and one clear call . . . [47]

    The speaker in the poem listens as her mother calms herself and her children by reciting snatches of classic poetry, from Milton and Tennyson, as the storm rages outside. Without mentioning the tornadoes that may accompany the severe thunderstorm, Muske cleverly introduces their presence by mentioning the washing machine that has stopped spinning and her mother's spinning motions and excited recitation. In the following stanzas her mother quotes more poetry, from the blind poet Milton, Wordsworth's "getting and spending we lay waste our powers," and the Bible. The storm passes, and the speaker reflects in the concluding stanzas on the importance of even the most fragmentary knowledge of poetry in moments of "extremity," in the sense of extreme need or peril. She matches the sublime in nature with the sublime in art:

Here were the words of the Blind Poet—
crumpled like wash for the line, to be
dried, pressed flat. Upstairs, someone called
        my name. What sense would it ever

make to them, the unread world, the getters and spenders,
if they could not hear what I heard,

nor feel what I felt
nothing ruined poetry, a voice revived it,
extremity.[48]

Another Minnesota poet, Robert Hedin (b. 1949), takes a lighter approach to tornadoes:

The last time any of us saw Gustafson's prize sow
She was rising over the floodlights
Of the poultry barns, pedalling off into the rain
And wreckage.[49]

The poem ends the next morning with the pig's owner sitting still dazed in the sty with "the empty swill pail / Vibrating in his hands." The poem is a variation on the popular image of livestock caught up in tornadoes and the stories of remarkable escapes. Why did the tornado take the sow and not Gustafson? Hundreds of books about tornadoes contain similar stories, none offers an answer.

Although most scientists deny any distinction between a tornado and a waterspout, traditional usage preserves the latter term. James Merrill (1926–1995), known for his poems about language and communication, describes an outbreak of waterspouts as both nature on a drunken spree and a moment of enlightenment for a man on the beach watching the storm. The poem begins:

Where foam-white openwork
Rumples over slate,
Flash of a fork, the first
Wild syllables in flight,
The massive misty forces
Here to be faced are not
Of wind and water quite
So much as thought uptwisted
Helplessly by thought,
A fullblown argument
Sucked racing through whose veins
Whitebait and jellyfish
Repeat the lacy helices
Threaded into the stem
Of a Murano—

Ending a few lines later:

> A bright-eyed reveler
> Looks down on cloth outspread,
> Strewn silver, fruits de mer,
> The lighthouse salt-cellar—
> A world exhausted, drained
> But, like his word, unbroken;
> Looks down and keeps his head.[50]

The poem twists like the waterspouts, which emerge as much from the speaker's mind as he thinks about the emptiness of life, as from the unstable air. The waterspouts resemble the twisted stems of Venetian glass goblets (Murano) and twirl like drunks. The speaker's brain reels, too, seeing the herring gulls as "syllables in flight," but looking at his picnic blanket, a kind of miniature world, he sees everything has order, meaning. His world, like nature's, is guided by design. Meteorologists speak of "organized" storms, knowing from experience and models that these are more likely to produce tornadoes. Poets, looking for the way the mind produces poetry, also find organization in the storm.

Hurricanes in American poetry do not begin with Hart Crane's (1899–1932) "The Hurricane," although it may be the best known. The short poem is written in unrhymed couplets, with a ragged rhythm mimicking that of the storm, and with an image of God mounted on a horse. Crane uses obscure words and odd mental associations, which now seem prescient when compared with eyewitness accounts of Katrina. The poem ends:

> . . . Battered,
> Lord, e'en boulders now out-leap
>
> Rock sockets, levin-lathered!
> Nor, Lord, may worm out-deep
>
> Thy drum's gambade, its plunge abscond!
> Lord God, while summits crashing
>
> Whip sea-kelp screaming on blond
> Sky-seethe, high heaven dashing—
>
> Thou ridest to the door, Lord!
> Thou bidest wall nor floor, Lord![51]

The use of words such as "levin" for lightning, and "gambade," which carries multiple meanings—the spring of a horse, a fantastic movement in dance, or, simply, antic motion—intentionally disorients the reader, as he would be confused by the storm, but these words also conjure fantastic images of a disaster in which nothing remains, door, wall, or floor.

A poet from a more recent generation, Victor Hernández Cruz (b. 1949), born in Puerto Rico and raised in New York City, takes a lighter tone but finds in hurricanes serious lessons about the danger of the meretricious in life. "Problems with Hurricanes" is worth quoting in its entirety:

>A campesino looked at the air
>And told me:
>With hurricanes it's not the wind
>or the noise or the water.
>I'll tell you he said:
>it's the mangoes, avocados
>Green plantains and bananas
>flying into town like projectiles.
>
>How would your family
>feel if they had to tell
>The generations that you
>got killed by a flying
>Banana.
>
>Death by drowning has honor
>If the wind picked you up
>and slammed you
>Against a mountain boulder
>This would not carry shame
>But
>to suffer a mango smashing
>Your skull
>or a plantain hitting your
>Temple at 70 miles per hour
>is the ultimate disgrace.
>
>The campesino takes off his hat—
>As a sign of respect
>toward the fury of the wind
>And says:
>Don't worry about the noise
>Don't worry about the water
>Don't worry about the wind—

If you are going out
beware of mangoes
And all such beautiful
sweet things.[52]

While the tone of the poem is playful, Cruz employs the device of putting the warning in the voice of a wise Caribbean farmworker, presumably one who has survived many hurricanes. Death by flying fruit is, of course, more than faintly ridiculous, but so is worrying about what you cannot control. Respect the fury of the storm, keep your head down, and keep your sense of humor is good advice for any dangerous situation.

The sun and heat are often used in poems for various effects, but less often as subjects. Climatologically it is fitting that one of the great poems about the sun comes from the Mexican poet and critic Alfonso Reyes (1889–1959). "Monterrey Sun" is a hymn in praise of the sun as a source of his youthful imagination and a lament that the sun is such a powerful source of creative energy that it exhausts him. At the heart of the poem are these vivid lines, translated by Samuel Beckett:

Indigo all the sky,
all the house of gold.
How it poured into me,
the sun, through my eyes!
A sea inside my skull,
go where I may,
and though the clouds be drawn,
oh what weight of sun
upon me, oh what hurt
within me of that cistern
of sun that journeys with me!
No shadow in my childhood
but was red with sun.[53]

Reyes's only peer is his contemporary Wallace Stevens, who personifies the sun as a flamboyant mythical character in "The News and the Weather," published in July 1941, which begins: "The blue sun in his red cockade / Walked the United States today." The sun is a shaper of personal and national identity; it can also signal the passing of time and the fragility of life, as in "This Amber Sunstream" by Columbia University professor Mark Van Doren (1894–1972). Like Reyes, Van Doren compares sunlight to a sea, flowing through his mind and body. The poet is awed by nature's indifference to human concerns. Unlike Reyes's outdoor sun, Van Doren's is the sun as it penetrates "the sunk hull of houses." The first stanza and the last two lines convey the essential message:

This amber sunstream, with an hour to live,
Flows carelessly, and does not save itself;
Nor recognizes any entered room—
This room; nor hears the clock upon a shelf,
Declaring the lone hour; for where it goes
All space in a great silence ever flows.
. . . .
No living man in any western room
But sits at amber sunset round a tomb.[54]

A poet and artist from Brooklyn, David Colosi (b. 1967), offers a playful poem in which the sun addresses the reader directly in the form of a rap song, because it is a "Sun with Issues." The poem is prefaced with a quotation not from T. S. Eliot but from the R & B group Martha and the Vandellas: "I got a heat wave burning in my heart. I can't keep from crying, tearing me apart." The sun thanks earthlings for the many sun songs—"Sunshine of Your Love," "Sunny Side of the Street," and others—but warns that it will "melt your radio in the street, burn the flip-flops off your feet, / Rub that SPF 30 on your skin, only then will I begin / to lick it off like bar-b-que sauce." The sun continues to enumerate the bad things it will do, including cause melanoma and heat stroke. Finally,

> And at the end of the day, when I go down,
> I'll be waiting in your concrete,
> in your asphalt seeping heat.
> Go ahead, run your air all night, I'll be sure to turn out the light.
> And at the stroke of dawn, when I'm coming on strong,
> I'll find you in your bathtub where I've left you cool and nice
> sunstroked and drowned in a pool of melted ice.
> "It's a sunshine day,
> everybody's smiling, (sunshine day),
> everybody's laughing, (sunshine day)."[55]

Colosi's sun is mischievous and malevolent, reveling in its power. In an era of planetary warming, the message is ominous, but is anyone listening?

Forecasters and forecasting are the subject of many poems, some witty, some dryly humorous, from Ambrose Bierce's literally damming portrait of the "Chief Forecaster" in hell, still predicting, "Cloudy; variable winds, with local showers; cooler; snow," to another recent poet laureate, Ted Kooser's (b. 1939) elegant elaboration of the weather as horse metaphor in "Weather Central":

> Each evening at six-fifteen, the weatherman
> turns a shoulder to us, extends his hand,

and talking softly as a groom, cautiously
smooths and strokes the massive, dappled flank
of the continent, touching the cloudy whorls
that drift like galaxies across its hide,
tracing the loops of harness with their barbs
and bells and pennants; then, with a horsefly's touch,
he brushes a mountain range and sets a shudder
running just under the skin. His bearing
is cavalier from years of success and he laughs
at the science, yet makes no sudden moves
that might startle that splendid order
or loosen the physics. One would not want to wake
the enormous Appaloosa mare of weather,
asleep in her stall on a peaceful moonlit night.[56]

The humor of Kooser's poem lies in his juxtaposition of the presumed authority of the TV weatherman and the power and unpredictability of the weather, which reduces the meteorologist to the role of "groom," who, although knowledgeable in equine/weather behavior, respects the "enormous" beast and wants to keep her peaceful. Even as a "cavalier," the weatherman is a knight errant, wandering in search of truth, but often led astray. He knows the physics can be loosened by "sudden moves" in the chaos of global circulation.

NOAA's History Web Site offers a selection of weather poems, mostly from the late 1930s when George W. Mindling, official in charge, Weather Bureau Office, Atlanta, Georgia, was an active versifier; and the mid-1940s, when the Southern Regional Headquarters published a newsletter, the *Breeze,* containing poems by Weather Bureau employees. Typical of Mindling's doggerel is "Soliloquy of the Weather Man," which begins:

> If I should say, "It's going to snow
> And folks won't need a fan,"
> They'd smile and say, "He does not know;
> He's just the Weather Man."

After two more stanzas giving examples of lack of respect for weathermen, the speaker lists some of the problems of making accurate forecasts, including the gathering and interpretation of data. Weather is generally predictable, he concludes, but the exact timing of the arrival of storms is not:

> No storm comes rolling on a track
> Like that of railroad trains;
> And yet there is no utter lack

Of order in the course of rains.
My greatest trouble's in their speed;
Sometimes they come too late.
A rigid schedule's what they need,
A bit more steady gait.
Then I could tell just when they'll come,
How long they're going to last,
And I would not appear so dumb
As often in the past.[57]

Mindling churned out dozens of similar verses on scientific instruments, Groundhog Day, cranks who call the Weather Bureau with inane questions, and writing an article for the American Meteorological Society. Like the poems in the *Breeze,* his poems provide a glimpse of the everyday life of meteorologists unavailable elsewhere.

As more women were trained as observers during World War II, some playfully expressed the "battle of the sexes" in verse. Joy Whiteside's "Weather Woman" is a good example:

Have you noticed it's the weather man
Whom people talk about?
I once believed implicitly
But I've begun to doubt.

I think it is a lady
Up there above the sky,
Who causes heavy rainstorms
Or makes warm breezes sigh.

One day she feels so happy,
The sun begins to shine,
We think that spring has really come
For that's a well-known sign.

And then without a warning
She changes overnight,
The skies are dark and gloomy,
No ray of sun in sight.

That's why I feel the way I do,
'Tis known since time began
That such an imp of fickleness
Could never be a man![58]

Whiteside's poem elicited comment in verse from both male and female observers who joked about what came to be called gender stereotypes in a later generation. The good-humored exchange masked the difficulties women faced in retaining professional employment after the war, but in 1945 women found satisfaction in sharing the rigors of weather observation. In "The Adventures of Annie Mometer," Lorena Pepper in Ketchikan, Alaska, wrote of Annie Mometer, whose fingers froze as she "plotted the pibal" (pilot-balloon observation), and presumably consulted her anemometer, in the snow. Other contributors commented on the challenge of studying physics and mathematics to qualify as observers. Limited though they are by conventional ideas about how poetry should look and sound, amateur poets who write to express their feelings about their work have two advantages over more skilled writers—a deep knowledge of their subject and an unself-conscious desire to communicate their thoughts to others. Weather poems by meteorologists, like cloud paintings by atmospheric scientists, are an imagining of weather with a double consciousness.

Pepper's Atlanta-based colleague Mindling, writing in 1939 in praise of the "raymette," the radio-transmitted instrument package recently adopted by the Weather Bureau, provides a sense of the excitement and optimism in the bureau as he speculates that soon:

> In the coming perpetual visiontone show
> We shall see the full action of storms as they go,
> We shall watch them develop on far away seas,
> And we'll plot out their courses with much greater ease.[59]

Weather in the most general sense—earth's aura, general circulation, climate—has been the subject of some of the most interesting American poetry by poets who have found in it ideas about the meaning of place, the nature of language, and the forces of history. For Archibald MacLeish (1892–1982), weather is the wind that blew from four directions in his childhood in northern Illinois on the shores of Lake Michigan. Northeast, southwest, west, and north winds each blew different scents and sounds, and each stirred unique dreams, particularly the north wind that brought images of the "Tyrrhenian sea where the hills saw / Once the long oars and the helmsman." But it is the northeast wind, "the wind over the lake / Blowing the oak leaves pale side out," that made the deepest impression as the speaker in the poem recalls the smell of the earth and the lake. MacLeish celebrates his "country," his Midwest with oaks, apple trees, skulls of buffalo, cicadas, and the lake, throwing in a brief glimpse of another sea, the Mediterranean with its long-oared triremes, by way of making the weather seem both familiar and exotic.[60]

Once one of the country's best-known poets, MacLeish wrote on many subjects and in many styles, leading one critic to call him "our poetic weathercock. A glance at his work in any decade will tell us which way the wind of thought and feeling and poetic fashion was blowing." "Weather" is an early poem, perhaps influenced by Carl Sandburg or MacLeish's own experiences as an artilleryman during World War I, as much as by the fashions of modern poetry. In "Memorial Rain," a poem that may be considered a companion piece to "Weather," MacLeish describes the dedication of a cemetery in Belgium where a friend is buried. The poet is disturbed by what he feels is the perfunctory ceremony and the interring of Americans on foreign soil. He remembers the "lake winds in Illinois" and finds the wind off the English Channel "strange," commenting with bitterness how "these happy, happy dead / Have made America." Later, as an editor of Henry Luce's *Fortune* magazine, MacLeish helped create public awareness of a vernacular American art and literature thriving alongside the new scientific and industrial civilization. More than a metaphoric "weathercock," MacLeish found in weather an appropriate symbol for the powerful forces shaping people and nations in the early twentieth century.[61]

A poet of the next generation, Reed Whittemore (b. 1919), builds his poem, "The Weather This Spring," around the idea of "weathering," the survival of the soul in an amoral world, in the cold war era with its threat of nuclear winter. The weather reflects language; both are rotten, both need "something so trivial / As a warm day in the sun." Whittemore's image of the conflict between nature and culture is striking and amusing:

> The weather this spring has been rotten, I don't know when I have
> Ever before set such store by something so trivial
> As a warm day in the sun.
> Past March, past April the cold has clung to us.
> Stalks are still limp in the gardens, trees are still bare,
> Of the color of gray not anything, persons,
> Places or things are spared, and if souls
> Live at all they do so in gray, gray holes.
>
> But words?
> At office and bank, in classroom and parlor,
> They saunter around us in shirt-sleeves, tanned and toothy,
> Greeting each other sociably through the overcast
> As if some sun of theirs, not ours, still glinted,
> Making meaning bearable.

The words are ungodly, the poem tells us, because they say they have their own souls and can ignore the seasons, "Leaving the rest of us, gray and envious, wishing / We too were fools, / And could weather it through with our own set of rules."[62] The words are "tanned and toothy" because they are fools, ignoring the reality of the unseasonable weather, which surely portends some calamity or reckoning. Weather is ontological truth; there is only one sun.

This is still true even

> Now that the cameras zero in from space
> To view the earth entire, we know the whole
> Of the weather of the world, the atmosphere,
> As though it were a great sensorium,
> The vast enfolding cortex of the globe,
> Containing contradictions, tempers, moods,
> Able to be serene, gloomy or mad,
> Liable to huge explosions, brooding in
> Depressions over several thousand miles
> In length and trailing tears in floods of sorrow
> That drown the counties as the towns. What power
> There is in feeling! We are witness to,
> Enslaved beneath, the passions of a beast
> Of water and air, a shaman shifting shape
> At the mercy of his moods, trying to bend,
> Maybe, but under pressure like to break.[63]

Thus begins Howard Nemerov's (1920–1991) poem "The Weather of the World," which celebrates the age of weather satellites but foresees that knowledge does not necessarily mean control. In contrast to Mindling's poem in praise of "raymette," Nemerov asks what is lost as well as gained by a global perspective on weather. In the second half of the poem, weather is personified as a human who, like all of us, "seeks balances that are / Inherently unstable." Weather and humans struggle for order in a chaotic world, the "god / Of this world, the apparent devil of the will / Whom God has given power over us." While the satellites send back pictures of "the weather of the world," we are left searching, "In windy eloquence and rainy light / As in the brilliant stillness of the sun," for ways to understand and control our inner weather.

In the poem that provides the title of this section, Richard Hugo links weather and time, specifically the passage of almost a century since the U.S. Army fought the Nez Percé in western Montana. "The wind is 95," the poem, "Bear Paw," begins, and the speaker laments the killing of the Indians so that a century later tourists can speed past the historical markers commemorating the events. "The wind / is infantile and cruel," and like Colonel Nelson A. Miles, "It cries 'give in' 'give in,'" to Chief Joseph and Chief Looking Glass. The poet instructs visitors to the site to "Learn what you can," "Marked stakes tell you / where they fell." "The wind / takes all you learn away to reservation graves." This is deeply felt political protest poetry. Written in the midst of the Vietnam War, Hugo's poem reminds the reader that the nation's past atrocities have become little more than dust blowing in the wind. He concludes with what might be called a plea for weather as an aide-mémoire:

> . . . One more historian
> is on the way, his cloud on the horizon.
> Five years from now the wind will be 100,
> full of Joseph's words and dusting plaques,
> Pray hard to weather, that lone surviving god,
> that in some sudden wisdom we surrender.[64]

The brilliance of this poem lies in its subtle association of event, place, and weather. The wind is understood as an artifact from the battle, like a rifle or a uniform, which historians must include in trying to tell the historical story. Contemplating the site of the final defeat of the Nez Percé in a weeklong siege in October 1877, and their subsequent confinement in Oklahoma, the poet feels the cold winds of early winter in the mountains and realizes that weather and a few dusty plaques are all we will be left with a century after any human tragedy. Weather remembered as context might keep us from repeating injustices.

More recently, Mark Strand (b. 1934), a Canadian-born, U.S.-educated poet, who followed Nemerov as laureate, imagines a weather-obsessed president of the United States giving a farewell address in which he reviews his "proposals, petitions uttered on behalf of those who labored in the great cause of weather—measuring wind, predicting rain, giving themselves to whole generations of days—whose attention was ever riveted to the invisible wheel that turns the stars and to the stars themselves? How like poetry, said my enemies. They were right. For it was my wish to make nothing happen." The president's platform was stasis, and he achieved it. "How lovely the mind is," he continues after applause, "when overcast or clouded with indecision, when it goes nowhere, when it is conscious, radiantly conscious, of its own secret motions." This shift, from seeking stillness to the pleasure in the secret motions of the mind, is the first hint that this seemingly witless chief executive may be taunting his audience. A bit later in his speech he asks, "How can I tell you what the weather has meant? The blue sky, its variations and repetitions, is what I look back on: the blues of my first day in office, the blues of my fifth, the porcelain blues, the monotonous blues, the stately blues, the ideal blues, and the slightly less than ideal blues, the yellow blues on certain winter days." After recalling "the weather of night," in which he "sailed [his] whole life," he concludes: "The blessings of weather shall always exceed the office of our calling and turn our words, without warning, into the petals of a huge and inexhaustible rose."

What do we make of this curious prose poem? First, and most obviously, it is a witty satire on political speeches empty of substance. Inaugural addresses, State of the Union messages, and campaign speeches are almost always formulaic and cliché-ridden. Change weather to "family values," or a "war on terrorism," and you will find similar empty rhetoric. There is also a broad hint that it is a satire on conventional verse. "How like poetry, said my enemies," about proposals on behalf of those who labor in the great cause of weather, and "they were right." Weather, poetry, politics, and the media are all solipsistic when viewed narrowly and to excess. Strand's president is resigning before his full term is up because, having accomplished nothing, he has achieved his goal and can leave office triumphant. But this is not quite true. In a brief introduction to the president's address, an anonymous observer lists the president's accomplishments: "a National Museum of Weather, in whose rooms one could experience the climate of any day anywhere in the history of man. His war on fluorocarbons, known as the 'gas crusade,' is still talked about with astonishment." Apparently, the president is both a memorialist and an environmentalist. His museum exhibits suggest that he believes that past climates have continuing relevance. History is important. Moreover, from the titles of those attending the speech, we learn that the president must have reorganized the cabinet. There is a "Minister of Potential Clearness," a "Warden of Inner and Outer Darkness," and an "Undersecretary for Devices Appropriate to Conditions Unspecified." The "Chief Poet Laureate and Keeper of Glosses for Unwritten Texts" also attended. Strand's "The President's Resignation" was published in his book *The Weather of Words: Poetic Invention,* in 2001. Whether his satire was meant to skewer any particular sitting or recent president is unclear. The poet's humor is broad, but timely, a neat reminder of the connections of weather and poetry as systems of signs.[65]

Wallace Stevens is America's preeminent poet of weather and atmospherics. Wind, clouds, the sun, the seasons, especially autumn, can be found in many of his poems because they work so well to illustrate one of his principal concerns, the relation of appearance to reality and the ways in which the artist's imagination responds to reality. Stevens was an uncompromising modernist who felt it was the role of the poet to ponder the effects of science on traditional beliefs. In a lecture at the University of Chicago in 1951, in which he invokes the ideas of physicist Max Planck on causality—asserting that it is neither true nor false, but a working hypothesis—Stevens submits "that poetry is to a large extent an art of perception" and that the imagination of the poet helps humanity "to believe beyond belief," to accept that reality exists, but is unknowable, and that the human mind can create beauty and order from the ever-changing hypotheses of science.[66]

In a 240-line poem, "The Auroras of Autumn," Stevens uses the poet/speaker's perception of the northern lights from his home in Connecticut to explore the passage of time and the end of life, both individual and global. The poem is written in unrhymed stanzas of three lines each, with eight stanzas in each of ten cantos or sections. The rigorous form of the poem contrasts with Stevens's sometimes unusual diction and syntax, making the total effect edgy, accentuating the processes of memory and discovery. Written in 1947, with the threat of nuclear war growing in public awareness, "Auroras" can be read as a comment on the ultimate terror facing the world, but it is also a poem about the ways in which the poet's perceptions of the aurora borealis provide insights into atmospheric physics.[67]

The poem is difficult because Stevens constantly shifts perspective, and his imagination challenges the reader to reassemble the insights he offers. A reader today benefits from information on autumnal auroras provided by a NASA Web site, which explains why there are usually more auroras in the fall, and the science behind the encounter of the sun's magnetic fields and the earth's stratosphere. NASA does not mention Stevens, but its official recognition of an aurora season provides a partial explanation of the poem's title. The increase in autumnal auroras is a predictable, if not well-understood, phenomenon. Robert Eather's comprehensive and beautiful book, *Majestic Lights: The Aurora in Science, History, and the Arts,* is also helpful in placing Stevens's poem in both scientific and literary contexts.[68]

Canto VI is meteorologically the heart of the poem. Stevens realizes that nature has a penchant for gaudy display and that the aurora is like the poet's own imagination, drifting idly through half-thought-of forms. Stevens's source for the line, "a theater floating through the clouds, / Itself a cloud," could be Napier Shaw, Eric Sloane, or any number of writers who used the theater analogy to describe cloud formations. Stevens takes this image in a different direction, discovering in the apparently random changes in the sky clues to a hidden order:

> It is a theater floating through the clouds,
> Itself a cloud, although of misted rock
> And mountains running like water, wave on wave,
>
> Through waves of light. It is of cloud transformed
> To cloud transformed again, idly, the way
> A season changes color to no end,
>
> Except the lavishing of itself in change,
> As light changes yellow into gold and gold
> To its opal elements and fire's delight,
>
> Splashed wide-wise because it likes magnificence
> And the solemn pleasures of magnificent space.
> The cloud drifts idly through half-thought-of forms.[69]

In the poem's final canto, the speaker plays with the four permutations of the phrase "An unhappy people in a happy world," rejecting "A happy people in a happy world" as "opera," which recalls the theater of the clouds and the observation in canto I that "the master of the maze" is "Relentlessly in possession of happiness." The speaker calls on a "rabbi," an authority standing for all priests, philosophers, scientists, and magicians, to tell his congregation about "This contrivance of the spectre of the spheres, / Contriving balance to contrive a whole." The speaker rejects the homilies of such philosophers because their contrivances lead them to experience the unhappiness, the doubts, of the people.

> In these unhappy he mediates a whole,
> The full of fortune and the full of fate,
> As if he lived all lives, that he might know,
>
> In hall harridan, not hushful paradise,
> To a haggling of wind and weather, by these lights
> Like a blaze of summer straw, in winter's nick.[70]

A haggling of wind and weather end the poem, a reminder of constant change and disorder in life. The aurora blazes like summer in the months preceding winter's reckoning, another deception. Stevens struggles to give us a glimpse of reality through the poet's imagination, using the spectacular displays of the aurora borealis to meditate on nature's purpose and meaning, always reserving the possibility that there is no meaning, nothing but the poet's dreams.

It is now generally accepted that Stevens was a nature poet who created a new, posttranscendental understanding of nature. "Wallace Stevens," as critic Gyorgi Voros argues, "sought to write a new Nature poetry that answered to what he perceived to be a great lack in American consciousness—a sense of the immediacy and profound presence of earth itself, rock-bottom foundation of human thought and experience." Further, Voros proposes, the science of ecology, with its focus on interrelationships within an ecosystem, "can be usefully applied to systems and patterns within a linguistic system. Language, like Nature, is prior to the individual use of it; as such it is (as Stevens recognized) itself a natural force and a process within which the individual dwells and comes into expression."[71]

Stevens's use of the nexus between language and weather exceeds in range that of other poets, I think. He has what can be called "a mind of weather," the ability to hold contradictory thoughts at the same time, specifically ideas about order and randomness. As an executive with the Hartford Insurance Company, Stevens surely followed developments in meteorology that affected his business. Although he did not live to see the successes of numerical weather predictions, he created what might be called "alphabetical climate forecasting," the building of poems (models) based on observed conditions and following the logic of language. Quoting literary critic Joseph Meeker, Voros concludes: "'We apply human mentality to the earth according to the requirements of the model we have adopted to explain it to ourselves.'" A final aphorism by Stevens, "words are everything else in the world," sums up the poet's meteorological mind—the world is weather, and everything else is a depiction of it.[72]

To know the weather of a place, consult the local meteorologist and the nearest poet. Meteorologists and poets search for order in a chaotic world, one with numbers and imagination, the other with words and imagination. Why should we "pray hard to weather"? Stevens, Ammons, Whittemore, and Strand show us that such worship may bring us closer to understanding language and improve communication. Stevens and Nemerov glimpse cosmic order in the weather. Stevens, Strand, Roberts, Shihab Nye, and others offer the heretical notion that order is constant change, and weather and sky are the essence of the unknowable future. Stevens anticipates in the solar storms that create the aurora the idea of space weather, an order behind disorder in the atmosphere. For Stevens, MacLeish, Hugo, and Strand, the answer is that weather binds us to history and to a place. Any one of these reasons should be enough to explain the godlike power of weather on the human mind.

Plotting the Weather

Brevity and elliptical language make poetry a suitable medium for talking about the elusive nature of weather, but novelists, too, have found in weather fruitful themes, symbols, and experiences. In twentieth-century America we could begin with almost any writer, especially the realists and naturalists, who used weather as a backdrop or as evidence of the forces of nature toying with human desires. Jack London comes to mind, as do Ole Rölvaag and Willa Cather. Later generations produced notable imaginative treatments of weather, drought in John Steinbeck's *Grapes of Wrath* (1939), the wind and cold of Chicago in Richard Wright's *Native Son* (1942), a blizzard in New York City in Henry Morton Robinson's *The Great Snow* (1947), and the hurricane in John D. MacDonald's *Murder in the Wind* (1956). Saul Bellow's *Henderson the Rain King* (1959) might qualify for its brief but crucial passages on physical and spiritual renewal through rain. Kurt Vonnegut's *Cat's Cradle* (1963), with its satire of his brother Bernard's snowmaking experiments, is one of the most amusing fictional creations involving allotropes of ice; and Toni Morrison's descriptions of winter along the Ohio River in *Beloved* (1987) are powerfully imagined and help confirm her heroine's connections to the natural world. Science fiction, whether hard-to-classify works such as Charles Fort's bizarre meteorology in *The Book of the Damned* (1972) and Scientology founder L. Ron Hubbard's *Wind-Gone Mad* (1992), or currently popular fantasy novels such as those of Kim Stanley Robinson and Bruce Sterling, provides numerous examples of weather as a last frontier in both a literal and a metaphoric sense.

To keep my sample of weather novels manageable, I will consider briefly three written before and immediately after World War II that use weather as major elements of plot and theme: Elizabeth Madox Roberts's *Jingling in the Wind* (1928), a fantasy of weather control; George R. Stewart's *Storm* (1941), a chronicle of a Pacific front that brings snow and flooding to California; and Gore Vidal's *Williwaw* (1946), an account of a small navy freighter caught in a violent windstorm in the Aleutian Islands. These imaginings of weather, each important, if neglected, in American literary history, will establish a baseline, a map of American weather, with which to compare four recent novels that find in weather keys to understanding some of the spiritual and moral conundrums of contemporary life: Paul Quarrington's *Storm Chasers* (2005), Rick Moody's *The Ice Storm* (1994), Jean Thompson's *Wide Blue Yonder* (2002), and Clint McCown's *The Weatherman* (2004).

Elizabeth Madox Roberts was born in Kentucky in 1886 and died in 1941. Educated at the University of Kentucky and the University of Chicago, she taught music and wrote poetry and fiction. Her early novels, *The Time of Man* (1926) and *The Great Meadow* (1930), earned her critical acclaim and a popular audience. *The Great Meadow,* a historical novel about pioneers in Kentucky, was made into a motion picture in 1931. *Jingling in the Wind* was her third novel and a departure from conventional realism and regionalism, leaving her admirers puzzled. Briefly, the novel focuses on Jeremy, a professional rainmaker in a rural county in what seems to be the United States in the not-too-distant future. A friend returns from a visit to a nearby city and describes some of the exotic people he has seen there, including Tulip-Tree McAfee, also a rainmaker. Jeremy falls in love with Tulip sight unseen and decides to visit the city and find her. His journey is interrupted many times, and each chapter introduces new characters who complicate Jeremy's quest in some way. The novel climaxes with a rain festival at which Jeremy is honored and named the "Rain-Bat." He wins the love of Tulip, and the book ends as it began with a spider on a web in the rain-soaked grass and the author commenting enigmatically: "Life is from within, and thus the noise outside is a wind blowing in a mirror."[73] Roberts seems to use the weather, and rainmaking in particular, as a metaphor for the artist's mind as she creates an alternative world that she uses to comment on and satirize America of the 1920s, with its liberated women, fondness for psychoanalysis, conflicts between science and religion, and worship of technology.

Early in the novel we are told that Jeremy follows the "Winkle System" of rain production. Winkle was not the inventor; the real inventor's name has been forgotten because he spent his life in his laboratory trying to perfect his system, while Winkle popularized the work and bought out his competitors. The inventor's name, Roberts writes, "was unknown to ninety-eight per cent of his countrymen, and of the two per cent who knew of it, fifty per cent had it wrong by an entire syllable, and fifty per cent of the remaining persons could not pronounce it and confused it with Einstein, Nietzsche, and Prometheus."[74] If this passage echoes the sentiments of Roberts's contemporary the cynical editor Henry L. Mencken, it is intentional. She was part of the revolt of midwestern and southern writers who came of age after World War 1, the domestic equivalent of Hemingway's "Lost Generation." Roberts's loss of innocence was not the result of the physical and psychological trauma of war, however, but the social and technological changes that the war helped to accelerate. The objects of her scorn go well beyond "the ignorant booboisie," to include both Christian fundamentalists and academic and scientific elites. Opponents of the Winkle System included both "evangelists of many sects," who condemned rainmaking as blasphemous and pagan, and scientists, who found the system deficient in theoretical sophistication. Supporters of rainmaking were further divided politically between those who favored private enterprise and those who supported a government-controlled "Rain Bureau." Shortly after describing the controversies among rainmakers, and pointing out the similarity to the conflicts over Darwinian natural selection, Roberts discusses psychology: "Having destroyed friendship and exhausted classical myth, they [the psychologists] were investigating the dreams of the happily married: Did they dream? How often? What? The psychologists were almost without occupation. The world had dwindled. They sat twiddling their thumbs."[75]

In preparation for the great rain fair, a committee decides that the event should be commemorated by a poet. To their dismay they discover that there were no poets available, because

philanthropists and other agents of similar sort had so rewarded the poets for staying out of America, had given prizes and traveling scholarships to those persons with the stipulations that they should keep beyond American shores, or had rewarded those who had chosen to stay afar, that scarcely a one was left. . . . But finally one was discovered, one who had been masquerading as something else, pretending that he had a living from some more genteel calling when actually he had no living whatever. Discovered in a small town far in the South, he was granted security, dragged out of hiding, and promised the fee.[76]

The great day finally arrives, with Jeremy presiding, and rain falls after an epic struggle between Jeremy and the clouds. Roberts describes Jeremy's rainmaking apparatus as a combination of electrical and chemical processes that together trigger precipitation. The essential element is an acid known as "$xkygrt_2$," also called "xk" and "exkay," a kind of elixir that is kept in a box "precious as a pyx." It is an "ointment to purge any cloud and instrument of catharsis." Roberts's fun with the idea of a chemically induced catharsis, or purification through purgation, in rainmaking, classical tragedy, or psychoanalysis anticipates a host of technological quick fixes developed over the next fifty years. Jeremy's synthesis of chemical and natural sublime works in this case, as "a thunder shook the city and there was a great cry of pleased terror from the women and the men."[77]

A sympathetic critic of Roberts's work, the historian Allan Nevins, found the novel "a gentle, clouded form of satire, sometimes rather wistful, and seldom more than reproachful. It is a mockery that shifts and changes in color and form from page to page, usually defying analysis."[78] Like a meteorologist whose forecast calls for rain but gets sleet, Nevins just misses the essence of Roberts's novel, which, like the weather itself, shifts and changes in color; a "clouded form of satire," indeed.

George R. Stewart (1895–1980) was the author of several popular novels and works of nonfiction and taught in the English department of the University of California, Berkeley. He began his career with a history about the Donner party, *Ordeal by Hunger,* and books on American place-names. The idea for a book about a weather system, Stewart recalled in a later edition of the book, occurred to him in the winter of 1937–1938 when he was living in Mexico and reading about a series of severe storms that struck California. His home was in the Berkeley Hills with a view, such as photographer Richard Misrach would have a half century later, of San Francisco Bay and the Pacific horizon through the Golden Gate Bridge. "No observing person," Stewart writes, "can live in such a spot without becoming storm conscious." Beginning his research with Sir Napier Shaw's *Manual of Meteorology,* Stewart recalled reading that "a certain meteorologist had even felt storms to be so personal that he had given them names." Thus was born Maria, a twelve-day January storm that is the main character of the novel. Stewart, reflecting his antiquarian interest in names, instructed readers to put "the accent on the second syllable, and pronounce it 'rye.'"

Returning to Berkeley to complete the book, Stewart consulted with the U.S. Weather Bureau, met with transportation and power company maintenance crews, and rode with the California Highway Patrol during storms. The result is a book in which the human characters are virtually anonymous while the storm emerges as a living force, challenging humans to use their intelligence to survive its fury. Stewart reminds his readers that the book was written during "those grim and terrible months of Dunkirk and the fall of France," inviting them to read the novel partially as an allegory of the coming struggle between democratic order and irrational force. Anticipating the so-called nonfiction novel and new journalism of the 1960s, *Storm* is a compelling account of the ways in which individuals and institutions respond to what they know and think they know about nature.[79]

Storm begins with a description of the dynamics of ocean currents and the atmosphere as they create weather. The storm's birth occurs in the North Pacific when a cold air mass descends from Siberia, pushing beneath the warm air over the ocean southeast of Japan. The storm is first detected by a ship bound for China when the crew notices a rapid change in temperature and barometric pressure, and the smell of dust in the air hundreds of miles from land. In the Weather Bureau office in San Francisco, a junior meteorologist enters the temperature and barometric data he receives from weather stations across the continent and ships in the Pacific. He correlates the data from the ship and the Japanese weather stations and begins to draw the isobars that represent a developing storm. He names this storm "Maria."[80] By the second day, the chief meteorologist begins to take notice of his assistant's projections. Stewart provides a detailed description of the making of a weather map in the 1930s.

The highway superintendent responsible for U.S. 40 over Donner Pass checks to make sure his men and equipment are ready to plow the roads. Stewart begins the third day with a reference to weather and the analogy of combat: "In the Northern Hemisphere the opponents are the Arctic and the Tropics, North against South. Uncertain ally to the South—and bringing, now withdrawing aid—the sun shifts among the signs of the zodiac. And the chief battle-line is known as the Polar Front." In January, with the Northern Hemisphere tilted away from the sun, the action is "in that No Man's Land which is the Temperate Zone [where] the storms raid and harry."[81]

On the third day the sequence of accidents and consequences of the storm begin to unfold. An owl touches a power line and causes concern at the power company as the linemen prepare for the storm. The junior meteorologist does his "public service" by talking to students from a local school about the weather. The general manager of the railroad, the chief service officer of the Bay Area airport, and officials of the telephone company also set their emergency procedures in motion. The airport manager, Stewart wryly observes, "was fond of the expression 'Acts of God,' but since he confined it to disasters he must apparently imagine God as practicing sabotage." Rejecting the storm as terrorist analogy, Stewart has the chief meteorologist say,

> "Storms and men—they're all different, and yet they're all the same. Each little storm starts out hopefully, but until it's all over, you can't say whether it was better than the ones that went before it—or as good." . . . "And in the end," the Chief went on, "it doesn't seem to make much difference. Every storm mixes up the air a bit. Sometimes it raises quite a hullabaloo. Then it's gone, and there you are in a high pressure area just like where you were before, with maybe another storm showing up on the edge of the map. Month in, month out, a lot of wind blows, but at the end of the year everything is just about where it was before."[82]

Weather, ultimately, is neither war nor disaster, but everyday life.

Saturday, the fifth day, brings one part of the storm across the Canadian border, and the rain begins in northern California. The flood-control coordinator for the Sacramento Valley checks the levees. A blizzard strikes the Midwest, while snow begins to fall in the Sierra Nevada. By the sixth day, wind and rain begin to disrupt air travel. Halfway through the storm the novel's pace quickens. More men and agencies go into action. Stewart shows the reader how vulnerable modern civilization is to weather, electric and telephone lines come down, roads and tracks are closed, farmers lose their crops, there are fewer shoppers on the city's streets. The seventh and eighth days of the storm bring a series of small disasters and a few fatalities. Day 9 includes a multicar pileup in a flooded underpass. During the tenth and eleventh days there is more serious flooding in the valley. By the final day there have been sixteen deaths directly attributable to the storm and millions of dollars in property damage; some crops are saved by the rain, others lost. Stewart's conclusion about the storm is simple. It is nature's challenge to humans; weathering, both its popular sense of surviving and its technical definition of mechanical, chemical, and biological erosion of exposed material, is at the core of life itself. Toward the end of the novel Stewart speculates on the effect of improved weather forecasts. He is pleased by the possibilities of greater international cooperation—good weather forecasts, like good weather, mean stability and peace—but he is troubled by the possible psychological effects:

> A century hence, Siberia and Patagonia alike, Arctic and Antarctic, the islands of the sea and the ships that cross the sea, may all report to one great bureau. Then the published tables of next year's weather may be as accurate as the published tables of next year's tides.
> Only man's quarrelsomeness seems likely to prevent this consummation. To master and apply the laws of the air without a world-wide co-operation is like trying to predict tides with an imperfect knowledge of the motions of the sun and moon.
> If the final success is attained, what will be the effect upon man? Will he at last have to stop talking and speculating about the weather? Will the foreknowledge that he must prepare against a tornado upon a given day be more strain than grasshopper-like ignorance and sudden disaster? Will the removal of the daily mystery only serve perhaps to make life at once safer and more boring?[83]

Published on the eve of World War II, Stewart's Book-of-the-Month Club best seller of science in the service of humanity, coping with the storm with minimal casualties, brought hope to readers. His and his readers' faith quickly faded, however, and it is no surprise that Stewart's next novel was a postapocalypse science fiction tale of survival, *Earth Abides*.

A weather novel with direct links to the war, *Williwaw,* by Gore Vidal (b. 1925), appeared in 1946. It is a first novel by a writer who is known today for more than forty works of fiction, history, and autobiography. Vidal served as first mate on a supply ship in the Aleutian Islands in World War II, an experience that provides the details for *Williwaw.* The title, Vidal tells us in a prefatory note, is an Indian word for a big wind that sweeps suddenly down from the mountain peaks that make up the Aleutian chain. This explanation of the word's origin is placed in doubt by other authorities, notably the *Oxford English Dictionary,* which defines "williwaw" as "a sudden violent squall, originally in the Straits of Magellan," and cites an 1842 source. In northwestern Australia, according to Marjory Stoneman Douglas, hurricanes are called "willy-willies." A sailor's term, "williwaw" has obviously crossed the equator many times, and whether its origin is Patagonian, Athabascan, or pure imagination is less important than its aptness. The very sound conveys a sense of howling winds in exotic places.[84]

Vidal's williwaw mirrors the emotions of six men who travel together on a small army freighter from one military base to another late in the war. Evans, the ship's skipper, although still in his twenties, was a fishing boat captain in Alaskan waters before the war. Duval, the chief engineer, is an older man from New Orleans. Bervick, the second mate, is an experienced seaman about thirty years old. Their passengers are Major Barkison, his assistant Lieutenant Hodges, and a Catholic chaplain, O'Mahoney. The major is anxious to get to a base from which he can fly with a report he has written on the closing of the Aleutian bases. Barkison is a West Point graduate and career officer, but he suffers from seasickness, as does the chaplain. The passengers are as garrulous as the crew is taciturn. In addition to the usual tension between men of different ranks and occupations, there is hostility nearing hatred between Duval and Bervick because of a waitress in one of the harbor cafés. Duval pays her for sex, while Bervick thinks he wants to marry her. The novel's action takes place over a little more than three days, during which the ship is driven onto a rocky coast and damaged during the williwaw. At the height of the storm Bervick angrily but accidentally causes Duval to fall overboard and drown in the icy water.

References to weather appear on almost every page of the short novel. Vidal uses the weather and the men's reaction to it as indicators of the rising and falling stress they are feeling. The tone of the book and the dialogue of the men are low-key, making the repetitions of statements such as "This is funny weather," "I think I'll go check with Weather," "We can't tell until we know the weather," and "The weather reports are likely to be pretty lousy" seem freighted with meaning. The major and the chaplain fear rough seas, while the lieutenant is eager to experience a storm. Evans, despite his experience, is keenly aware of his lack of education and scant knowledge of meteorology. As it becomes clear that the ship will encounter a storm and as the conflict between Bervick and Duval grows public, Vidal has Evans think "of the falling barometer and the stormy sky. For some reason, as he thought, the word 'avunculus' kept going through his head. He had no idea what it meant but he must have heard or read it somewhere. The desire to say the word was almost overpowering. Softly he muttered to himself, 'avunculus.'" This wonderful conflation of "avuncular" and "cumulus" seems to help Evans resolve some of his internal conflicts. He reprimands Bervick for quarreling and decides to proceed cautiously. He returns to his cabin, looks in the mirror, and says, "quite loudly, 'Avunculus.'" The major, despite his education, is no less guided by idiosyncratic behaviors. Vidal writes, "Major Barkison had a sure method of foretelling weather, or anything else for that matter. He would, for instance, select a certain patch of sky and then count slowly to three; if, during that time, no sea gull crossed the patch of sky, the thing he wanted would come true. This method could be applied to everything and the Major had great faith in it."[85]

Vidal's description of the williwaw is brief and vivid. Powerful hurricane-force winds and fifty-foot waves toss the ship violently. With luck and skill Evans manages to get the ship lodged between two rocks in a small, calm harbor, where they ride out the remainder of the storm. Everyone is relieved, and the major suggests a commendation for Evans. After they manage to get off the rocks and continue their voyage, Bervick tosses Duval a hammer, causing him to fall overboard. Knowing that the ship could not stop in time to save Duval from hypothermia, Bervick does not report the incident and lies when asked if he knows what happened to the chief engineer. Others have their suspicions because the bad feelings between the two men had been so obvious. Instead of a commendation, Evans may receive a reprimand, and an investigation may ensue. The major, newly promoted to lieutenant colonel, feels generous and files an accident report on Duval without further inquiry. The skies clear, and Evans looks forward to time off while the ship is repaired in Seward or even Seattle. The williwaw clearly suits the mood of men who have been at sea and at war too long. All the characters, their nerves frayed, are capable of violence, or at least unchecked anger. They know that justice is not being done, but they no longer care. They want out of the situation, away from williwas, real and imagined.

Weather in the novels of Roberts, Stewart, and Vidal is managed to some degree by magic, hard work, and good luck, but knowledge of the weather is basically limited to temperature and barometric pressure. A half century later, weather is better understood, forecasts are more widely available, and storms are managed by an ever-expanding network of public and private agencies. Weather novels of the past two decades reflect these changes in several ways, chiefly by emphasizing the manner in which mediated weather rather than real weather creates the climate in which most of us live. The exception, significantly, is a Canadian novel in which the characters seek extreme weather to fill spaces in their lives that neither TV nor daily weather can.

Paul Quarrington (b. 1953), author of *Storm Chasers* (*Galveston* in the Canadian edition), was born and lives in Toronto. The author of several novels and books on hockey and fishing, Quarrington is also a musician, artist, and filmmaker. *Storm Chaser* is a taut and beautifully written story of tragic personal loss and the search for meaning in life. There are three main characters. Caldwell is a former physical education and science teacher in Toronto whose wife and son were killed in a storm-caused accident on the day he wins $16 million in a lottery. Beverly was raised by her alcoholic grandfather in southern Ontario after her parents were murdered in a drug deal; her own daughter was killed when she was caught in the cyclonic suction of a swimming pool intake drain. Caldwell and Beverly have lost conventional religious faith and now seek redemption for the guilt they feel by placing themselves in the path of destructive storms. The third protagonist is Jimmy Newton, known to television viewers and Internet weather weenies as "Mr. Weather" for his well-publicized adventures photographing tornadoes and hurricanes. "The National Oceanic and Atmospheric Administration," Quarrington tells the reader, made no secret of the fact that they checked Jimmy Newton's website daily, that they at least factored his thinking into their data. . . . He had a reputation for bravery, although it was more truly an utter recklessness."[86]

The novel begins as these three, and two young women seeking an inexpensive Caribbean holiday, board a plane in Florida for "Dampier Cay," a small island southwest of Jamaica, where Hurricane Claire is predicted to make landfall. Newton, who recognizes Caldwell from previous storm chases, tries to engage him in a discussion of weather. Caldwell claims he was in Seattle on a fishing trip, to which Newton responds:

> "Bullfuck. Ingshit. You went there looking for lightning."
> Caldwell nodded. "Maybe so, I like lightning."
> "I don't." said Newton. "Lightning is like foreplay. I'm interested in getting fucked."
> . . . Jimmy Newton droned on, the flat pitch of his voice sitting a quarter tone above the hum of the airplane's twin engines. It made Caldwell long to plug his ears. Newton was explaining something to do with the quantification of chaos, which, if successful, would enhance the dynamic modeling of weather systems a hundredfold. . . .
> Newton abandoned the subject of chaos abruptly and pointed out the window. "You ever do this, Caldwell?"
> "Do what?"
> "We're in a cloud now, right?"
> Caldwell turned and saw whiteness swimming by outside the airplane. "Right."
> "Okay, so we wait, and . . . look, we're not in the cloud any more."
> "Uh-huh?"
> "But when did we leave? You know? You can never tell when you leave a cloud, you just know when you're out of it."[87]

Though both men are intensely interested in experiencing the hurricane, their motives are far apart. Near the end of the book, as the hurricane destroys the island, and one of the local residents is killed, Quarrington returns to Newton's observation: "There had been no clear demarcation between being and non-being. It was like the cloud game, the heaven game. Polly slipped out of life, into death, and no one could say when, exactly, it happened."[88] Caldwell's family and Beverly's daughter also became nonbeings in the fraction of a second that a cloud forms or disappears.

"Weather is God," Jimmy Newton announces to Caldwell, Beverly, and the young women in the bar of the hotel as they await "Claire's" arrival. Take all your personal problems "outside when a category three is passing through. That's clean, baby. That's all your problems blown away." Beverly demurs:
"I don't think I agree with you, Mr. Newton," said Beverly. "I think God *is* the little human problems. All that other stuff is flash and filigree, you know. A cheap trick. *Pay no attention to the man behind the curtain.*"

Then, recalling her tornado chase with Newton in Oklahoma, she adds:

"But what struck me about the tornado—when we finally found one, thanks to you, Mr. Newton—was that it was composed mostly of the, um, *detritus* of human lives. Condoms, candy wrappers. Nails and wedding rings."
Caldwell opened his mouth to say, "I know what you mean," but he choked on the first word and reached instead for his whisky. [89]

Beverly is obsessed by the Galveston hurricane of 1900, in which more than 6,000 people died. The tropical depressions of hurricanes seem to match her personal depression from which she seeks release. For her, a hurricane is "the man behind the curtains" distraction, making us miss the real magic in small disasters. To know this for sure, she chases dangerous storms. Weather disasters raise many philosophical problems, not the least of which is, what are we worshiping when we seek God in the weather?

Caldwell and Beverly, having discovered that they are looking for the same thing in the storm, go into a small church as the full fury of the hurricane strikes. The church and most of the buildings on the island are destroyed; the only survivors are the husband of the hotel owner and the two young women who have had a memorable holiday and are dubbed the "Hurricane Party Girls" by the media. In the final paragraph, Quarrington, in the spirit of his sardonic tale of salvation through weather tourism, challenges the reader to suspend scientific belief and imagine that Caldwell and Beverly survived by floating away on a raft made of the church floor and that they live in peace on some deserted island. And in fact they do. Detritus is, after all, what remains after weathering—faith and dreams, as well as candy wrappers and wedding rings.

Weather in *The Ice Storm,* by Rick Moody (b. 1961), comes close to being as powerful as it is in *Storm Chasers.* Halfway through this mordant story of dysfunctional families in New Canaan, Connecticut, Paul, the sixteen-year-old son of one of those families, calls from Manhattan, where he is visiting friends from boarding school, to ask if he should come home early because of forecasts of bad weather. His younger sister, who is home alone, advises him to stay because their parents are at a party and won't miss him. To which Paul responds, "No one believes in the weather anymore."[90]

Weather in Moody's novel is a system of signs, cruel and contradictory, such as children receive from parents. Set just after Thanksgiving, 1973, with the Vietnam War dragging on, Nixon under suspicion for the Watergate break-in, and OPEC's oil embargo driving up the price of gas, Moody's novel is about two families, the Hoods (Ben and Elena and their children, Paul and Wendy) and the Williamses (Jim and Janey and their sons, Sandy and Mike), who are too absorbed in their own lives to understand their children's need for love and guidance. Ben, who works for a public relations firm and specializes in media and entertainment, is having a desultory affair with Janey, whose husband seems to be a venture capitalist who works from home. Elena, ignored by Ben, has retreated into alcohol and self-help books. Janey is just bored, as is fourteen-year-old Wendy, whose observations and judgments provide the moral core of the book.

Wendy, in rebellion against the stifling suburb, experiments with sex with both Sandy, a dreamy underachiever her own age, and Mike, a slightly sociopathic thirteen-year-old. From the beginning of the book, the weather reflects her mood:

> Rain. Some fat, smiling weatherman would say it was *raw*. New Canaan was maybe a single degree above freezing. Surfaces contracted. There had been hail, too. Her poncho didn't keep out the cold, but she withstood it, shivering, because she was precociously brilliant—everyone said so—and impractical. Anything was better than the homely, pink ski jacket her mom had bought her.

She goes to Mike's house and they bike to the Silver Meadow, an isolated residential psychiatric facility, where they can be alone. Weather enhances the experience:

> The light was failing. The precipitation had turned to snow. Or something close to it, fierce nuggets of precipitation. Precipitation like an insult. But the anticipation of licentiousness thrilled Wendy, worked that tantric magic on her. Winter didn't trouble her. She could have walked miles in the slush and ice, like a superhero.[91]

Superheroes, specifically Stan Lee's "Fantastic Four," function as an alternative family for Wendy and Paul, who is especially addicted to the comics. Need you be reminded, one of the four superheroes is "Storm," a girl who can make herself invisible. Ben puts aside his comics to drink and smoke pot with his school friends. In this haze he half hears the weather report and thinks: "No matter how many times the weather repeated its four symphonic movements, the specifics of rainfall and wind direction and velocity and barometric pressure seemed new to Paul."[92] New, because the weather remakes the conditions in which we live each day, repetitive, but subtly and importantly different. This storm will precipitate enormous changes in Paul's life.

Meanwhile, the Hoods and the Williamses are attending a "key party" in which couples place their car keys in a bowl from which the husbands will later select them, going home with the wife of the owner of the car. Slightly disgusted with themselves and each other, and in various stages of drunkenness, Elena goes off with Jim, while Ben searches for Janey. In Manhattan Paul struggles to get away from his stoned friends, one of whom vomits on his sneakers. The parties degenerate, as does the weather:

> Outside, everything had changed. Meteorologically, the phenomenon, which occurred rarely in that part of the Northeast, went like this: rain, sleet, and snow, propelled by subfreezing winds—warmer temperatures aloft and freezing temperatures at ground level—began to harden instantly on trees, rooftops, power lines, and other surfaces. The ice built up on every surface. . . . Abandonment was in the parlors of America, in the clubs, in the weather.

After describing Ben passed out on the floor of his host's bathroom, Moody speaks in the author's voice about how life is composed of coincidences and lapses of coincidence, how events in our past and future are linked by topography and history, and the analogy between the novelist and God: "Though metaphors of the mind of God are characterized by coincidence and repetition, examples from nature aren't as tidy. Nature is senseless and violent. So this part of the story is violent, and because it's senseless, too, it's not from the point of view of any of the protagonists."[93]

The senseless violence is the death of Mike Williams, who, wandering the streets of New Canaan in the early morning after the storm, sits on a guardrail that has been electrified by a downed power line and is killed instantly. It is Ben, staggering home, who finds Mike's body, carries him home, and calls the paramedics, who take the body to the Williams house for identification. Gradually both families assemble from different homes where they have spent the night. Parents berate their children; Jim Williams gives a self-justifying, transparently false excuse to Wendy and Sandy to explain why he and Mrs. Hood spent the night together.

The ambulance arrives, and both families realize the enormity of the tragedy that has happened, partly because of their indifference to one another. The Hoods suddenly remember that their son, Paul, might be waiting at the station for a ride home. Ben and Elena suspend their bickering and drive through the historic streets of New Canaan—"Here John Gruelle had first drawn his famous Raggedy Ann; landscape artist D. Putnam Bradley had painted his sweet pastorals; here Hamilton Hamilton had pen-and-inked, and Childe Hassam had Gallicized a little verdant scene." Moody's list of earlier Canaanite accomplishments contrasts with the Hoodses' failures, while at the same time implying that much that we consider significant is trivial after all. In the station Paul has spent the time contrasting the ideal family of "The Fantastic Four" with his own pathetic family, but when they arrive and his father tearfully tells him about Mike's death, they are amazed to see a sign in the sky: "An actual sign in the sky. The conversation stopped and there was a sign in the sky and it knotted together everything in that twenty-four hours. Above the parking lot. A flaming figure four. And it wasn't only over the parking lot. They saw it all over the country."[94]

The flaming four may be a validation of the superheroes or of the Hood family. For Paul looking back on his life, it marks the end of the story that he has carried with him for twenty years and needs to close. The flaming four is a phenomenon worthy of inclusion in some future edition of Charles Fort's *Book of the Damned*, an unverified sign, a truth that has been damned, as Fort would say.

The curious interplay of family and atmospheric dynamics is explored in a similar but happier way in Jean Thompson's *Wide Blue Yonder*, a novel about a single mother, Elaine, her rebellious seventeen-year-old daughter, Josie, and her ex-husband's almost mute uncle, Harvey, who spends his days watching The Weather Channel. The novel, set in Springfield, Illinois, is divided into four parts, the months of the hurricane season of June through September 1999. As the book begins, the reader sees The Weather Channel through Harvey's eyes. We learn later that his apparent mental slowness is partly the result of poor eyesight, but Elaine's concern for him stems from the fact that her ex-husband, Frank, is ashamed of his uncle and ignores him. Thompson's opening paragraphs evoke Harvey's joy and confusion:

> Beige Woman was saying Strong Storms. She brushed her hand over the map and drew bands of color in her wake. All of Illinois was angry red. A cold front currently draped across Oklahoma, a raggedy spiderweb thing, was going to scuttle eastward and slam up against your basic Warm Moist Air From the Gulf. The whole witch's brew had been bubbling up for the last few days and now it was right on the doorstep. Beige Woman dropped her voice half an octave to indicate the serious nature of the situation, and Local Forecaster nodded to show he understood.
>
> . . . Local Forecaster stood on the back porch steps and turned his nose to the southwest, where all the trouble would come from later. He sniffed and squinted and tried to tease the front out of the unhelpful sky, but it stayed as shut as any door.

Back inside it was Man In A Suit's turn. He stood, palm up, balancing
Texas by its tip. Texas was green today. Florida was full of orange suns. Man
In A Suit tickled the Atlantic coastline and whorls of ridged white sprang up,
high pressure. The jet stream arrows came leaping out of the northwest,
blue and swift, full of icy glittering.

There was always Weather. And every minute there was a new miracle.[95]

The Weather Channel gives Harvey's life meaning. He calls himself "Local
Forecaster" and limits his conversations to summaries of what he has heard on the
weather reports. When he learns that his name, "Harvey," will be used for the
eighth hurricane of the season, he is delighted. He will become something other
than what he is; he will, for a time, be free from his limitations. In the meantime,
he must endure the well-meaning efforts of Elaine to clean him up, get him to the
dentist, and prevent Frank from putting him in a nursing home. Two subplots develop, butterflies on the horizon. Jodie falls in love with a twenty-five-year-old policeman, Mitchell, who takes her home after she narrowly escapes from some of
her classmates who have a near-violent encounter with a drug dealer. Her pursuit
of Mitchell is part of her rebellion against her mother and the tedium of Springfield. Jodie, called "Sunshine" by her mother, is a zephyr compared with the tempest of Rolando, a young petty criminal whose desire to escape his life of poverty
in Los Angeles is so strong that he commits a series of crimes, including murder,
in his flight across the country to the inevitable appointment in Springfield. As he
engages in baggage theft at LAX, Rolando senses the weather anxiety of the travelers: "Although the weather in Los Angeles was summer-perfect, there was fog
in Seattle, there were high winds in Phoenix and thunderstorms over the Great
Lakes. There were delays and cancellations, everything backing up, people getting fretful."[96] Weather and some people, the author implies, only come to our
attention when they are an inconvenience or a threat.

Elaine hires a Mexican immigrant woman, Rosa, to clean and cook for Harvey
once a week. As if in anticipation of the arrival of these disparate characters,
The Weather Channel is running commercials for "off-brand automobile insurance, chain saws, Magic Mops, Shoe Away, Zim's Crack Creme," and other products Elaine had never heard of. "Where did they get these commercials?" she
asks herself.[97] But the odd and obscure goods and services offered are part of
the abnormal and baffling weather, as is the bizarre and mysterious behavior of
the elderly, the middle-aged, and the young. Josie's relationship with her mother
becomes more strained; her infatuation with Mitchell interferes with school and
work. She takes him to visit Harvey, and they agree that constantly watching
The Weather Channel would drive anyone crazy. July melts into August, Harvey
and Rosa fall in love, Josie and Mitchell sneak into her father's house while he
is away and make love, but leave evidence that Frank's wife passes on to Elaine.
Before any confrontation, however, Josie discovers that Mitchell has another

girlfriend and, heartbroken, she hides in Uncle Harvey's house. Meanwhile, Rolando continues his crime spree across the country and, high on drugs, shows up in Springfield.

September arrives. It has been a disappointing hurricane season for Harvey and The Weather Channel:

> Arlene never got past a Tropical Storm. Bert was a hurricane. He made landfall along the south Texas coast and pooped out. Cindy stayed out in the Atlantic, past Bermuda. None of them anything to write home about. But low-pressure masses kept firing themselves from the Cape Verde Islands, from the coast of Africa, like baseballs. An ocean full of storms. Dennis was lumbering toward Florida. Emily was a whirling blob somewhere east of the Windward Islands. Floyd was stacked up behind them, a giant saucer of wind and rain. Floyd was the one everyone at the Weather was keeping an eye on, the one getting all the attention. If there was going to be a Harvey, it would simply have to wait its turn.[98]

Frantic, Elaine begins to search for Josie. She meets with a policeman, Kellerman, from the juvenile division, who tells her a story about chance and order in daily life, to reassure her that Josie's behavior is "normal." He explains "cellular automations":

> "It's a computer simulation that tries to mirror the way the laws of nature work. It has to do with how complex results derive from simple beginnings. Seemingly random events are actually predictable. The result of repeating a sequence of possible combinations."
>
> Elaine shook her head. "Sorry. Not a computer person."
>
> "It's not really about computers. I'm not explaining it very well. Imagine water vapor freezing into snowflakes. The flakes can take any number of shapes, an infinite number. But you start with just a few basic combinations of crystals. . . . Replicate the sequence according to a regular set of rules, over and over and over, like a computer would, and you get, guess what?"
>
> "No two snowflakes alike."
>
> Kellerman nodded, pleased. "So let's say you wake up on an ordinary day and you make a series of choices. Coffee or tea. Shower or bath. Read the paper starting with the front page or the funnies. Every one of those choices is a sequence. A simple beginning that generates a potentially complex result. Something you read in the paper alarms you and you spill the coffeepot on yourself so you have to go to the emergency room where you're given either the right or the wrong treatment; if it's the right kind you get to go home and the sequence of your life continues, but if it's the wrong kind you're admitted to the hospital or maybe you die, and each of those events

generates its own set of results. Just like the snowflakes, except the rules get more complicated."[99]

That this awkward attempt to explain chaos theory contains a clue to the weather and the events that are about to unfold in the lives of Elaine and her loved ones hardly needs to be stated.

Within the next few hours, as Hurricane Floyd spins destructively on the television and Josie learns about the childhood sexual abuse that has caused Harvey to withdraw from the world, Rolando invades Harvey's home and holds Josie and Harvey hostage. Although Harvey the storm never makes it to the hurricane category, Harvey the man gathers energy, bringing the destructive force of a frying pan onto Rolando's head. Elaine, Frank, Mitchell, and other policemen gather in front of Harvey's house, unaware of what is happening inside. When Rosa slips in to make sure Harvey is all right, he proposes to her. Josie explains to her family, police, and television crews what has happened, while Rolando, seeking redemption, escapes. In the final chapter, Harvey wonders what kind of weather they have in Mexico: "You never heard that much about it, like it was too far down on the screen for anyone to see." Josie encourages her uncle to fly to Mexico on a honeymoon: "You could fly through a cloud. In the real sky, not the one on the dumb TV." Harvey responds with, "Off we go into the wide blue yonder," and realizes that "everyone he loved was underneath the same sky."[100] Thompson's cleverly constructed tale of love, madness, and weather tells us a lot about the perception of weather in the early twenty-first century, the role of the media in teaching people about weather, and our need to identify with nature, including weather. Like Quarrington, Thompson is fascinated by weather pathologies and by storm seekers, some of whom are in pursuit of answers about themselves, and others in avoidance of them.

Taylor Wakefield, the protagonist in Clint McCown's novel *The Weatherman* (2004), is a young man drifting through his twenties with two burdensome memories. When he was eleven years old, he witnessed his delinquent cousin, Billy Hatcher, kill a black man outside an illegal gambling den in Montgomery, Alabama. Billy tries to kill Taylor and makes it clear that he will if Taylor turns him in. Taylor's other childhood trauma was losing the state spelling bee to a Seminole girl, Alissa Powell, when he misspelled "responsibility." A precocious child, he soon becomes a "Super Puff Kid" on a call-in radio quiz show in Birmingham.

From these seemingly unrelated incidents, McCown constructs a compelling story and a fascinating exploration of the parallels between the behavior of the atmosphere and of history. McCown is interested in causality. Can we discover either the laws that determine weather or those that shape behavior—and if we can, what happens to chance and free will? His novel is a series of set pieces on the capriciousness of weather and human conduct. Taylor Wakefield becomes a weatherman by accident, but he seizes the opportunity to explore the chaos of complex systems.

When the novel begins, Taylor has graduated with a degree in history from the University of Alabama, Tuscaloosa, remaining at the university to work as a reference librarian. The butterfly that stirs his tranquil life and sets in motion an emotional storm that will culminate in bloodshed is a small notice in a newspaper announcing that William R. Hatcher is running for attorney general and that, as a prosecutor, he has just indicted a well-known racist for the murder in Montgomery on August 26, 1963, of Arvin "Blackie" Wilson, a small-time gambler. Guilt-ridden but still afraid of his cousin, Taylor goes to Birmingham and contacts a reporter covering the election. The reporter refuses to pursue the story because the prosecutor has five witnesses against the accused, but he sends Taylor to Channel 11, Alacast Network, a tiny cable television station whose owner, Rusty Wilderman, is also running for attorney general. Rusty offers Taylor a job as weatherman.

Mulling over the accidents and coincidences in his life, Taylor sees the connections between weather and history in the origin of the universe, the big bang theory. (His own life began, in a sense, with the big bang of his cousin's pistol.) "The comforting thing about the Big Bang," he muses, "is that it binds everything together as part of a single chain reaction. Cause and effect from start to finish, and everything's a part of it, right down to the most unexpected and obscure detail in the show. . . . But that doesn't mean predictions come easy. Hail can still fall from a clear blue sky. Every weatherman knows what it's like to be wrong."[101] The weatherman at Channel 11 is expected to be a storyteller, not a reporter, as Taylor learns in a conversation with the station manager:

> "Events aren't random," I said. "Every fact leads to another." I felt confident of this. The cause-and-effect streamline of history was clear proof.
>
> "Wrong," he said curtly. "Every fact leads to a *million* others. But we can't include a million facts, so we pick the most interesting handful."
>
> "What if they're the wrong ones?" I asked.
>
> He shook his head, and I could see he was annoyed by my slow learning curve.
>
> "It doesn't matter which ones you pick," he said, the impatience creeping into his voice. "Facts are just the starting point. Imagination is what shapes the news. The weather, too." He grabbed the brass knob and rattled the door in the frame. "If you want to be a weatherman, that's the first thing you'll have to learn." . . . "Weather isn't a story you make up in the newsroom," I objected.
>
> "The forecast is. Nobody's got the facts on tomorrow's weather—too many variables. Every prediction you've ever heard was just some guy's calculated guesswork. Fiction. We could probably have a chimp pick forecasts out of a hat and nobody would know the difference."[102]

Told he can say anything he wants to after he has read the wire bulletin forecast of the weather, Taylor decides to alert his viewers to his cousin's criminal past and to search for the girl who humiliated him in the spelling bee fourteen years earlier. Linking the random events of his past becomes an obsession.

In his on-air debut he gives a very tentative weather forecast, adds some weather trivia, and offers his first paycheck and his weatherman's blazer to the person who can tell him where Alissa Powell is living. He then concludes his three minutes with a warning: "Be careful how you vote in this next election, because one of the candidates killed somebody." His forecasts continue in this vein. On Wednesday he tells his audience that the accused man did not kill Arvin Wilson. On Thursday he admits to witnessing the murder. On Friday he predicts an end to weather as we know it and promises a revelation come election day.

Taylor's life becomes seriously complicated when he finds Alissa, who has become a novice nun, convinced her life was ruined by winning the spelling bee because it did not bring her family the fame and fortune they expected, and her father was sent to prison when he turned to crime. Releasing her pent-up rage, she slugs Taylor, and the nuns suggest she leave the convent. She goes with Taylor to try to resolve her anger. Meanwhile, Billy attempts to convince Taylor that he has reformed, gotten religion, and now opposes the death penalty, but Taylor persists in trying to expose Billy's crime, and Billy threatens him.

The human tempest gathers when the reporter working on the election finds evidence that Billy bribed the five witnesses against the man he is prosecuting. Billy promptly announces that another man, Wilkie Smith, has confessed to the crime, but before Taylor can respond, the reporter is murdered and Smith kills himself in jail. Billy wins the primary election, prompting Taylor to observe:

> Not many people know it, but sometimes general elections are decided by the weather. If it rains, the Republicans have an edge, while the sunny days bring out the Democrats. . . . So in some instances, in the tighter races, the winning candidate is quite literally determined by the heavens.
>
> But heaven stays out of the primaries. . . .
> Things can go terribly wrong in a primary.[103]

After the election, Billy demands to meet with Taylor and Alissa. As he nears this rendezvous, Taylor thinks of himself as a catalyst, bringing Alissa, Billy, and himself together for a confrontation that may have unintended consequences. In this sense, the dreaded meeting reminds him of "seeding the clouds. That's a sore spot with some weathermen, those purists who view it as a form of cheating. If the sky itself can be manipulated to suit our temporary needs, prediction becomes a different game entirely. Forecasts become mere self-fulfilling prophecies, divinations with no basis in the divine."

At the meeting, Billy menaces them with a pistol and tells Taylor that the only reason he did not kill him the day he murdered the gambler was that his weapon jammed. Taylor begins to realize just how complicated the chain of events in a predetermined universe may be. Billy, denying inevitability, wants the past to go away. "The past is like a Third World country—nothing but filthy hotels and a bad exchange rate. Nobody goes there anymore." Taylor wants to know why things happen the way they do. "I still believe in free will," he thinks while facing Billy's gun,

> But I've also come to see the tapestry of choices that binds us in a common framework, the way one thing leads to another, the way threads that started miles and years apart will someday inevitably intertwine, weaving and knotting themselves together in a way that might have been predictable all along. . . . Because philosophically, I want to have it both ways. I want to say I'm caught up in determined movements larger than my own and so absolve myself of any personal responsibility. But I also want to believe I'm more than just a puppet of predestination, more than a domino knocked from behind, forever falling forward with an assigned momentum. But here's the catch, if there's such a thing as free will, there's also such a thing as accident, a moment not dictated, a wild thread unraveling in the tapestry. Accident is an argument against belief, against faith, because accident, by definition, falls outside the scope of any plan, even God's. If we acknowledge accident, we open the door to chaos. And chaos, like its negative twin, the providential universe, frees us from personal responsibility. If disorder rules the world, there are no pawns, no queens, no bishops, no game pieces at all. No rules to play by.[104]

Just before the violent climax, Taylor wonders if his misspelling of the word "responsibility" was an accident, a choice, or the result of the development of language at the very dawn of time, "when nothing yet existed but the weather."[105]

All seven of the novelists discussed in this chapter find in the weather parallels to human behaviors, but for writers of the past twenty years weather is more authority than adversary. Beyond simple analogies are larger questions addressed by science, religion, and literature. What is the nature of the universe? To what extent is behavior, atmospheric and human, governed by laws? Why do accidents happen? What role does chance play in life? As McCown points out, concurring with Quarrington, Moody, and Thompson, belief in accidents precludes faith, in either science or religion. Reducing weather or human behavior to laws and regularities comes with a cost to free will and its congener, accident. Novelists, looking at the weather, delight in the paradoxes of the chaotic atmosphere because it resembles human behavior. Weather may not have free will, but if it remains partially unpredictable it has something like it—surprise. Humans may not have free will either, but they also remain unpredictable.

Weather in American poetry and fiction of the past 100 years is too diverse to be easily summarized. Like weather talk in general, poems and novels about weather serve multiple purposes. Meteorology and the atmospheric sciences expanded the language of weather in the twentieth century but lost some of the older vocabulary that continues to enrich poems and stories about clouds, sky, sunshine, and snowfall. In the twenty-first century we translate the weather from physical sensation, electronic images, or oral tradition to the printed page, to the languages of science, literature, philosophy, or sundry combinations. Some things can be explained mathematically, some only by narrative. Howard Nemerov even draws a distinction between weather that is inherently poetic and weather that is inescapably prosaic. Snow is poetry, he tells us, because it "flew," while rain is prose because it "fell."[106] This is not a trivial distinction when transcribing weather.

5

Suffering Weather

Windshield Meteorology (A Storm Chase Diary)

Day 1: Monday, June 23, 2003

I arrived at Denver International Airport just before noon to join Tempest Tours for an eight-day lecture tour to discover, in the words of the company's press release, "the science and romance of storm chasing, and to pursue nature's most spectacular weather."[1] I note that Denver's airport looks to me not like tents or snow-capped mountains but like a cluster of inverted tornadoes. I discover that the lavatories are also designated as tornado shelters. The implications of this spin sublimely in my mind as I hurry to catch an airport bus.

The tour was scheduled to leave at 3:00 P.M. from a nearby motel, but Bill Reid, leader of the storm chase, called me before I left Washington, D.C., to ask if I could meet him immediately upon landing. Reid, who shares the name of the British engineer who pioneered hurricane research in Barbados in the 1830s, explained that meteorological conditions were favorable Monday for "supercells," rotating thunderstorms with internal areas of rotation called "mesocyclones," and there was no time to waste.[2] Supercells are the best indicator of tornadic thunderstorms, and the possibility of a supercell sets storm chasers in motion. When the bus drops me at the meeting place, about 12:30, the chase van is idling, and we are off as soon as I have climbed in.

Although people have been watching thunderstorms for aeons, an understanding of how clouds form and how they behave is relatively recent. Supercells, identified and named by a British graduate student, Keith Browning, in 1962, are a kind of cumulonimbus, or thunderhead (an Americanism traced to 1843). For me, the tour is an opportunity to learn more about the science of meteorology and the appeal of storm chasing. I have chosen this tour because Charles "Chuck" Doswell, a prominent meteorologist who specializes in severe storm and tornado formation, will be on the chase.

We make hasty introductions in the van, as Bill drives toward Nebraska. Bill looks to be in his late thirties, with tornado-tousled dark hair. Later I learn he has a master's degree in climatology from Cal State Northridge and has been storm chasing for several years. Chuck is in his late fifties, tall, with white hair and moustache. He has a Ph.D. in meteorology from the University of Oklahoma and began storm chasing as a graduate student in 1972. My fellow participants, each of whom paid $2,300 for eight days of storm chasing, are a ninth-grade science teacher from New Jersey, a computer programmer from northern Virginia, and a photographer from Detroit. The teacher hopes to interest his students in meteorology, the computer guy wants to start storm chasing in Virginia, and the photographer is trying out different kinds of cameras. We don't bond. In a week's time we share less personal information than agents from the CIA and the FBI.

Chuck immediately begins to regale us with stories about past storms and information about storm chasing. He talks about the need for tornado spotters to warn communities in a tornado's path. The relative infrequency of tornadoes (between 800 and 1,400 annually in the United States) compared with thunderstorms (fewer than 1 percent of thunderstorms produce tornadoes) makes emergency management difficult. Moreover, 86 percent of all tornadoes are characterized as "weak," 13 percent are "strong," and 1 percent are "violent."[3] I think people correctly surmise that the odds are in their favor when a storm develops, but the consequences of being wrong can be fatal. Storm chasing is a kind of "deep play" along the lines of the Balinese cockfight described by the anthropologist Clifford Geertz. Geertz borrowed the term from Jeremy Bentham, who coined it to characterize gambling in which the stakes are so high that from a utilitarian point of view it is irrational. Geertz explains how the seemingly reckless wagers of the Balinese actually help to hold their culture together. A similar argument might be made for storm chasing, which creates a community of risk takers who briefly reconcile modern scientific experiment and primitive struggle with nature.[4]

People complain that their sense of security is shattered after a destructive tornado, but their sense of security is an illusion, Chuck observes. Forecasting and disaster mitigation are only as good as what citizens are willing to pay for, he continues. The movie *Twister* grossed $300 million, Chuck points out, while the annual per capita expenditure for weather is $7, about the same as a movie ticket. What is a human life worth? This is one of the questions at the heart of weather hazard and all disaster mitigation—indeed, of all health care. How much are we willing to pay for safety, and who pays for it? To what extent can life be risk-free? How free are we to risk our own and others' lives? Doswell's point is that the National Weather Service (NWS) saves a lot of lives at $7 per capita. He notes that the manufactured home association fights legislation that would require mobile home parks to have tornado shelters. Chuck's passion for using meteorological knowledge to serve the public good is apparent. He wants us to see the public value of storm chasing. We are being initiated into the culture of serious weather observers. Chuck, seeming to gather energy from the atmosphere, lectures on supercells and their formation as warm moist air flows into western Nebraska from the south and cold air blows east from Colorado.

About 2:30 we stop in Sterling, Colorado, for fuel. I have a chance to check out the van. It is equipped with a laptop connected to the Internet by cell phone, various radios, and a video screen mounted on the ceiling of the van behind the driver and visible to the three rows of passenger seats. On the video we can watch what Bill and Chuck are accessing on the laptop, or, depending on where we are, The Weather Channel and local TV news and weather programs. The van is also equipped with a Global Positioning System and DeLorme software for maps and motel guides. Before we enter Nebraska, we stop, and Bill and Chuck consult NWS and Storm Prediction Center atmospheric data. One screen gives the convective available potential energy (CAPE), a combination of temperature and humidity at various altitudes, which indicates how much energy is available for thunderstorms. Another set of screens provides the lifted index (LI), a more complicated reading of temperature and humidity measurements that provides an estimate of potentially unstable air. Other measurements taken by NWS balloons provide wind speed and direction in the air up to 50,000 feet. Satellites and radar provide additional information. All this information is turned into models predicting changes in the atmosphere when no real-time data are available. Looking out the window of the van at the bottoms of the clouds, then seeing their tops from a satellite photo on the TV screen, is disorienting yet exhilarating.

Chuck grows more excited, explaining, "We're chasing air and energy and motion; nothing that exists now or later. Strange really." His point is that the visible storm, the cumulus supercell, is a stage in the birth, growth, and death of a storm system. What we can see is just a moment in an unending process. We are literally chasing chaos. Chuck points to the clouds and announces, "the fist of God," with only a hint of the sarcasm he often expresses for clichés. We speed past Buffalo Bill's ranch, and I can't help but compare our storm chasing to that of the buffalo hunters who harassed the shaggy beasts from Pullman cars 130 years ago. We won't be hunting tornadoes into extinction, of course, but we are using resources—gas, highways, space—in our pursuit. The buffalo are reduced to a few herds, the aquifers are drying up, small farms are disappearing, so the tourist gaze shifts from the Great Plains to the sky above. Strange, really.

6:00 P.M. We stop on a small hill near North Platte, Nebraska. Meteorological science says that supercells should be forming; the atmosphere says no. Chuck looks at the cloudless sky and labels it "invisible convection." We drive northeast toward Broken Bow on a township road. Who knew in 1787 when the Northwest Ordinance was enacted that the grid system of land surveying would provide storm chasers with a fine network of roads?

9:00 P.M. We stop again to photograph the clouds that are beginning to produce thunderstorms and lightning. It's a good show. Later we drive through a violent rainstorm to Kearney. Dinner at midnight.

Day 2: Tuesday, June 24

From the *Lincoln Journal Star* I learn that last night's storms caused flooding, record hail, and tornadoes in several Nebraska towns. A man was killed when his garage collapsed in Deshler.[5] We meet in Bill's room at about 10:00 A.M. for the orientation that we skipped yesterday and a briefing on today's storm prospects. The models are not encouraging. No clear indications of supercells. Although there are many variations, a classic supercell is described by meteorologist Thomas Grazulis as "more than just a severe thunderstorm. A supercell tends to have a unique structure, with many recognizable features. It is an isolated thunderstorm dominated by its overall rotation and the presence of a unique area of rotation called a mesocyclone. That 'meso,' as it is often called, would be positioned as a vertical column between the wall cloud and the overshooting top." A supercell can rise to 60,000 feet, more than twice the height of Mount Everest.[6] We stop at a Walgreen's drugstore for supplies. A farmer spots our multiantennaed van and tells us about the winds that took his shed and two trees; he and other locals are amused by out-of-state storm chasers. Tornadoes are something to be avoided in this area. I ask Chuck about what the media called the "unusual" number of tornadoes in Oklahoma in early May. The question evokes a short lecture on the meaning of "normal." Variability is the only constant.[7]

On the road, Bill calls attention to the shadow of a cloud moving at about our speed, that is, 50 knots or 60 miles per hour, at an estimated altitude of 2,000 feet. This observation causes Chuck to comment on a storm chaser's need to read signs on the ground and in the sky more than relying on instruments. Storm chasers do not have to be scientists to make a contribution to understanding severe storms. Their photos, videos, and descriptions help complete the description of each unique storm. Science, Chuck believes, is observing nature, and then offering explanations that can be verified. "The atmosphere shows me that my scientific explanations are incomplete," he avers. "If this were all cut-and-dried, it wouldn't be fun." Chuck admits to being a romantic about storms. He is also totally devoted to his chosen field. When the tornado season—April through June in the southern plains—ends, he does armchair chasing using boxes full of old data.

In *The Tornado: Nature's Ultimate Windstorm,* Thomas Grazulis goes further to suggest that "storm chasing can be a religious experience—the power of the Creator revealed in the spectacle of the thunderstorm." The origins and current popularity of storm chasing are more mundane. Grazulis dates scientific storm chasing to May 4, 1961, when Neil Ward of the National Severe Storms Project (the forerunner of the Storm Prediction Center) chased a storm in Oklahoma using radar. Amateur storm chasing is said to have begun with Dave Hoadley's chase on July 6, 1962, in North Dakota. Hoadley founded *Stormtrack*, a newsletter for chasers, in 1977.[8] Others attribute the first "hobbyist" chase to Roger Jensen,

Storm chasers looking for a tornado in western Kansas, June 2003. Photo by author.

also in North Dakota, in the 1950s. It took another ten years for storm chasing to develop to the point where chasers were making systematic observations of many storms. Chuck and his fellow graduate students at the University of Oklahoma began chasing in April 1972, before the development of portable computers and cell phone communication. Professional chasing began in the mid-1980s when Warren Faidley, a journalist, began selling photographs of lightning, tornadoes, and other storms to national magazines. He incorporated in 1989 as Weatherstocks, relentlessly promoting himself as the "Cyclone Cowboy."[9] He is most likely a model for Jimmy Newton in Paul Quarrington's novel *Storm Chasers.*

Storm chasing tours began in the 1990s. Storm chasing may be a victim of its own success as more and more chasers converge on the same spot, clogging the roads and obstructing emergency vehicles and the view. Storm chasing as adventure tourism is gaining national recognition. Like an African safari, a storm chase involves a certain amount of risk (mostly on the highways), the pursuit of savage elusive creatures, and their capture on film. Sampling exotic local cuisine and buying souvenirs from friendly natives are also part of both experiences. There is also an aesthetic component; the clouds, storms, and sunsets are beautiful.[10]

As we scan the sky, the radio announces a tornado watch for Custer County, about 30 miles southwest of our location. Bill and Chuck confer, expressing doubts about the NWS model. It's 3:00 P.M. "The time of day when a decision can make or break the chase," observes Bill. Chuck elaborates on the process of decision making based on visible clouds. Chuck laughs about The Weather Channel prediction of no severe weather in Nebraska today, complaining that The Weather Channel has become a "weather magazine," doing last year's weather. Bill and Chuck banter about what kind of tornadoes they want to see as the cumulus begin building rapidly to the northeast. They complain again about *Twister,* a movie they hate but which seems to set the terms for any discussion of tornadoes. The major objections to the movie from storm chasers are the unrealistic depictions of tornado behavior and the portrayal of chasers as reckless thrill seekers. In reality, responsible storm chasers maintain that the chances of actually seeing a tornado are slim and that the goal of a chase is to observe a tornado from a safe distance, not to become part of it.[11]

Back in the van about 4:00. Excitement is rising. The chase is really on as we speed toward Burke, South Dakota. Bill drives *and* operates the computer, often getting weather updates by cell phone at the same time! "There's a moisture convergence point," Chuck exclaims. "It's a Darwinian survival situation, one of the towering cumulus clouds will take over and kill the rest!" At 4:45 we descend toward the Missouri River. We need to be north of Mitchell, South Dakota, about 100 miles northeast. Will we make it in time? Now we spot a domed cloud to the northwest. Where are we heading? From the backseat of the van the sky isn't really visible. I try to read highway markers. We swing north on Route 11, then east to 281, then north again. We're about 80 miles southwest of where the radio says the storm is most intense. A tornado warning has been issued for Mitchell until 5:45. We're thirty minutes away.

The radio reports a tornado near Mount Vernon and others forming. At 5:30 we cross the Missouri River at Fort Randall. I think of Lewis and Clark, who passed this way in July and August 1804. Did they see any tornadoes? (Later I check their journals and do not find any reports of tornadoes.)[12] We shift our target area to Yankton and Vermillion, 40 miles southeast. Bill videotapes while driving! Beautiful storm clouds about 20 miles ahead. Chuck calls it a "left mover." He gets NWS Sioux Falls radar on the computer after several tries. We stop for photos and then drive into the rain and hail.

Near Centerville, South Dakota, we stop and within minutes see tornadoes forming, funnels "touching down"! This, I learn from Chuck's Web site, is the wrong way to describe tornadoes in contact with the earth. There is nothing coming down from the cloud; rather, a vortex at the surface becomes visible when it gains strength, decreases in scale, and picks up debris, dust, and moisture. Nor do tornadoes "skip," in the sense of lifting and descending, but the circulation at the surface may strengthen and weaken every few seconds.[13]

I don't think anyone can miss the erotic movement of tornadoes. It's easy to understand why chasers get excited and scream orgasmically at the sight. Our group is quiet and seemingly calm, although I felt my heart rate increase, I'm sure.

Everyone is busy photographing. I watch and listen and feel the winds shift and the temperature drop as we experience what Chuck calls the RFD, or rear flank downdraft. Frogs are croaking, birds twittering. The air seems alive. We see cloud-to-ground and cloud-to-cloud lightning. I measure wind gusts of 22 miles per hour on my Kestrel, a handheld weather instrument, but Bill estimates 30 miles per hour, and I trust his experience. Over a two-hour period we count at least eight, maybe as many as ten, tornadoes—often, three funnels from one cloud at the same time. Bill and Chuck agree it is one of the most spectacular displays they have ever seen.

Back in the van we watch TV news of the tornado damage in Manchester, South Dakota, a small town that was totally destroyed. Fortunately, there were no fatalities, although some people were injured. None of us on this tour wants to talk about the irony of desiring to see an event that may bring other people harm. We are not engaged in "disaster tourism," but we have joined the voyeurs watching disaster video on the evening news. The realization that the object of our quest is made visible by the debris of someone's field, or barn, or home chills us, like a "gust front" in a tornadic thunderstorm, and sends us scurrying in denial to the silence of our individual mental storm cellars.

NWS radio warnings of tornadoes all around us add urgency to the moment. Chuck does not think we're in danger. My ignorance is bliss. My faith is in his judgment and the knowledge that neither Tempest Tours nor any other storm chasing organization has ever lost a customer to a tornado. We reach Sioux Falls about 11:00 and eat a celebratory steak (a Tempest Tours' tradition following a chase in which a tornado has been spotted) at the Fryin' Pan. Heavy rains have left the streets flooded. We share the restaurant with a few Native Americans who look like they have seen worse storms.

Day 3: Wednesday, June 25

The *Sioux Falls Argus Leader* provides some photos and details of yesterday's storm. It was a big one.[14] Briefing in Bill's room at 9:30—low chance of tornadoes today. Thunderstorms have stabilized the air. Chuck thinks a squall line is our only hope for severe weather. The small forest of tripods in the back of the van reminds me of trees flattened by a tornado. We drive east into Minnesota. Talk in the van is about cameras and sports. Chuck says he may write a paper about yesterday's storm because its behavior was unusual, moving north but spinning tornadoes from east to west. He'll look at the data gathered by instruments and combine them with his own observations. Outside, a few fair-weather cumulus begin to form. "Post convective debris," Chuck calls them.

Noon. We pull off I-90 and go into Fairmont, Minnesota, for lunch and to use the Internet at the public library because Bill is having difficulty with the cell phone connection. During lunch Bill regales us with earthquake and tornado statistics. His memory for dates and events is amazing. He seems able to recall in detail not only his own storm chases but those of other chasers. Based on the weather models, we decide to drive south into Iowa. We're all feeling the emotional fatigue from yesterday's spectacular storm. Two of my fellow chasers are dozing; I'm feeling sleepy. We stop to look at some clouds that Bill and Chuck pronounce "uninteresting." Out of boredom Chuck jokes about taking hostages in a farmhouse and making them watch The Weather Channel until they beg for mercy.

We leave for Sioux City, Chuck driving and pointing out low (2,000 feet) postfrontal stratocumulus in parallel north-south lines. They stretch away to the west like sheets of corrugated iron. We arrive about 7:00 P.M. and find a motel in South Sioux City, Nebraska. At dinner Chuck entertains us with his comments on the special qualities of tornadoes. They are neither too large nor too small. You can see all the elements develop and feel part of the storm. Every storm is a process, not a thing. We're clearly getting into deep questions here. For Chuck, science is experience and acknowledging what you don't know. Every storm is a new challenge. Chuck talks about the three revolutions in the study of tornadoes: radar, numerical modeling, and storm chasing.

Day 4: Thursday, June 26

Beautiful clear morning. The sky is a pale blue with the "vault of heaven" look. Only a storm chaser could be disappointed in a sky as calm as this. The *Sioux City Journal* reports on the destruction of Manchester, South Dakota, by Tuesday night's tornado, the F4 tornado that hit Coleridge, Nebraska, and the "40–60 tornadoes" that "touched the ground" in the Sioux Falls area. The Centerville storm that we witnessed flipped over a garage and moved a modular home off its foundations.[15] The Weather Channel replays footage of Tuesday's tornadoes. Bill looks at the NWS predictions and concludes that Saturday will be the next appearance of supercells in northwest Kansas and northeast Colorado. Bill favors North Dakota as a target, but dew points are too low. Chuck explains how models shift with slight fluctuations. He doubts the NWS estimate of thirty to forty tornadoes Tuesday.

Chuck lectures on the mesocyclone as we drive to Omaha. Chuck thinks all storms are multicell, a series of mesocyclones. The change of velocity in tornadoes is different from what causes change in large-scale systems. Models of the atmosphere are made from quantitative data; there are no similar data for tornadoes because there are no measurements inside tornadoes. There isn't even a peer-reviewed definition of tornado. Doswell prefers to emphasize process rather than form, modifying the standard definition—"a small mass of air (whirlwind) that spins rapidly about an almost vertical axis and forms a funnel cloud that contacts the ground; appears as a pendant from a cumulonimbus cloud and is potentially the most destructive of all weather systems"—by emphasizing the necessity for deep, moist convection in the clouds and intensity sufficient to cause surface damage. Measuring tornadoes by intensity and damage is a problem because any storm damage is often erroneously attributed to tornadoes when actually caused by straight, high winds.[16]

If we actually could put plastic balls with sensors inside a tornado, as in *Twister,* how would it produce better forecasts? I note again that, although Chuck and Bill hate the movie, it dominates their discussions. *Twister* needs to be refuted at every point. When I check my e-mails, I find a news item concerning an attempt by *National Geographic* to take pictures inside the Manchester tornado by placing cameras in an armored "device," the remains of which were found more than 430 feet away on Wednesday. Although all the cameras were smashed, "the film was being sent back for processing just in case, but officials said they doubted it captured an image." Is this legit, or tornado lore? Chuck will have an opinion, I'm sure. Describing the hazards of being near a sand and gravel pit in a tornado, he says the experience "blew him away." How much of our language is weather-dominated?

The full story of *National Geographic*'s project to place cameras encased in a 95-pound container called "Tinman" inside a tornado is reported in the magazine's April 2004 issue. Storm chaser Tim Samaras and photographers Gene Rhoden and Carsten Peter succeeded in dropping the unit in the path of one of the Manchester tornadoes; the resulting pictures were not very revealing, but they mark another advance in tornado research. No one has come this close to photographing a tornado from inside the swirling debris. Photographs of tornadoes and the damage they cause have been an important part of meteorological research and assessment since April 26, 1884, when the first documented photograph of a tornado was taken. This photograph, by A. A. Adams of Garnett, Kansas, shows a ropelike funnel about 13 miles southwest of the town. Taking 120 years to get 13 miles closer to the vortex of a tornado may seem like progress by inches, 6,864 inches a year, but when someone finally gets a video from inside a tornado funnel, the results will be spectacular.[17]

The view from the weather office is so different from that of the chaser that the office meteorologists can't imagine the uniqueness of each storm, Chuck argues. They see similarities where chasers see differences. Is this a variation on the old lumper versus splitter dichotomy in science? Is it a significant difference? I think so. Science in general looks for regularities and laws; the arts seek difference and novelty. Weather challenges both perspectives because it is constantly in motion, revealing patterns only in the numbers of a computer model or the retinas of an observer's eyes. Chuck complains about computerized forecasts, which provide a minimally acceptable forecast, neither very good nor very bad, in contrast to human forecasts, which can be outstanding or terrible. Society is unwilling to pay for the recruitment and training of human forecasters. Forecasting and science are very different, but Chuck tries to bridge the gap.

4:15. We're back in the van rolling past Boys' Town, Nebraska. What did Father

Flanagan tell the orphans about tornadoes? We stay in Fremont. We discuss NWS forecasts, which are not good for chasing—low dew points, weak winds, fuzzballs. Fuzzball clouds have fuzzy edges instead of the well-defined cauliflower shape. Chuck has even more colorful terms for storms that don't develop to his satisfaction, such as "bird farts." From 6:00 to 8:30 P.M. we watch Chuck and Bill's video of Tuesday's tornadoes and some other storms filmed by Bill. Great footage!

Day 5: Friday, June 27

The *Omaha World-Herald* headline reads, "Federal Aid Iffy after Twisters." It seems the damage in Nebraska did not meet the $1.8 million required by the Federal Emergency Management Agency. A second story, with photo, touts a hailstone with a circumference of 18¾ inches that fell in Aurora, Nebraska, that exceeds the world record claimed by Coffeyville, Kansas, in 1970 by 1.3 inches.[18] I'm sure this "record" will not go unchallenged. Weather as source of local pride is not confined to the Midwest but may be more obvious here because of its extremes. TWC shows locally severe storms in North Dakota, while the weekend forecast calls for storms in Colorado and Nebraska—too great a distance for us to cover both sites, making this a tough call for Bill. He and Chuck think the CAPE looks good in northwest Kansas. The Storm Prediction Center in Norman, Oklahoma, indicates isolated thunderstorms with good structures. The forecast for Saturday and Sunday is for hail and supercells in Colorado. As we depart at 11:00 A.M., Chuck talks about the atmosphere as a heat engine. He gives a good overview of the general circulation, the Coriolis effect, and other phenomena.

As we zip through Wahoo—"Home Office of Late Night with David Letterman"—Chuck describes the importance of the polar front boundary and how the strength of the jet stream depends on the contrast of temperatures—how it breaks into waves, troughs, and ridges. Each wave has a life cycle. Unequal heating is felt strongest at the earth's surface. Clouds are common over islands in the ocean because of the temperature difference between the water and the land. William Churchill's "Weather Words of Polynesia" lists 267 names for different kinds of clouds and 435 names for winds in various Polynesian dialects. In Hawaiian, Churchill writes, a waterspout is called *waipuilani*—water (*wai*), to force (*pui*), heaven (*lani*). Tornadoes in the continental United States, if named at all, are often called by the names of the towns they destroy, or the day they occur, if it's a holiday.

Chuck convinces me that American weather, particularly in the Great Plains, is more diverse and severe than anywhere else in the world. The unique geography—no physical barrier between the plains and the pole—makes it the supercell center of the world. His lecture gets more complex as he explains vertical wind shear. I'm getting a whole year, or at least a semester's course, of basic meteorology crammed into a few hours. It's wonderful, but exhausting. It's not clear why the atmosphere needs supercells, Chuck says, adding that more research needs to be done on non-supercell storms that produce tornadoes. Next he talks about scale and Mesoscale Convective Systems (MCS). A MCS is a multicell convective complex leading a squall line followed by showers.

He continues, explaining squall lines and derechos—geostrophic winds and hydrostatic balance. He then responds to my request for a brief history of the Storm Prediction Center (SPC) and its predecessors. Some of this history is in Grazulis and Marlene Bradford's *Scanning the Skies: A History of Tornado Forecasting,* but I want Chuck's version.[19] Doswell begins with the Tinker AFB forecast of March 1948, which everyone agrees was the first "official" tornado warning, although John Park Finley of the Army Signal Corps began making forecasts in 1884. After the U.S. Weather Bureau was created in the Department of Agriculture in 1891, the civilians in charge terminated what they saw as a "military" project. World War II forced the Weather Bureau to cooperate with the military services. A Weather Bureau–Air Force–Navy Analysis Center was created in 1952, and the Weather Bureau created a separate Severe Weather Unit (SWU) in Washington, D.C. The following year, the chief of the Weather Bureau, Francis Reichelderfer, changed the name of SWU to Severe Local Storms Center (SELS). Nineteen fifty-three was the year of the great Waco, Texas, tornado and other deadly tornadoes, and the chief of SELS, Don House, succeeded in moving operations to Kansas City, Missouri, in 1954. Chuck says that in 1962 or 1963 the research division of SELS split from the forecasting division and moved to Norman, Oklahoma, to become the National Severe Storms Laboratory (NSSL).

Failure to issue warnings for the Palm Sunday 1965 tornadoes led to Allen Pearson replacing House as chief of SELS that year. Chuck says House was a scientist; Pearson, a PR guy. The SELS forecasting division was renamed the National Severe Storms Forecast Service (NSSFC) in 1966. Chuck went to work there briefly in 1967 after receiving his B.S. in meteorology from the University of Wisconsin. The big Tri-State Tornado outbreak of April 3–4, 1974, led to another investigation, and SELS formed a new research unit. In the meantime, Chuck got his Ph.D. and returned to spend six years with SELS. Joseph Schaefer became chief of SELS in 1995, moved the NSSFC to Norman, and renamed it the Storm Prediction Center (SPC), where it is now part of the National Centers for Environmental Prediction. As Chuck sees it, all the moves and reorganization illustrate the influence of personalities in institutions. He has expressed himself forcefully on the subject in an essay on his Web site, "A Jaundiced Look at NOAA and NWS Management," in which he traces some of the current problems in the bureaucracy to Robert White's success in creating the Environmental Science Services Administration (ESSA) in 1965.

2:30. Lunch in Kearney. At 3:20 we pull off on a dirt road to watch a small thunderstorm, then resume the drive west and watch rain to the northwest and southwest as we roll toward McCook along the Republican River. We turn south and see what may be a supercell on our right and ahead. NWS issues a severe weather warning for northwest Kansas. Chuck is cheerful and recites tornado clichés like a mantra: "It struck without warning!" "It sounded like a freight train." "It looked like a war zone!" Bill stops to throw dust in the air like an actor in *Twister*. Everyone is giddy with fatigue and anticipation. At 5:20 P.M., we enter Kansas and stop outside Colby, but are forced to move when a combine comes down the road. It's harvest time here, and farm equipment is one more hazard for storm chasers. We continue south ahead of the storm, then east out of Oakley. At 6:45 we stop to watch but move fifteen minutes later because of lightning. We go west and stop under the leading edge of the storm. A small lightning fire burns about a half mile away. South again at 7:15 P.M. to stay ahead of the storm. Chuck warns of hail. We stop 4 miles north of Monument Rocks National Landmark and feel the wind change from inflow to outflow. My Kestrel tells me that the temperature has dropped from the 80s to 71°F and the wind is 10 to 12 miles per hour. Horseflies are biting.

Rain intensifies. Chuck points out "whale-mouth" clouds. We move to find a place with fewer power and telephone lines blocking the view, stopping outside Scott City near a field that smells like a feedlot. The sunset is not great, but the tail cloud is sucking moisture, and the storm continues its rotation. A sheriff pauses to talk about the weather. The clouds look like Chinese dragons. Chuck says he cannot see any shapes in the clouds, only meteorological processes. Back in the van, we drive through Scott City just as the tornado warning alarm is activated. We're hit by pea-size hail. On to Garden City for the night.

Day 6: Saturday, June 28

The morning weather report is not encouraging for severe storms. We drive toward Goodland on our way to Colorado, stopping to look at hail damage to fields of corn. We pause in Scott City for snacks and continue toward Colby, Kansas, where we have lunch. I photograph the Colby welcome sign ("The Oasis on the Plains"), look at the metal palm trees near a shopping mall (are they camouflaged cell phone towers or just roadside attractions? are they a lightning hazard?), and buy a "tornado-in-a-bottle" souvenir decorated with figures from *The Wizard of Oz*. Chuck does a W. C. Fields routine about a dog he had that was afraid of thunderstorms. We go on to Colorado on I-70 to Burlington, and then turn north.

Chuck explains problems with data gathering related to wind speed and direction. Some observation points in Oklahoma are in river valleys running east-west, excluding north-south winds. I ask about the AgriMet systems used in Idaho and Oregon that give farmers real time on temperature, winds, humidity, soil moisture, sunshine, and so on. He says there is a similar mesonet system in Oklahoma. Chuck goes on to say that local environments are not well understood, meteorologically. The new Denver airport was built in an area of frequent non-super-cell tornadoes. There seems to be insufficient communication between the Federal Aviation Administration and the NWS. We stop in Wray, "The All American City." Forecasts not promising for tornadoes. About 5:00 P.M., we stop to watch a storm about 20 miles away. Bill expects it to develop by 7:30 P.M. We can hear thunder. Chuck describes the storm in anthropomorphic terms: "The little guy's got to move fast if he's going to survive. He needs to get his air from the other side. He's anticyclonic, a 'left-mover.' He wants to be pushed from west to east. Storms don't know where they are. The right-mover is spinning off another left-mover." We watch in silence.

Returning to the van, I find that Bill wants to drive south. About 7:30 we stop to watch storms 180° from northwest to southwest. The clouds are beautifully layered, and the sunset behind the squall line highlights the cumulus clouds between the storms. We feel the outflow wind and see many lightning flashes. Everyone takes photographs. Finally we drive back to Burlington for dinner about 11:00 P.M.

Day 7: Sunday, June 29

Again the weather forecasts are not good for severe storms. Chuck likes the Limon area. Bill yawns as he looks at the computer screen. It obvious we're all tired. We look at Monday's forecast showing the summer pattern of weak upper air. We drive toward Limon watching Chuck's 1995 highlights. (Potential storm chasers prone to motion sickness need to consider the effect of watching shaky handheld video while driving on bumpy roads with the landscape flashing past the windows.) In addition to Chuck's voice on the video, we hear Rush Limbaugh between storm warnings. Even Rush can't heat up the air enough to produce a storm.

Lunch in Limon, then south through Colorado Springs on U.S. 25. Chuck says he's suffering from SDS, supercell deprivation syndrome, because he hasn't seen one in so long. He talks about Bill Paxton of *Twister*. "He's a nice guy," Chuck says laconically. We watch a thunderstorm over Colorado Springs and then head north. Chuck mentions the "open mouth" experience of his first storm chase, and I am reminded of the story Charles F. Brooks tells in his book *Why the Weather?* (1935) about the natural tendency of the mouth to open when you look up at the sky. Chuck relates the history of storm chasing in the 1990s, the VORTEX project, Howard Bluestein's TOTO (Totable Tornado Observatory), which is now in the NOAA museum in Silver Spring, and the work of Thomas Grazulis. We drive through Denver and go northeast on U.S. 76 toward Fort Morgan. A towering storm cloud looks like a chess piece. We stop north of Fort Morgan to watch the lightning and then have dinner at an agreeable and appropriately named restaurant, Memories.

Day 8: Monday, June 30

At our morning weather briefing, Bill thinks northeast Colorado is best for today. We return to Memories for a big breakfast, then to the public library for Internet forecasts. While Bill and Chuck copy their videos, I explore the Fort Morgan museum, which has just opened in the same building as the library. Good local history collection, including a Glenn Miller corner for Fort Morgan's most famous resident. Chuck gives us two PowerPoint lectures, one on terminology and one for advanced spotters. Both are highly informative and sum up nicely what he has been pointing out during the week. I am amazed by how things have worked out on this tour. It would be very different if we had these lectures first, then seen the tornadoes, and different still if no tornadoes had developed. Like the weather, the tour has been determined by small initial conditions with tremendous consequences.

In his advanced spotter presentation, Chuck notes the importance of being able to recognize storms that do not develop tornadoes. The number of reported tornadoes has increased greatly between 1915 and 2003, but he thinks at least half are missed annually. Reported nontornadic severe thunderstorms increased from 5,000 in 1980 to 25,000 in 2000, a change attributable in part to better observation. Chuck has examined NWS reports from many states and found that what gets reported depends on who's reporting. The three most important priorities for storm spotting are (1) safety, (2) service to community, and (3) ground truth for NWS and science. NWS warnings are too broad at the county level, and radar is not a replacement for human eyes. His list of things spotters need to know includes location and speed of storm, escape routes, and how to take photographs for research on storms. He feels that about 10 percent of spotters are serious.

At 3:30 we head north on Colorado Route 52. Not much going on, but Tempest Tours' policy is that as long as one person wants to chase, the chase will go on. Chuck talks about his work with the Spanish government. He will be in Mallorca for five and a half weeks beginning in September. University research is comparable there, but there is no tradition of government storm warnings. About 6:30 we are in the Pawnee National Grassland, where we watch some thunderstorms. I learn that the ratio of male to female chasers is 10:1, but 30 percent of Tempest Tours' guests are female. Bill tells us that we drove more than 2,600 miles. We head back toward Fort Morgan, and I see a coyote in a field. The sign of the trickster! The weather gods are having their fun. There's a small but beautiful sunset. We reach Denver about 9:30 P.M., have our final meal together, and receive our Tempest Tours T-shirts. They are tan with a lightning bolt encircled by the words "Tempest Tours—Storm Chasing Expedition" on the front, and a red tornado watch box on the back with the words "Project Red Box 2003 Storm Chase Team." We will be able to purchase videos later.

The tornadoes I saw on June 24, 2003, were sensual and sensuous; black against dark gray clouds, they spun and undulated, swelled and shrank, teased and threatened. A tornado, the meteorologist on our team kept repeating, is not a thing but a process; air, energy, motion momentarily visible. It was a great show, the kind of natural sublime that tourists sought in the Romantic era. The commercial storm chase is the twenty-first-century equivalent of a visit to Niagara Falls or a hike into the Alps to see the Brockenspector.

Tornadoes: The Era of Description

Tornadoes are the most American of disasters. This is not to say that tornadoes are the most deadly or costly weather disasters, or that they are unique to the United States, only that their frequency, compared with that in other parts of the world, and the place they hold in the public imagination make them more central to the understanding of the cultural meanings of disaster in American history. Tornado stories are an essential part of regional identity in what is often called "America's Heartland." These stories juxtapose the commonplace (volatile weather throughout the year) with the extraordinary (people and objects blown for hundreds of yards without harm) to convey the distinctive quality of life on the Great Plains.[20] In the various American Edens weather is a spoiled child, proudly displayed to visitors, but needing discipline when wild.

How have tornadoes been described, defined, and explained over the past 150 years? How have the experiences with tornadoes shaped attitudes toward disasters in general? Do the attitudes toward disasters reflect broader values in American society? Do the representations of tornadoes in art, film, and literature contribute in any way to our understanding of either tornadoes or their cultural meanings? Has our quest for the sublime in nature been replaced by something more extreme—a desire to experience terror more than awe, tumult more than splendor?

These are broad and complicated questions to which there are no simple answers, but one thing is very clear: there is a large and growing popular literature on tornado disasters. There are dozens of detailed accounts of the major tornadoes of the past century. Add the ever-growing number of books on hurricanes, blizzards, floods, and other weather events, and a substantial genre of disaster narratives emerges. The tornado narratives, popular and technical, are the products of two eras of tornado awareness. In the first half of the twentieth century, tornadoes were awesome displays of nature's power, poorly understood scientifically but perceived by residents of the Great Plains and other tornado-prone regions as inescapable hazards of life and as a defining element in their regional identities. Beginning in the 1950s a major change in attitude toward weather hazards emerged. Radar gave weather forecasters a crucial new tool in predicting

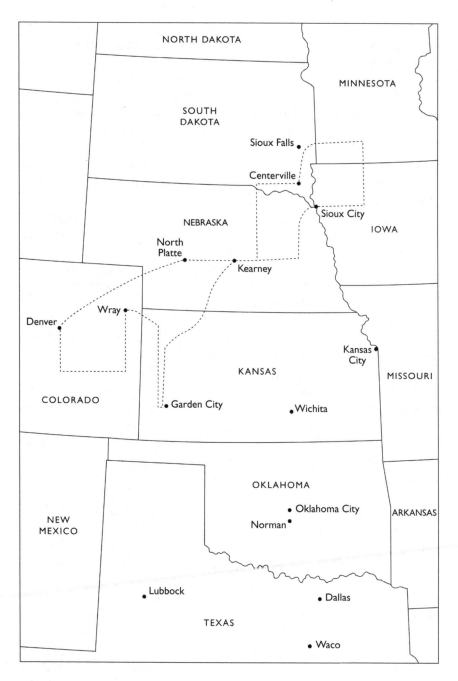

Map of tornado alley and author's chase.

tornado outbreaks, and the Weather Bureau became proactive in educating and warning the public about severe weather. Equally important, the federal government, anxious about civil defense in the nuclear era, began to create institutions to study and manage disasters, and the public, prepared by rapid technological and social changes, began to believe that science and government could greatly reduce life's risks, if not eliminate them altogether. The first era can be illustrated briefly by accounts from 1887, the 1920s, and the 1960s, the second by tornado studies of the 1950s, 1970s, and 1990s.

Tornadoes: What They Are and How to Observe Them; With Practical Suggestions for the Protection for Life and Property (1887) by John Park Finley, a sergeant, later a lieutenant, with the U.S. Army Signal Corps, is one of the first and most comprehensive of tornado books. Over several years, Finley had recruited and trained more than 2,400 volunteer tornado observers and analyzed data on 600 tornadoes. A colorful character with a wide range of interests, Finley remained with the army after Congress created a new weather service in the Department of Agriculture in 1891. He later served in the Philippines, where he did linguistic and ethnographic research with William Churchill among the Subanu on the island of Mindanao.[21]

Finley's tornado book explains the topographic reasons why the Mississippi Valley and Great Plains are especially prone to tornadoes—the lack of mountain barriers between the warm, humid air from the Gulf of Mexico and the arctic air from Canada that meet in the trough between the Rockies and the Appalachians—and clarifies the differences among cyclones, hurricanes, whirlwinds, waterspouts, hailstorms, thunderstorms, and tornadoes. Relying heavily on the experience and vocabulary of his farmer-observers, Finley lists four "premonitory signs" of a tornado, all of which "are within the comprehension of any ordinary observer and can readily be detected by him."[22]

The signs, similar to some still used by observers in the field, are (1) sultry, oppressive atmosphere; (2) peculiar formations of clouds; (3) peculiar actions of clouds; and (4) a roaring noise. Finley quotes extensively from his observers to document these phenomena. "It seemed as if the lightest garments I could put on were a burden to me," remarked one in reference to the sultry atmosphere. "The clouds seemed to be boiling up like muddy water, the upper surface of the cloud reminding me of the incessant eddies or whirls seen in the muddiest portions of the Missouri River," recalled another concerning the appearance and movement of the storm clouds. The approaching sound was described as the "rush of 10,000 trains of cars." Although this is one of a litany of clichés that the media often report after a tornado, the choice of railroad sonance is not unreasonable given the lack of other sources of roaring sounds on the Plains in the 1880s.[23] Clearly it was not lack of imagination that caused the consensus on the sound of a tornado; once the twister was sighted, the observers' visual imaginations ran wild. They described tornados as "balloon shaped," "basket shaped," "egg shaped," "trailing on the ground like the tail of an enormous kite," "bulbous form," "like an elephant's trunk," as well as "funnel, "hour glass," "acorn," "serpent," "turnip," and "pear."[24]

Finley condensed these observations into "General Instructions to Volunteer Tornado Reporters," which he appended to his text. He also offered advice on the location and construction of farm buildings, the protection of livestock, and the design of "tornado-caves" or storm cellars, for which Finley had been a judge in a design contest sponsored by the Burlington Insurance Company of Iowa. From 122 submissions, Finley chose the work of John R. Church of Rochester, New York. Church's design for a 7-by-12-foot chamber could be adjacent to a house cellar or freestanding. It had a 6-foot 6-inch ceiling, a brick roof supported by iron beams and covered by earth sown with bluegrass, a cement floor, 18-inch-thick stone walls, air vents, and oak doors reinforced with sheet iron. The estimated cost of this "cave" ranged between $154 and $345, depending on materials used. Although the primary purpose of the structure was to provide shelter for the few minutes during which the tornado passed by, Finley reminded his readers that they may be "compelled to use it as a home for a considerable time while repairing or building anew from the ruins of the storm."[25]

Public safety campaigns seem to have had mixed success. Almost a century after Finley's report, a survey of one rural area of Cleburne County, in north central Arkansas, found a storm cellar for every four houses, some built as early as 1914, but none built to Church's specifications. Another study, of suburban Sallisaw, Oklahoma, in the early 1990s found that 30.6 percent of the older houses had "hidey holes," whereas only 18.2 percent of the houses in the newer area had such shelters. The older cellars of Grassey Township in Cleburne County were similar to root cellars, generally 4 by 6 feet, with sandstone walls, a dirt floor, and a log roof covered with corrugated sheet metal.[26] In the 1960s, the use of concrete blocks became popular, and in 1966 prefabricated storm cellars using septic tank molds were installed throughout the area. The tanks were 6 feet 4 inches in diameter, 5 feet high, and set only 15 inches into the ground. Owners customized the entries with stone and concrete. Even a well-made storm cellar could evoke traumatic memories, however, as a folklorist wrote in 1992:

> From my childhood experience it [a storm cellar] can be described as a dark, partially underground cement box with damp concrete walls covered with the slimy tracks of slugs, a floor gritted with dead black roaches and rapidly filling with water, an entrance festooned with spidery cobwebs, and a tin-metal door drumming from the pounding rain in symphonic rhythm with the clanging metal chain and the thunder. This ghastly scene was lit by a smoking kerosene lamp and frequent blue arcs of lightning.[27]

Remains of old cellars can conjure the terror of living year after year in the path of tornadoes. Exploring Chase County, Kansas, William Least Heat-Moon found a cellar of rough-cut stone with wooden shelves and broken Mason jars. "These cellars," he observes, "once kept cool home-canned food (and rat snakes) and, when a tornado struck like a fang from some cloud-beast, they kept families that mocked their own timorousness by calling them 'fraidy holes, and it did take nerve to go into the dim recesses with their spidered corners and dark, reptilian coils." Heat-Moon found the cellar "dismal and haunted," but the local residents were fatalistic, happy to tell stories of loss and survival. They call tornadoes "Old Nell," and those who have been under her skirt describe the brilliant colors—cherry red, yellow, electric blue—of the sky at the center of the vortex. Like Finley's volunteers a century earlier, they name the shapes of individual twisters—cones, cylinders, tubes, ribbons, thrashing hoses, dangling lariats, writhing snakes, elephant trunks.[28]

Finley was a pioneer not only in tornado research, forecasting, and hazard mitigation through shelter but also in disaster management and public safety education. Moreover, his remarks on tornado preparedness show an appreciation of human behavior in disasters. Noting that individual cellar-caves may not be possible in cities and that city dwellers "are not inclined to give attention to the face of the sky," he recommends appointing "weather-guards" in prairie towns who will issue warnings and conduct people to places of safety. Not unexpectedly, he links severe weather to a kind of nationalism: "There is no country on the face of the globe where meteorology can be studied with so much advantage, practically and scientifically, as in North America. The elementary principles of meteorology, especially in regard to storms, should be taught in every school, country [sic] town, and city. In the colleges and universities an advanced course should be prescribed."[29]

With this promising start, why did the Signal Service terminate tornado research, and why didn't the Weather Bureau resume it? As Joseph Galway and Allan Murphy make clear in their studies of Finley and his work, there were at least three factors behind the decision of Signal Corps Weather Service chief General Adolphus W. Greely to suspend tornado research and forecasting in 1887. First, the struggle between the military and civilians for control of the Weather Service was reaching a critical point, culminating four years later with the transfer of the service to the Department of Agriculture, and Greely was faced with deep budget cuts and strong political pressures. Midwestern congressmen were not happy to have their states described as sites of frequent tornadoes. It was bad for real estate values. The wording of Greely's 1887 report—"it is believed that the harm done by such a prediction would eventually be greater than that which results from the tornado itself"—does not specify what the particular harm may be, but the credibility of the Service and the loss of political support were legitimate concerns.

A dozen years after Greely's dictum, the *Monthly Weather Review* published a number of articles on tornadoes and their impacts on society. Nevertheless, Cleveland Abbe, editor of the journal, defended the ban on forecasting by citing the lack of a telegraph and telephone grid sufficiently fine to detect storms less than 5 miles wide, and by arguing that the destructive areas of tornadoes are so small that the chance of being injured is one in 10,000 years: "We have no right to issue erroneous alarms. The stoppage of business and the unnecessary fright would in its summation during a year be worse than the storms themselves, so few and so small are they." He recommended insurance to compensate victims and further research to improve warnings. Two months later he printed a letter from a Weather Bureau section director in Oklahoma who asserted, somewhat facetiously perhaps, that more deaths were caused by damp tornado cellars than by the twisters, adding that the citizens of Kansas and Iowa and the newspapers of the Midwest had themselves shunned the word "tornado" for fear that it would drive settlers away.[30]

In addition, the *Review* was concerned about newspapers that published unconfirmed reports of tornado control by cannon fire and retouched photographs purporting to be taken during tornado outbreaks. The chief of the Rivers and Floods Division published identical photographs that had been used to illustrate stories about tornadoes in Oklahoma and Missouri in 1898 and 1899, concluding with an observation that would be endlessly repeated by media critics: "The argument seems to be: 'If there was a disaster, it must have been a tornado; if a tornado, it must have had a funnel; if a funnel, there must be a picture; this is a photograph, therefore it will do.'"[31]

A second factor inhibiting tornado forecasting was personal rivalry. Finley won first prize from the *American Meteorological Journal* for an essay on the relation of tornadoes to regions of low barometric pressure. Third prize went to a civilian employee, H. A. Hazen, who became Finley's implacable enemy and who was, in 1890, placed in charge of tornado research by General Greely. Hazen held this position until 1897, when Chief Willis L. Moore of the now civilian U.S. Weather Bureau suspended all research on tornadoes. Moore faced budget cuts and chose to place his emphasis on public education and promotion of the bureau rather than on research. This hiatus in tornado and other scientific work lasted until 1916. Nevertheless, individual observers continued to contribute articles on tornadoes to the *Monthly Weather Review,* and a considerable body of knowledge existed by 1925.[32]

Finally, although Weather Bureau regulations in 1905 extended Hazen's ban on the word "tornado" in forecasts, it allowed the substitution of "severe thunderstorms" and "severe local storms."[33] These euphemisms for tornado were well understood in the affected regions, and residents responded accordingly. As cities grew in "Tornado Alley" (an area extending north from Texas through Oklahoma, Kansas, and Nebraska, into South Dakota), the potential for fatalities and property loss increased.[34] The years 1910–1930 saw rising numbers of tornado-caused deaths. The Weather Bureau resumed the systematic collection of tornado statistics in 1916 and by 1925 had compiled data that revealed not only an increase in the number of these storms but the problems of defining them. Noting the discrepancies among regional directors in reporting tornadoes, and "a greater tendency on the part of most newspapers to brand as a 'tornado' a violent storm which results in loss of human life than to apply the term to a similar storm which does not involve loss of life," one meteorologist remarked that despite the official definition of tornado, "in practice, the information a section director can secure is very often inconclusive."[35]

The Weather Bureau lifted the prohibition on the word "tornado" in 1938 and redefined it in 1959 as "a violently rotating column of air, in contact with the ground, either pendant from a cumuliform cloud or underneath a cumuliform cloud, and often (but not always) visible as a funnel cloud."[36] This definition did not require a funnel cloud, the essential mark of vernacular definitions, nor did it mention damage. "Old Nell" became an abstract thing composed of numbers.

The most destructive tornado ever recorded in American history struck east-central Missouri and southern Illinois and Indiana on March 18, 1925. Unusual in its size, strength, and duration, it began at 1:01 P.M. near Annapolis, Missouri, and ended at 4:30 P.M. in Princeton, Indiana, a distance of 219 miles. Moving at 50 to 70 miles per hour in a path about a mile wide, it killed more than 700 and injured thousands more. The death toll would have been even higher if the area had been more heavily populated, but most of the storm's path was through farms and villages. Murphysboro and West Frankfort, Illinois, coal-mining communities with populations of 12,000 and 18,000, respectively, were the largest towns in the tornado's path. A total of 284 people died in Murphysboro, where 40 percent of the town was destroyed, and 148 in East Frankfort, which experienced 20 percent destruction. Some villages were totally destroyed, and most of the area never recovered economically.

There are two full-length books on the disaster. Peter S. Felknor's *The Tri-State Tornado: The Story of America's Greatest Tornado Disaster* is a compilation of newspaper accounts and thirteen interviews with survivors, while Wallace Akin's *The Forgotten Storm: The Great Tri-State Tornado of 1925* is a more synthetic account based on additional newspaper sources. Both authors claim personal connections to the event—Felknor, because he grew up in St. Louis in the 1960s, and Akin, because he was raised in Murphysboro and was two years old at the time of the tornado. Both have meteorological training.[37]

Felknor juxtaposes long quotations from newspapers with shorter excerpts from his interviews with survivors seventy years later. Since most of the survivors were children and were in school when the tornado struck, their memories of their journeys home through the wreckage of their communities are especially vivid. Newspaper accounts, gathered chiefly from adults, tend to focus on the moment of impact, narrow escapes, and acts of heroism. Both the press and the survivors invoked the analogy to a "war zone."[38] Although comparisons between human and natural disasters are obvious, Americans were becoming more aware of tornadoes when memories of World War I were still vivid. Like the sound of a train, "war zone" was once fresh, not the cliché it has become.

Both Felknor and Akin devote a chapter to the long-term consequences of the Tri-State Tornado. Akin's personal memories include playing in a marsh just outside Murphysboro that had survived the tornado but was drained by the Works Progress Administration during the 1930s. The tornado, he implies, was just the beginning of Murphysboro's decline. Ill-advised geoengineering projects, a decline in coal production, and the Depression shut more businesses. The Mobile and Ohio Railroad never reopened its repair shops. Felknor provides a detailed picture of the relief effort that combined Red Cross, Salvation Army, National Guard, church, and individual contributions. A firm in Nashua, New Hampshire, sent sixty cases of blankets, and the Sears-Roebuck Agricultural Foundation, using its rural radio station WLS, raised $35,000 for the victims. Felknor's interviewees recalled living in tent cities through the spring and summer; some were grateful for the help their families received, whereas others were critical, especially of the Red Cross because it seemed to neglect the working poor in favor of the unemployed. That postdisaster public opinion swings between the euphoria of spontaneous cooperation and the paranoia of charges of political favoritism seems inescapable.[39]

Felknor found that many survivors of the tornado of March 18, 1925, never fully recovered from the trauma it caused. When subsequently threatened by tornado weather, they admitted to becoming nervous, checking the sky, or retreating to their basements. In contrast, others claimed that they no longer worried, and some who did worry still refused to move. The memories of those who experienced the storm conflict on whether people built storm cellars after the disaster. One respondent remembered one or two being built in Griffin, Indiana, and then abandoned after a few years; another survivor recalled "many storm cellars were built in the back yards [of West Frankfort], and one can still see them there today."[40] Griffin, a small town, was totally destroyed. The differences may be due to individual memory or to the size of the communities.

The Tri-State Tornado highlighted the Weather Bureau's shortcomings, but it also vindicated the work that had been continued after Finley's, both the scientific research and the disaster mitigation efforts of shelter builders and insurers. The evolution toward an understanding that a tornado is wind, not a funnel, and that, lacking any actual measurements of wind speeds in the vortex, the only way to estimate wind strength is by the damage left behind, was well under way in the 1920s.[41] Contrary to most accounts of tornado disasters, therefore, the Weather Bureau did not ignore tornadoes but worked within the institutional restraints inherited from its turbulent origins. Indeed, as Charles Doswell suggests, "It is likely that the tornado disasters of the 1920s in the United States, including the famous 'Tri-State' tornado of 18 March 1925, were responsible for the recognition that more national infrastructure was needed to respond to the tornado threat."[42]

That infrastructure now includes extensive private and public research programs, including research sponsored by the insurance industry, into ways to reduce the destructiveness of tornadoes through improvements in home and office construction. Tornado insurance, available since the 1880s, allows individuals and institutions to protect themselves from extreme losses. Federal relief, available since 1974 at the discretion of the president and Congress, may provide affected communities low-interest loans as well as temporary food, shelter, and medical assistance. The Federal Emergency Management Agency (FEMA), created in 1979 to work with private and state disaster relief agencies, has had a mixed record due to mismanagement and political interference.[43]

Because of the large number of fatalities, the Tri-State Tornado raised tornado awareness nationally and created a market for tornado stories. One of the more interesting, for several reasons, is Laura Kirkwood Plumb's "My Personal Experience with a Texas Twister," published in the popular magazine *Scribner's* just three months after the Midwest disaster. She describes her home, 90 miles northeast of Amarillo in the Texas Panhandle, in terms instantly recognizable to readers of Dorothy Scarbough's book *The Wind*. The desolation and loneliness of the place drive Plumb "to turn to the only thing left for companionship—the sky. I became a connoisseur of sunsets." She also learns the native Texan's "habit of keeping one eye on the horizon and the other on the storm-cave for six months of the year. If no other refuge is available when a cyclone strikes, he lies down flat on the ground and lets the world whiz by!" When an evening comes that she sees a tornado approaching, her husband refuses to take shelter because there's no funnel. Their house blows away as he stands in the doorway.[44]

Laura Plumb and her son were inside, tumbling over and over, but they also survive with a few bruises. She remembers the sound of the tornado not as a freight train but the roar of a mountain avalanche. Apparently, her husband's only comment was, "Had he not seen Harold Lloyd chasing a house in the movies, he would have known the scene was faked." She, too, invokes popular culture to explain her feelings: "Song-writers, columnists, jokers, and all that frothy fringe of the litterateurs tell us things are not the same from the outside looking in as they are from the inside looking out. This holds true in cyclones!" Plumb's most revealing comment comes at the end of her essay when she writes, "It seems impossible that such a thing could happen in the midst of this twentieth-century civilization, but it can and does."[45] Her article suggests, I think, that Americans were, by 1925, expecting a level of protection from natural hazards that the Weather Service could not yet provide. Her house was in a remote location, without a telephone or radio, though apparently she had electricity, since she mentions a refrigerator. Rural electrification was certainly crucial to bringing farms and villages into the network of emergency services. Her sense of outrage at the tornado's power to destroy her life and property signals a new feeling of entitlement, which includes, in her case, a trip to Seattle for a summer's rest. Storm hazard mitigation by moving to a safer place is the oldest and probably remains the most reliable course of action.

Preparations for the country's entrance into World War II revived interest in storm spotting. Many munitions plants and supply depots were located in the Midwest and vulnerable to lightning storms and tornadoes. By December 1942, the Weather Bureau had organized more than 100 networks of voluntary observers in areas near military facilities to provide warnings for use in the evacuation of workers. These spotters were distinct from the existing network of cooperative observers that the bureau had relied on since the 1890s for climatological data. As Doswell notes, the spotter network grew to more than 200 by 1945, and spotters were retained after the war, partly because of the continuing importance of military bases, but more because of their proven utility. Despite wartime censorship, tornado warnings were extended to civilian populations in Kansas and Missouri in 1943, and the bureau published a pamphlet of instructions for network observers in June 1944.[46]

Postwar developments in tornado research and forecasting were rapid and coincided with the beginnings of federal and state disaster management programs and weather-hazard mitigation. Partly at the urging of the airline industry, Congress appropriated funds in July 1945 for the Thunderstorm Project, a multiagency research program involving several universities. Three years later, on March 25, 1948, two air force meteorologists, Colonel Ernest Fawbush and Major Robert Miller, issued the first tornado warning from their office at Tinker Air Base near Oklahoma City. Although their forecasts were distributed only within the air force weather offices, news of their success leaked to the public, and by 1950 the Weather Service had created the Tornado Project, which was issuing tornado forecasts throughout the Midwest by 1952.[47] Principally in response to the increasing nuclear threats of the cold war, but with some attention to natural disasters, the military medical services, in conjunction with the National Academy of Sciences and the National Research Council, created the Committee on Disaster Studies (reorganized in 1957 as the Disaster Research Group) to fund research on and planning for large-scale disasters.[48]

As noted earlier, in 1953 the Weather Bureau created the Severe Weather Unit, which was renamed the Severe Local Storm Warning Service (SELS) and moved from Washington, D.C., to Kansas City, Missouri, the following year, partly in response to the unusual number of extremely destructive tornadoes in 1953. Waco, Texas, was struck on May 11; Acadia, Nebraska, on June 6; Flint, Michigan, on June 8; and Worcester, Massachusetts, on June 9. Several hundred lives were lost, and there was widespread fear that atomic-weapons tests in Nevada were triggering the tornadoes. Although most meteorologists rejected the link, the public perception was fueled by extravagant claims by advocates of weather modification, such as Nobel Prize–winning chemist Irving Langmuir and New Mexico senator Clinton Anderson. Howard T. Orville, a navy meteorologist and chairman of the congressionally created Advisory Committee on Weather Control, was asked in 1954 if Americans can have "weather made to order." His answer was a resounding yes.[49]

Texas began using volunteer spotters after deadly tornadoes struck there on April 9, 1947, and other states followed suit. In the late 1950s the pace of severe storm research quickened. *It "Looks" Like a Tornado,* a handbook for tornado network observers, came out in 1959; the following year a pamphlet for the public, *Tornadoes: What They Are and What to Do about Them,* appeared. The next ten years witnessed the rapid growth of both tornado and natural disaster research. Although civil defense programs expanded during the 1950s, offering one kind of disaster planning, it was a series of earthquakes and hurricanes that led to the creation of the Office of Emergency Preparedness in the Kennedy White House in 1961. An unusual outbreak of more than thirty tornadoes in Iowa, Wisconsin, Illinois, Michigan, and Ohio on Palm Sunday, April 11, 1965, with almost 300 fatalities, stimulated the establishment of the Natural Disaster Warning System (NADWARN), which included a tornado-specific plan, SKYWARN. SKYWARN was also the beginning of NOAA Weather Radio and an integrated system that includes spotter information, amateur radio operators, and Weather Bureau broadcasts over a frequency that requires a special radio.[50]

Tornado research continued apace through the last decades of the twentieth century, revealing interesting social and cultural dimensions. A study of the effects of tornadoes in the 1950s and 1960s in five midwestern states and five southern states found that the southern states had five times more deaths per 100,000 inhabitants. Moreover, the difference in mortality continued after the Weather Bureau began issuing tornado warnings in 1953. After interviewing thirty-three tornado survivors in Illinois and twenty-four in Alabama, the authors of the study concluded that there were significant local differences in the perception of natural disasters. A majority of southern respondents believed in God's will and in luck. When tornadoes were imminent, residents of Illinois turned to radio and television, or their own weather instruments, whereas Alabamians watched the sky. Forty-six percent of those interviewed in Illinois rated Weather Bureau forecasts as excellent; only 12 percent of those in Alabama agreed. "To be effective," the authors write, "a warning must be heeded"; people themselves are part of the problem of natural disasters.[51] As much as any other characteristic, the tension between expert opinion and individual belief marks the Americanness of tornado and other natural disasters.

From the prairies of Kansas in the 1880s to the fields of Illinois and Alabama in the 1960s, witnesses and survivors of tornadoes agreed on two things: the twisters were unpredictable, but with watchfulness and luck you could probably escape from them with your life. The emphasis in the descriptions of tornadoes was on their unusual and distinctive features. Tornadoes and the severe storms that spawned them were spectacles, hypnotic in their dramatic displays. Fear often gave way to ecstasy, and the undulations of the distinctive funnel cloud became iconic, photographed, filmed, and painted to represent the turbulence and destructive power of nature.

Tornadoes were a test of courage and resolve. They were certainly not sought out, but neither were they a reason to flee the region. Few Americans were critical of the Weather Bureau in the years 1887–1950 for its refusal to issue tornado forecasts; they did not expect such service until they became aware of the technological advances that made such forecasts feasible. By the early 1950s, newspapers, popular magazines, and television were making the public aware of threats—political and environmental—that existed on a scale heretofore unimagined. The drumbeat for civil defense rumbled like thunder, and the storm cellar morphed into the fallout shelter. The North American Air Defense Command (NORAD) and NADWARN promised protection with radar and sirens. Made-to-order weather came with bravado made-to-order. The promise of improved tornado forecasts and the creation of emergency preparedness offices around the country began to change the way Americans perceived weather hazards. Attitudes toward tornadoes swung from fascination to fear; the response to rescue swirled from gratitude to blame.

Tornadoes: The Era of Management

For residents of Waco, Texas; Flint, Michigan; and Worcester, Massachusetts, in the spring of 1953, the promise of better tornado warnings must have seemed hollow. The tornado that hit Waco killed at least 114 and caused millions of dollars in damage; 115 died in Flint. The following day, 94 died in a tornado that swept through downtown Worcester. The attention given the Worcester storm by the emerging profession of disaster management adds to its significance. Anthropologist Anthony F. C. Wallace, well known later for his studies of culture and personality, ethnohistories of the Iroquois, and historical studies of textile and coal-mining communities in Pennsylvania, produced a report for the newly organized Committee on Disaster Studies of the National Academy of Sciences and the National Research Council, in which he created a model for analyzing the stages of a disaster.

Wallace's contribution to disaster studies is twofold. First, he offers a model that deals with both the space and time dimensions of disasters. His spatial model consists of five concentric rings based on the role that the populations in each area play. The core area is divided into Total Impact and Fringe Impact areas. The third ring is designated Filter Area, and the two outer rings are labeled Organized

Community and Organized Regional Aid. The time dimensions are (1) Steady State (the city before impact), (2) Warning, (3) Impact, (4) Isolation (the impact area before rescue can begin), (5) Rescue, (6) Rehabilitation (the attempt to restore the steady state), and (7) Irreversible Change (the city achieves new equilibrium). Although it assumes an urban disaster, the model also is applicable to rural catastrophes.[52]

Sketchy as it is, Wallace's model continues to be cited.[53] Some of his key points are as follows: First, equilibrium in a cultural system simply means that the population is maintaining what it would define as normal daily life, handling the stresses of economic and social tensions without resorting to extraordinary measures. Equilibrium does not mean that the system is unchanging. Indeed, the momentary situation, affected by season, time of day, business trends, age of the population, and similar factors, may influence the degree and nature of a community's response to a disaster, but "to be a disaster, the impact must upset the total system."[54]

Second, Wallace's warning period is divided into "warning" and "threat," the former meaning the general probability of danger; the latter, a specific prediction of time and place. Clearly, Wallace did not consult the Weather Bureau terminology, since in its usage in 1953 a tornado "bulletin" or "forecast" was a prediction for a large area, and a "warning" was an imminent threat.[55] Such lack of coordination between the bureau charged with issuing watches and warnings and the agencies dealing with weather-related hazard mitigation continues into the twenty-first century.[56] Third, Wallace's concept of a Filter Area through which traffic and communication must pass, located between the Impact Area and the site of the first organized aid, presents interesting problems for disaster relief workers in tornadoes, which are notoriously erratic in their paths of destruction. Although he does not discuss the Worcester tornado specifically in connection with Filter Areas, he does note that in the Isolation period the proximity of emergency services is largely a matter of chance. Moreover, it is in this period that secondary impacts, such as live electrical wires, fires, and the leakage of poisonous or explosive gases, are most dangerous.[57]

Read as an artifact of the early cold war, Wallace's *Tornado in Worcester* is a scenario for dealing with the aftermath of a nuclear attack. The line between nature's and human terror had begun to erode, but the culture of fear and the dependence on federal relief and authority had yet to emerge. Tornadoes of the 1970s and 1990s (the 1980s were free of major tornadoes) contributed to the changing outlook toward weather disasters by revealing them as predictable, even if only by a few minutes, and manageable, given sufficient public and private resources. The terrorist attacks of September 11, 2001, shifted public concern back to security from foreign invasion. On August 29, 2005, Hurricane Katrina and the federal government's slow response revealed that security was an issue in both weather and man-made disasters. But not all citizens are equally protected, nor will they receive equal compensation for loss. All disasters teach this lesson, but acknowledging it is painful. Disaster planners are charged with restoring the pre-disaster social order, not improving it.

A devastating tornado struck Lubbock, Texas, on May 11, 1970, at about 9:30 P.M. It destroyed an area 9 miles long and 1.5 miles wide at its widest point. At least twenty-six people were killed, most in the poorer Mexican American neighborhoods near the downtown area. The storm developed rapidly after dark, which made it difficult for spotters to see. Based on radar and visual reports, the Weather Bureau issued a tornado warning for Lubbock and adjacent counties at 8:15. The city's disaster officials had been alerted at 7:30 and were in their office an hour later. The city's nine radio stations and three television stations repeated the Weather Bureau warnings for more than an hour before the tornado struck, but its sudden appearance in the center of the city meant that the sirens did not go off until the last minute. Thanks to the forecast and warning system, most residents had time to seek shelter.[58]

Immediately after the storm, sociologists from Texas Tech University began a study of the responses of the Anglo, African American, and Mexican American populations. Lubbock was a city of 150,000, located in the high plains of west Texas. At the time of the tornado, approximately 75 percent of the population was Anglo, 15 to 18 percent were Mexican American, and about 8 percent African American. Although there had been no major racial conflicts, juvenile delinquency and crime were issues in the community—hence, the focus on race in the postdisaster study. Three thousand people were left homeless, and many of the poor were moved into a Federal Housing Administration (FHA) development that had been built in the 1960s but largely abandoned by white residents when African Americans began moving in in 1966. By destroying the Mexican American barrio, the storm brought about a rapid integration of the housing project.

The researchers interviewed 200 families; 45 percent had lived in the housing project before the tornado, and 55 percent were resettled there. Forty-eight of the families were Anglo, 52 were African American, and 100 were Mexican American. There were no significant differences in response among the three groups. Poverty, not race, was the main variable in this storm because the tornado damage was greatest in a low-income housing project. Contrary to the usual clichés, survivors were not caught completely by surprise, and many had access to tornado cellars and other shelters, but they did describe the noise of the storm as "a train passing over" and "bees in a can." Asked what they would do differently if a tornado struck again, forty families admitted they would probably do nothing differently, twenty-nine families admitted they "didn't know," fifteen families said they would go to a storm cellar sooner, and ten families emphasized the need to build storm cellars, including public storm facilities.[59]

In the wake of the Lubbock disaster, Congress authorized another program on storm preparedness; the National Weather Service (the U.S. Weather Bureau was renamed that year) produced a spotter training film, *Twister*, in cooperation with the Civil Defense Preparedness Agency; and Theodore Fujita of the University of Chicago and Allen Pearson, director of the NSSFC, published their scale of tornado intensity, forever changing the way tornadoes are described and classified.[60]

Because no instrument has ever measured the speed of winds in a tornado, estimates of wind speed are based on the most destructive thing it does at any single point along its path. The Fujita-Pearson Scale (usually shortened to Fujita Scale or F-scale) was formulated in 1971 in response to growing dissatisfaction among meteorologists, insurance companies, and the public over the NWS's inability to distinguish between minor and major tornadoes. Meteorologists, civil engineers, and others familiar with the effect of wind pressure on different kinds of structures had long offered estimates of the wind speeds necessary to blow a roof off a house or lift an automobile, but little systematic assessment of tornado damage had been done before Fujita and Pearson.[61]

Fujita's biography is central to understanding what one writer has called "the golden age of tornado research."[62] Fujita brought a remarkable number of interests and skills to the study of tornadoes. Tetsuya (he renamed himself Theodore after settling in the United States in 1953) was born in Kitakyushu, Japan, in 1920. He graduated with a degree in mechanical engineering from Meiji College in 1943 and was appointed assistant professor of physics the same year. He worked on the measurement of impact forces, from bomb damage at Hiroshima and Nagasaki to typhoon damage on his home island of Kyushu. His typhoon studies became his doctoral dissertation in 1952. In addition, he was a cartographer and photographer, producing three-dimensional maps of volcanic calderas on Kyushu in 1940. In 1950 he began corresponding with Horace Byers, chairman of the meteorology department at the University of Chicago, and three years later Byers brought him to Chicago as a research assistant. In 1956 Fujita became research professor and senior meteorologist and established the Severe Local Storms Project. Ultimately this became the Wind Research Lab, reflecting his broad interests in microbursts, wind shear, mesoscale systems, satellite study of storm development, and global circulation, as well as tornadoes.[63] As one tribute to Fujita put it after his death in 1998, "In essence, Fujita gave us the language of severe storms."[64]

Fujita began his damage assessment surveys of American tornadoes with the Fargo, North Dakota, storm of 1957, but the Lubbock tornado of 1970 was the catalyst for finalizing the intensity scale. What is often ignored or forgotten in reporting tornado-strength estimates is that they are always made retroactively, based on damage to objects on the ground. The Fujita Scale does not estimate the wind speeds a few meters above the ground, which may be much higher, nor does it always distinguish between tornadic winds and equally strong straight winds. It does recognize differences in construction, but not siting and alignment to the force of the wind. The Pearson Scale adds a ranking for path width and length, which is seldom reported in media accounts. Finally, the Fujita Scale was designed to be consistent with the venerable Beaufort wind scale, with its twelve-part rating from calm to hurricane force, that is, above 72 miles per hour. The Fujita Scale begins with F0, winds 40 to 72 miles per hour, which may break tree branches and damage signs, then continues to F5, winds 261 to 318 miles per hour, which inflict "incredible" damage, lifting strong frame houses off foundations and throwing automobiles for 100 meters. F6 to F12 winds, 319 to 700 miles per hour or Mach 1, the speed of sound, are theoretically possible but not expected in tornadoes. Fujita was thinking of winds generated by nuclear weapons.[65]

Over the past thirty-five years, there have been many efforts to improve or replace the Fujita-Pearson Scale. The chief problem is that identically constructed buildings can be damaged differently depending on the angle of the wind or the location of the building in relation to other buildings and surrounding topography. Variation of wind speeds at different heights is also a factor. Wind speeds may be overestimated if many flimsy buildings are extensively damaged, underestimated if sturdy structures appear unscathed. In summarizing the problems with the F-scale, Grazulis notes the importance of knowing whether the determination of wind speed was made by an experienced observer or by an estimate based on damage. Some critics have suggested adding a confidence level to the calculation of wind speed. The media focus on top wind speeds, and a community assigned a high F-scale damage rating can reap the benefits of public and private assistance.[66]

In February 2007, the NWS replaced the old Fujita Scale with a new measurement of tornado damage, the Enhanced Fujita Scale. The new scale, developed by a team of meteorologists and wind engineers, uses twenty-eight damage indicators, from small barns and farm outbuildings through mobile homes to "big box" retail buildings and transmission-line towers. This is much more comprehensive than the old scale and requires a more careful evaluation of the degrees of damage. Each type of building now has its own set of damage indicators. Weather Service personnel and local damage assessment officers will receive thorough training in using the new scale. The original scale tended to overestimate wind speeds necessary to do significant damage. As with any change in procedures affecting disaster aid and insurance claims, the Enhanced Fujita Scale will face challenges and further revisions. Since there is still no way to measure winds in tornadoes, the Enhanced Fujita Scale remains an estimate of wind speeds. The ranking of tornadoes under the original Fujita Scale will not be changed; enhanced history began February 1, 2007. Once again, NWS's choice of terminology is revealing. While the Fujita Scale is obviously improved by adding indicators of damage, the connotations of "enhanced"—"increased," "exaggerated," "heightened"—all promise something more than the usual scene of gutted double-wides and toppled trees. The scale will surely enhance the storm.[67]

Popularization of the F-scale by TV weathercasters turned tornadoes into a kind of summer disaster game—F1 here, F5 there—with viewers sending in their videotapes of the storms. This focus on destruction obscured the advances in tornado research that led to a better understanding of these storms and that improved the warning systems. The successful launch in 1967 of the Applications and Technology Satellite (ATS), a geostationary platform allowing the study of clouds in tornado areas, was one sign of progress. Other developments include Bluestein's TOTO experiments in the early 1980s, Doswell's work on tornado formation and forecasting, and the deployment of NEXRAD, an improved Doppler radar system, all of which provided solid science for predicting tornadoes.[68] The impact of such disasters on people was also becoming better understood.

By the time of the Andover, Kansas, tornado of April 26, 1991, tornado research, forecasting, and warning had progressed significantly. The tornado that struck Andover, a suburb of Wichita, was not especially devastating in the context of the past century of tornado disasters. Nevertheless, it killed 13 people, 11 of them in the Golden Spur Mobile Home Park. More than a thousand survivors were left homeless from the power of the F5-rated storm. The town of 4,047 suffered an estimated $30 million in property damage, a significant part of the $200 million lost throughout Kansas by tornado damage that day. The most remarkable thing about the Andover tornado is the response to it. As has been noted, the systematic study of the impacts of disasters on individuals and communities began in the 1950s as part of civil defense programs activated by the threat of nuclear attacks.

Out of the early studies of bombing in World War II, weather disasters, and fires, new fields of sociology and psychology emerged. Ultimately the need to understand the causes and consequences of hazards caused by storms, air pollution, and earthquakes created the new professions of disaster emergency managers, risk analysts, and natural hazard mitigation planners—jobs that did not exist before the 1970s. Despite the passage of a series of disaster relief acts, federal aid was seldom available to victims of local weather disasters until, in 1974, Congress gave the president the power to make disaster declarations and, in 1988, vastly enlarged the scope of assistance with the Stafford Act. The past twenty years have witnessed the growth of both public and private disaster bureaucracies and rising public expectations about relief for property loss.[69]

The Andover tornado is a case study in the developing culture of disaster. The agencies responsible for warning, rescue, and recovery all worked relatively well. As part of the recovery phase, the community produced its own assessment of the process, an innovation that has been copied in subsequent weather disasters. *Tornado: Terror and Survival* is a hybrid book, combining brief summaries of eyewitness accounts of the storm as it moved through McConnell Air Force Base, south Wichita, Andover and its Golden Spur Mobile Home Park, and into the countryside, with a meteorological overview, criticism and praise of the warning and rescue and relief efforts, and an assessment of the process of physical and psychological recovery. The National Weather Service issued a tornado warning for all of Sedgwick County at 5:46 P.M., and a minute later all seventy-three sirens in the greater Wichita area began to sound. Nine minutes later the first of the tornadoes appeared southwest of the city. Sirens at McConnell Air Force Base sounded at 6:10, fourteen minutes before a tornado entered the base, destroying housing but missing the fueled, loaded, and ready-to-fly bombers and fighters. Later congressional hearings determined that the NWS warnings had provided a seven-minute lead time in most of the area, better than the three-minute national average. (By 2004 the NWS was claiming an average of ten to eleven minutes.) Some people ignored the sirens, however, thinking they were emergency vehicles. In Andover, the city-operated siren, purchased used to save money, malfunctioned, but a police officer drove through the mobile home park warning residents to seek shelter a few minutes before the twister appeared.[70]

Responses to the warnings, alarms, and the tornado itself varied widely. People sought shelter where they could, most in bathrooms or other interior rooms in their houses, some under cars or in ditches. One family took shelter beneath an overpass on the Kansas Turnpike with a television crew from KSNW in Wichita. The KSNW videotape was shown repeatedly on news broadcasts and gave the impression that overpasses were good tornado shelters, a misconception that meteorologists and emergency managers are still trying to correct. Called the "famous" and "infamous" Andover footage, the tape shows cars being tossed by an F2 tornado. Current thinking on tornado safety strongly advises against seeking shelter under overpasses unless they have external girders or there is no other alternative. Currently, the Sedgwick County, Kansas, Emergency Management Web site clearly warns travelers not to take shelter under overpasses. A fatality in Oklahoma in 1999 led to a reassessment of the overpass shelter theory, but the jury is still out on whether vehicles are safer places than ditches.

The National Weather Service and the American Red Cross recommend that people leave automobiles immediately if a tornado is approaching, but meteorologist Thomas Schmidlin has offered evidence that if the winds are less than 150 miles per hour (F1 and F2), vehicles are safer than even interior closets in mobile homes. Since the tornadoes and hurricanes of the 1990s, building codes have been improved in many states. New manufactured homes, as the industry prefers to call mobile homes, are now supposed to withstand 120-mile-per-hour wind gusts up to three seconds. About 8 percent of American houses are manufactured homes, but 49 percent of all tornado fatalities were residents of mobile homes. Schmidlin and his associates did extensive surveys of tornado damage to vehicles and some wind tunnel tests on models, but meteorologists and emergency managers have called for more studies before advising mobile home dwellers to seek shelter in their vehicles because windshields are easily shattered by flying debris. Legislation requiring mobile home parks to have storm shelters is the best solution. Indiana legislators recently passed a law requiring every manufactured home built after June 30, 2007, to be equipped with an emergency weather radio, ignoring the fact that without a shelter in the mobile home park, the warning may provide just enough time to kiss your pets and kids good-bye. If the radios save lives, they are well worth the cost, but it is obviously better to also require mobile home parks to build adequate shelters.[71]

The Golden Spur Mobile Home Park had a concrete bunker-style shelter, and sixty to seventy residents managed to get in before the tornado struck. But some lived too far away to walk there by the time they received the warning, and a few chose to stay with their pets, which were not allowed in the shelter. Others, suffering from claustrophobia, insisted on standing in the shelter's open doorway. Since only 33 percent of all mobile home parks in the United States have any kind of storm shelter, the residents of Golden Spur were, in that sense, fortunate. One family finally got to use the bomb shelter it had built in the 1960s in anticipation of nuclear war, and several people were sheltered by crowding into a walk-in cooler in an Andover restaurant. Witnesses to the tornado had the usual variety of impressions. A few compared the sound to a freight train, but more, perhaps because of their proximity to an air base, mentioned a roar like a jet engine or the launch of a space shuttle. Some were more imaginative. "It looked like thousands of black birds. It sounded like a huge vacuum cleaner," recalled one man; another thought the twister sounded like "hundreds of skateboards." Tornado comparisons change with the times.[72]

Rescue and recovery efforts generally went smoothly, but there were inevitable criticisms. The failure of the emergency siren was the most obvious flaw, but the mayor's decision to bar the Mennonite Disaster Service and residents from the mobile home site while search and rescue teams used front-end loaders on the wreckage of homes was controversial. The Mennonite crews specialize in collecting personal items before they are destroyed by search and cleanup teams. Residents accused the mayor of "bulldozing" their personal belongings. The mayor claimed that cordoning the site was necessary to prevent looting. There were also complaints from some of the injured that their rides to hospitals were slowed by sightseers. Apparently there are storm *disaster* chasers as well as storm chasers. A well-publicized weather disaster also draws politicians and celebrities. Senator Robert Dole and his wife, who was head of the American Red Cross at the time, flew into McConnell promising immediate relief. The senator got an earmark of $1.15 million through Congress to buy 154 new warning sirens for Kansas communities. Comedian Jerry Seinfeld helped raise $20,000 for the Salvation Army at a concert in Wichita. Three thousand "We'll Weather the Storm Together" T-shirts were sold to raise $15,000 for the Kansas Red Cross.[73] President George H. W. Bush signed a federal disaster declaration, making residents of Wichita and Andover eligible for grants and low-interest loans. FEMA offered $2.1 million to the affected areas of Kansas, with $189,375 going to Andover, and more than $350,000 was raised from private sources through the Andover Tornado Aid Fund.

While the money was crucial for the rebuilding effort, the impressive volunteer effort seems to have been even more important for the psychological recovery of the community. In addition to federal and state assistance, private relief agencies such as the Red Cross and the Salvation Army, churches, businesses, and individuals—some from as far away as North Dakota—joined in the cleanup and recovery effort. Thousands of volunteers spent weeks in Andover, making friends and sharing disaster experiences. Counseling began immediately, organized by the Counseling Center of Butler County, and was soon supplemented by the National Organization for Victims Assistance (NOVA), a private Washington, D.C.–based nonprofit, which helped to train volunteers from local hospitals to conduct "debriefings" with tornado victims, encouraging them to talk about their feelings.

Continuing the emphasis placed on children in disasters by the early research of the Committee on Disaster Studies in the 1950s, the Butler County center prepared special instructions for parents. Many teachers incorporated the disaster into writing assignments for their elementary school students. The essays, poems, and accompanying artwork give a good indication of children's resiliency. Some children played "tornado," building towers with blocks and furniture, and then tearing them down. Boy Scout troops helped plant more than 250 trees in the mobile home park and along the route of the tornado. They also took the lead in constructing a monument for the victims.[74]

Each weather disaster, be it tornado, hurricane, flood, blizzard, or heat wave, highlights recurring and changing patterns of American responses. Five patterns in particular stand out.

Volunteerism

Weather disasters stimulate American civil society. Each new catastrophe reveals the vast extent of both the voluntary and professional infrastructure for rescue and relief. Nobody bowls alone after a disaster. As the Andover tornado clearly demonstrates, a small community with damage confined to a limited area may actually renew itself and emerge stronger in the wake of a storm. Older elements of philanthropy from churches and fraternal organizations have been supplemented by more specialized services for children and the elderly, even rescue for pets.

Technological Fixes

Paradoxically, communities generally ignore the danger posed by severe weather until a disaster occurs; then they are motivated to make improvements in warning systems, but leaving specific responses up to individuals. Installing sirens, upgrading communication networks, and enforcing building codes are, of course, easier than changing human behavior or relocating. Trusting to machines and luck, while taking risks and ignoring expert advice, is deeply ingrained in the American psyche.

Responsibility

Although outside aid and government assistance are expected and welcomed, much of the onus for disaster prevention is assumed to be an individual's. Being prepared, knowing where to get weather information, building storm cellars, buying insurance, and having an evacuation plan are private matters. Even though personal responsibility is emphasized by FEMA and other disaster relief agencies, Americans today seem less confident of their ability to control their own destinies. Weather forecasts were once assumed to provide sufficient information for individuals to make sensible preparations for potential dangers. Forecasts now raise expectations about federal and local emergency plans that will protect residents of a threatened area from harm. Weather disaster movies and television programs contributed to the impression that severe weather is beyond an individual's ability to withstand "nature's wrath."

Blame

Related to the apparent shift of responsibility from individuals to "authorities" is the growing sense of outrage. The whine "Why wasn't I warned?" usually translates as, "Why didn't you make me listen?" The growth of what some have called "the nanny state" in the past twenty years has blurred the boundaries between private and public liability. But blame cuts two ways. Americans outside disaster sites increasingly question the foolishness of people who build in floodplains or on beachfronts, drive through swollen creeks, evade building codes, and fail to practice due diligence in weather awareness. Americans are still sympathetic and generous to those they believe are truly victims but scornful of those who seem to be waiting for "windfall" relief. Some have argued that governments want their citizens to live in fear because the frightened are more tractable. The century after the Weather Bureau had hesitated to cause panic by issuing tornado forecasts, to today's brightly colored warnings of polluted air, ultraviolet rays, excessive heat, and terrorist threats seems long indeed.

Missed Opportunities

The growth of national disaster management has not always led to efficient local action. Partly because weather forecasting is rooted in local conditions and atmospheric research looks at larger patterns, the two endeavors have not always meshed. Scientists understand tornado genesis much better than they did ten years ago, but the public conceptions of risk remain a mystery.

This brief review of a century of tornado experiences, scientific research, and disaster mitigation in the United States can only begin to reveal the complex interplay between humans and storms, scientists and the public, institutions and individuals, imagination and experience. As Charles Doswell points out, there are many "publics," many users of weather information; there are also many creators of weather information.[75] Cultural responses to weather disasters include all of these "communities."

Decade of Disasters

The 1990s mark a significant change in the ways in which weather disasters are perceived and addressed in American culture. Some of the changes began a few years earlier, but the devastation caused by Hurricanes Andrew and Iniki in 1992, the East Coast "Blizzard of the Century" in March 1993, the Midwest floods that summer, the Chicago heat wave of July 1995, the ongoing southeastern droughts, and dozens of slightly less deadly hurricanes, tornadoes, and floods created indelible memories of weather's lethal impact. Loss of life and property from weather was exacerbated by expanding population and continued building in vulnerable places. Presidential major disaster declarations rose from an average of twenty-five a year between 1984 and 1988, to forty-five annually in the years 1993–1997. Congressional earmarks added millions in disaster relief in districts and states whose representatives had political clout. The emergency management and natural hazard mitigation bureaucracies grew larger and more complex, but conventional models of risk management were challenged. Research on the causes and consequences of weather and other catastrophes questioned traditional definitions of disaster. The magnitude of weather disasters has expanded from localized storm damage to global climate change. The last twenty years have created a distinctive disaster culture in the United States, in which weather and other hazards are viewed not as "acts of God," or even the by-products of our technological society, but as a permanent part of our social and political systems. Tornadoes, hurricanes, blizzards, and heat waves may add stress to existing conditions, but they are only part of the uncertainties with which we live daily.[76]

Hurricanes grabbed and held the attention of the American public, from Andrew in 1992 through Katrina in 2005, displacing tornadoes as "America's storms"—at least in media coverage. Hurricanes are Super Bowl–like events that often strike cities with superdomes, not just villages in Tornado Alley with a high school gym. Hurricanes are bigger in every way—taking days, even weeks to develop, and destroying more property than even the biggest tornadoes. Tornadoes are like jazz quartets, improvising, surprising, picking up and dropping airs; hurricanes are like heavy metal bands, clanging and screeching, accompanied by pyrotechnics and surging fans. Katrina, whose horrors are still being tallied, has already spawned dozens of books—notably Douglas Brinkley's 716-page tome—a myriad of Web sites and government reports, at least one major photography exhibition (by Robert Polidori at the Museum of Modern Art in New York City), and songs and films. Andrew has three books and countless articles, but Hugo, a storm that occurred three years before Andrew that took more lives, awaits its Brinkley and Spike Lee.[77]

Older "superstorms" have received new looks from journalists and historians. The Galveston hurricane of 1900, still the deadliest with 6,000 to 12,000 fatalities, has had two full-length histories in the past twenty years, while the 1928 storm that killed more than 3,000 in the Caribbean and Florida has three new books, plus a technical reassessment in *Bulletin of the American Meteorological Society*. Not surprisingly, there are two recent books on the infamous 1935 Labor Day hurricane that killed several hundred World War I veterans working in Civilian Conservation Corps camps in the Florida Keys. Perhaps because of its more densely populated landfall, the New England Hurricane of September 1, 1938, has half a dozen books, two of which were published in the past five years. Hurricane Camille, said to be the most powerful storm (in terms of barometric pressure and wind speeds) ever to affect the mainland of the United States up to the time of its arrival on August 17, 1969, has three full-length accounts. Floyd, the 1999 hurricane that caused a massive evacuation of the coast and extensive flooding in interior North Carolina, rates just a single book as of 2007. Hurricane, tornado, flood, and blizzard books are tossed onto bookstore and library shelves after each storm surge. Individually they offer exciting stories of heroism and corruption; together they build a case for the normality of these events, but will they be a tocsin against storm amnesia?[78]

Hurricanes, from the mid-1980s until Katrina, killed fewer people each year than any other extreme weather event. The annual mean loss of life from hurricanes from 1986 to 1995 was twenty. On average, tornadoes claimed more than twice as many lives, lightning almost nine times more, extreme heat nineteen times more, and winter storms and extreme cold, more than forty times more than hurricanes. The death toll from hurricanes dropped each decade from 1910 on, except for the 1920s and 1930s with their four devastating storms. Property loss, on the other hand, rose dramatically, from a few million dollars annually in the first decades of the twentieth century to more than $30 billion a year in the 1990s. Estimates for the total cost of Katrina are still coming in, with NOAA estimating in early 2007 a total of more than $80 billion and a few billion more for Rita. This stark contrast between property loss, even when adjusted for inflation, and deaths has helped to frame what Roger Pielke Jr. calls the "U.S. hurricane problem." Better forecasting and emergency planning seem to have had great success in saving lives, but little or no impact in protecting property.[79] Why?

The answer, according to Pielke, is that the problem has been framed, that is, defined, too simply as one of saving lives and property. Better warnings and emergency management account for the decline in loss of life; "'normalizing' past hurricane damage to present-day values (using wealth, population, and inflation)" explains the apparent increase in property loss. The 1926 hurricane that killed 327 in Florida and Alabama would, Pielke believes, cause $129 billion in damage and probably kill thousands if it struck the same area today. A better definition of the weather problem involves focusing on vulnerability—seeing the connections between the incidence of hurricanes and other hazards and the populations and property at risk. The probability that a hurricane will make landfall is a matter of extrapolating from past events. We know that on average two hurricanes strike the United States every year, and two intense hurricanes inflict damage on either the Gulf or the Atlantic coast every three years. Using calculations made in 1997, the annual cost of hurricanes, averaged over the past century, is more than $5 billion. Given the extraordinary hurricane seasons of 2004 and 2005, this figure is certainly low.

The probability that New Orleans will be hit by a Category 3 or above hurricane is three in a century, while Miami's chances are 11.1, or more than once a decade, according to the National Hurricane Center. Although the city of New Orleans actually declined in population in the 1990s, surrounding parishes grew, some as much as 20 percent in a decade. Population growth along the Mississippi, Alabama, and Florida Gulf coasts has also exceeded 20 percent in many places. There have been similar population increases along the Atlantic coast from Florida to Maine. Florida's total coastal population has grown by 75 percent in the past twenty-five years. More than 45 million Americans, 16 percent of the total population, live in coastal communities from Texas to Maine, and 3.9 million more in Puerto Rico and the Virgin Islands. Hawaii, with 1.2 million people, is vulnerable to typhoons. Vulnerability is also a factor of wealth and health. In the city of New Orleans, more than 60,000 people, about 14 percent of the population, described themselves as disabled. With 17 percent of Miami's population and 21 percent of New Orleans's residents living in poverty as defined by the U.S. census, the number at greater risk is large and growing.[80]

With a better understanding of hurricane incidence and exposure, reducing vulnerability should be a simple matter of implementing the four phases of disaster as established by emergency managers—mitigation, preparation, response, and recovery. Mitigation, as used in academic studies and policy making, is the recognition and elimination of, or adaptation to, hazards; it includes such factors as choosing where to live, land-use planning, building codes, and improving forecasts. Preparation is an element of mitigation, but it is distinct in that it involves planning for specific impacts—training emergency workers, transportation for evacuation, providing shelters, education of the public—and creating the infrastructures to deal with the emergency and its aftermath. Response begins with individuals who have been affected by the disaster. The term "first responder" is a misnomer when applied to police or fire departments because survivors begin clearing debris, searching for other victims, and helping one another long before rescue crews arrive. In Katrina, members of the media were among the first responders. Studies have shown that response planning that does not include all citizens is incomplete and results in misunderstandings and conflicts between authorities and the public. Official response to an impending weather disaster begins with warnings, continues with specific instructions such as evacuation, and concludes with damage assessment and the initiation of recovery plans, created during the mitigation and preparation phases. Recovery should include the restoration of utilities, if necessary, short- and long-term medical care, maintenance of order, and the rebuilding of structures, roads, and other damaged infrastructure.

The four-phase model of disaster planning is meant to clarify the range of issues arising in any event that threatens lives and property and to help coordinate the dozens of public and private agencies involved in emergency planning and management. Natural hazard mitigation plans are required by law for most cities, counties, and states. FEMA and the Department of Homeland Security provide guidelines and work with other federal, state, and local authorities, although the shift from all hazards to terrorism has caused confusion, as Katrina dramatically revealed. The trouble with defining "hurricane problem" is that, although it recognizes that not all populations are vulnerable to the same degree, it assumes that the affected area is stable before the disaster and will return to its original condition once the recovery is complete. Consequently, this approach treats each extreme weather event as discrete, a named storm with a beginning and an end, even while acknowledging that the incidence of each storm is part of a recurring seasonal pattern. Larger historical trends affecting the social and economic health of a city or region are ignored. The vulnerability model is not really concerned with the variety of cultural responses to weather hazards.

Pielke and other natural disaster scholars adopt a rationalist view of human behavior. Presented with a problem, a threat to life and property, people will, they assume, act to minimize personal, if not community, loss. But will they? An anthropological model, which presumes that people perceive threats, and their options for avoiding them, very differently based on their age, sex, health, wealth, past experiences with disaster, and attitude toward risk, provides a more nuanced approach. Although all disaster plans recognize that there are people with special needs, no scenario that I have seen confronts the dynamic changes in individual attitudes toward compliance with emergency preparedness plans that take place during the course of an event. The cultures of weather disasters arise and evolve in part separate from the events themselves. These cultures are shaped by preexisting conditions and media interpretations, as well as the actual experience. Preexisting conditions, such as class conflict or a failing infrastructure, are magnified by disasters, as became obvious in New Orleans. The media provide stories, not always true, which help people make sense of events and place themselves in larger contexts.

Katrina, as Marita Sturken and participants in a Social Science Research Council-sponsored roundtable remind us, was both a weather event and a weather media event. Sturken argues that television coverage of the hurricane began with satellite images of the swirling clouds, remote from conditions on the ground, reassuring because prediction promises a measure of control. Once the media realized the extent of destruction, the focus shifted to the misery of the victims, illustrating for viewers the stark contrast between the organized storm and the disorganized communities on the ground. Russell Dynes, cofounder of the Disaster Research Center at the University of Delaware, extends Sturken's remarks by calling attention to how the media made New Orleans a synecdoche for the entire Gulf coast, and then turned the Crescent City's story into a series of searches for damage, bodies, rescue workers, "the person in charge," and "the bad guys." David Alexander, professor of disaster management at the University of Florence, Italy, completes the SSRC analysis by suggesting that the media employed two conceptual models of Katrina, one from Hollywood disaster movies in which a destructive event inevitably leads to a breakdown of the social order (and the emergence of a hero to restore it), and a civil defense model actually employed by many emergency planners in which the restoration of order has a higher priority than rescue and relief.[81]

These very preliminary assessments are helpful because they partially explain why media coverage, especially television, uses conventions familiar from detective fiction. Katrina, or any weather disaster, is like a crime. Reporters and cameramen go to the scene, record the evidence, interview eyewitnesses, count the victims, interrogate the authorities who have obviously failed to do their job, and round up the usual suspects—in this case, rioters, looters, and the New Orleans Police Department. The reporter as hero was explicit in news stories long before Dan Rather or Anderson Cooper whipped off their ties and jackets, rolled up their sleeves, and tousled their hair. Everybody loves a mystery. Reporting rumors feeds the drama. They are false clues. They can be denied in tomorrow's story.

Scientists also had difficulty explaining the enormity of Katrina. In a thoughtful report, "The Mourning after Katrina," Jeff Rosenfeld, editor in chief of the *Bulletin of the American Meteorological Society,* sought to understand why, despite repeated and detailed warnings from meteorologists and engineers, there was so much chaos and so many expressions of surprise. There were warnings about weaknesses in the levee system in both the popular press and the scientific literature. FEMA officials had concluded at the end of a weeklong simulation of a Category 3 storm, Hurricane Pam, in 2004 that a storm of this magnitude would cause "casualties not seen in the United States in the last century." Beginning in 2000, *Popular Mechanics, Scientific American, National Geographic,* and the *New Orleans Times-Picayune* all ran stories on potential flooding with casualties in the thousands.[82]

As long ago as 1989, John McPhee's book *The Control of Nature* clearly showed why the Corps of Engineers' system of levees would ultimately fail. Some scientists were indignant because government officials and the public had ignored their warnings, but most recognized the truth expressed by wildlife ecologist Scott Shalaway who wrote after Katrina that "for the last 35 years the people in charge of funding wetlands protection have bet their political careers that, because catastrophic events such as Katrina are so rare, none would occur on their watch. It's a safe bet, but this time Katrina beat the house." The New Orleans plan for a weather disaster was to hope that it would never arrive—regretfully the plan most of us employ.[83]

Post-Katrina emergency preparedness plans for New Orleans still place primary responsibility for shelter and evacuation on individuals, who are urged "to sit down with their families before the June 1 beginning of the hurricane season to make their own emergency plan." For the elderly and infirm, there is a new City Assisted Evacuation Plan (CAEP), but, citizens are warned, "CAEP is not intended to replace the individual's personal responsibility in preparing their [sic] own evacuation. It is meant to be an evacuation method of last resort." Special CAEP pickup points are identified, and residents are warned that "there will be no shelter of last resort." "Last resort" is a troubling term, suggesting last-minute departures after an individual's evacuation plan has failed, after power outages, and when pickup points may be inaccessible. Another class of special-needs residents seems to be getting almost as much attention as the disabled. Both the New Orleans and the FEMA Web sites have lengthy instructions for evacuating pets. Reports that some Katrina victims refused to evacuate because they could not take their pets probably prompted these government interventions, but they are further evidence of the way hazard mitigation filters into every crevice of daily life.[84]

Six months after its first reflections on Katrina, the American Meteorological Society again tried to assess the event. Gina Eosco and William Hooke, of the AMS Policy Program, call for expanding the focus of hazard mitigation beyond "management of evacuations and emergency response" to making potential disaster sites "safer, better, perhaps even more equitable, for the future." Using the analogy of the National Transportation Safety Board, they propose the creation of a National Disaster Reduction Board to analyze and report on disasters with the goal of correcting both technological and sociological problems. To achieve this they also urge greater public-private cooperation. Thomas Schmidlin, a meteorologist in the geography department of Kent State University, makes two heretical observations: first, that no government can protect its citizens completely from themselves; and, second, since the projected death toll from the simulated Hurricane Pam was 25,000 to 100,000, the actual number of fatalities in Katrina, now estimated by NOAA to be 1,577 in Louisiana, suggests that preparation and evacuation went better than expected.[85]

Most observers now agree that things could have been worse. Mayor Ray

Nagin claimed that 80 percent of New Orleanians left the city, and another estimate puts the total number of metropolitan region evacuees at 1 million. In seeking to explain why some residents in the hurricane's path chose to evacuate while others chose to stay, Schmidlin cites the work of Earl Baker and others who have studied the behavior of people living in the path of hurricanes, noting that the three most important factors determining their decision to evacuate are (1) the vulnerability of their location, (2) whether they believe that they have been told by authorities to evacuate, and (3) the predicted severity of the storm. There are other considerations, some of them paradoxical. Knowledge of the danger makes most people more cautious, but having survived earlier hurricanes causes many to think they are less vulnerable. People with children are more likely to evacuate, whereas households with elderly members are less likely to abandon their homes.[86]

Other studies have found even more obstacles to informed responses to hurricane warnings. Kirstin Dow and Susan Cutter, geographers at the University of South Carolina, interviewed several hundred residents of Hilton Head and Myrtle Beach, South Carolina, in the aftermath of Hurricanes Bertha and Fran, which had been forecast to make landfall in their area on July 12 and September 5, 1996, respectively. Both storms struck near Wilmington, North Carolina, providing the researchers the opportunity to study the impact of false alarms. The governor of South Carolina ordered mandatory evacuation for each event, limiting the order to areas east of the Intercoastal Waterway for Bertha but extending the order to all of coastal South Carolina in the case of Fran, predicted to be a Category 3 hurricane. A total of 41 percent (750,000 in both South and North Carolina) of the residents evacuated for Bertha, as did 59 percent for Fran. In both storms a significantly higher percentage of residents in the wealthier Hilton Head community evacuated, and in both storms more people cited the forecasts made by the NWS and The Weather Channel than the governor's orders for their decision to evacuate. The media in general and TWC in particular were rated more reliable than the governor or local officials. Cutter and Dow refrain from drawing sweeping conclusions for a single case study, but they suggest that public officials cooperate with the media to speak with a single voice. Providing detailed explanations for evacuation orders might also encourage greater compliance. Finally, they recommend study of "local disaster culture," attitudes and behaviors of those who choose to "ride out" the storm. Urging increased attention to individual differences in perceived threat, Cutter and Dow surmise that "while residents do not find that officials are 'crying wolf,' they are searching elsewhere for information to assess their own risk."[87]

Tim Frazier, a geographer from Pennsylvania State University, is only indirectly concerned with false alarms, but he recognizes the problem, since the overevacuation, or "shadow evacuation," of the coastal areas of the Carolinas for Hurricane Floyd, September 16, 1999, created what some labeled a "fiasco." He comments, "Over three million residents took flight during the largest peacetime evacuation ever witnessed in the United States." The evacuation also created the largest traffic jam in the nation's history. To discover what lessons were learned from this experience, Frazier interviewed nine state and local officials in South Carolina, including traffic safety engineers, police chiefs, emergency preparedness directors, and fire chiefs, asking them about their challenges and plans for future evacuations. Predictably, his informants pointed to the lack of bridges and roads from the coast to the interior, candidly admitting that the state could not afford to build more and that gridlock was inevitable. To alleviate the problem, transportation planners seem limited to technological fixes less expensive than bridges and highways, and to compliance from the public—neither being a certainty in a time of crisis. People in a panic to get away may or may not know or follow assigned evacuation routes, and the new technologies of Intelligent Traffic Systems (ITS), which consist of traffic cameras, signals, and signage designed to minimize bottlenecks, can easily be put out of commission by high winds, power failures, or human error.[88]

Roger Pielke Jr. mentions one example of technological failure that could have had dire consequences. For the past twenty years, the National Hurricane Center has used a computer model called SLOSH (Sea, Lake, and Overland Surges from Hurricanes) to estimate storm surge heights in more than thirty basins along the coast from Texas to Maine. Emergency planners, using the National Geodetic Vertical Datum, selected buildings they felt were well above the expected surge for Hurricane Hugo, September 21, 1989, for evacuation shelters. One of the shelters was a high school in McClellanville, South Carolina, which was thought to have an elevation of 23.53 feet above sea level. The surge broke through the doors of the school and reached a depth of 6 feet, sending those inside scrambling onto tables, bleachers, and even the rafters. After the waters receded, it was discovered that the base elevation of the school was only 10 feet. The more complex the technology, the more vulnerable it is to errors in crucial data. Jim Reed, looking at the aftermath of Floyd, concludes that "even if all needed lanes were opened promptly and all shadow evacuees stayed home, the possibility exists that gridlock may be unavoidable during huge evacuations." This was certainly the impression Americans received from the half-page color photograph in the *New York Times*, September 25, 2005, showing more than a hundred cars parked on ten lanes and the median of about 50 yards of I-45, in Texas, while attempting to evacuate before Rita. Storm fleeing, like storm chasing, may be less a weather event than a traffic event.[89]

Overcoming complacency, but not overalarming the public, is the emergency planner's most obvious conundrum. Less evident is what might be called the culture of hurricanes that encourages hurricane parties, hurricane trivia stories, and, most enduring of all, hurricane music. A Google search for "hurricane party" yields more than a million hits, among them something called "essortment," which is linked to other sites offering advice on party giving. "Essortment" offers the following in answer to the question, "What is a hurricane party?"

A hurricane party is any celebration that passes the time while waiting out a storm. Hurricane parties help everyone relax by bringing friends and family together to endure a stressful situation, and help conserve supplies by keeping everyone together in one location. The party is usually staged in a well-equipped home with abundant space and necessary emergency equipment such as a generator, hurricane shutters, and a portable television. Despite a reputation for heavy drinking and reckless behavior, hurricane parties are a unique facet of the disaster mentality of people who regularly persevere through dangerous storms.

A successful hurricane party keeps everyone occupied and free from stress or worry about the ongoing storm. Hurricanes last several hours, and power outages eliminate television, movies, video games, and other electronic entertainment. Board games and team competitions are perfect distractions for children and adults, and hurricane themed drawing or charade challenges contribute to a lighthearted atmosphere. . . . A party doesn't reach hurricane intensity without serving signature hurricane cocktails. First, combine gin, rum, and other liqueurs with grenadine, a heavy, blood red pomegranate syrup used for sweetening and coloration. Mix tropical juices such as orange, grapefruit, cranberry, raspberry, or pineapple and serve over crushed ice for an intense and delicious drink. . . . If possible, serve the drinks in hurricane glasses—tall, narrow glasses with exaggerated curves and flaring brims—for additional festive appeal.[90]

This insouciant advice goes on for several paragraphs, concluding that "a party is far less festive with injuries to tend, but with proper precautions, creative activities, and abundant comfort food, the merriment can outlast even the most violent hurricane." The author of "Why Do We Have Hurricane Parties?" is less relentlessly cheery but echoes the theme of fantasy play: "A hurricane is, for sure, a release from routine. It's a time of growing anticipation that something out of the ordinary is about to happen. It's a time for bracing against the elements, a time of survival. For some, the anticipation, the element of danger is exciting and, with businesses and schools shut down, it's the perfect excuse to celebrate—no work, no class, let's party!"[91]

Do hurricanes serve, for some, as an escape from ennui? Or are they just another excuse for drinking, as Brinkley implies in his description, provided by one of his graduate students, of Saturday, August 27, 2005, at Pat O'Brien's French Quarter bar. Hurricanes, like any disaster, can be considered liminal states, anthropologist Victor Turner's useful concept of the period of transition between two organized social systems. Characterized by ambiguity, rule breaking, even violence, liminal states can be regular ritual events or created in the collapse of order following disasters. By incapacitating the existing authority, a storm breaks down distinctions between rich and poor, foolish and wise, sacred and profane. The good hurricane party, as its advocates aver, allows its participants to be irresponsible. On another level, looting may be simply a necessity to survive, but it is also a ritual of reversal in which order is restored by forcing the "high" to experience what it is like to be "low." "Liminality," Turner writes, "implies that the high could not be high unless the low existed." Looting, most disaster experts believe, is almost always exaggerated by rumor and early media reports, inviting a police and military response to "restore order," but looting and an increase in crime have been documented for many disasters.[92]

To commemorate the liminal rituals of disaster in twenty-first-century American society, we market it (with T-shirts, bumper stickers, buttons, postcards, songs, exhibits, annual reunions, "Floodweiser" beer, "Chris Owens" bottled water) or personalize it (with graffiti, poems, tattoos, blogs). As the *Bulletin of the American Meteorological Society* noted,

> Shortly after the deadly hurricane, tattoo artists reported numerous requests for body art depicting aspects of the storm. JuJu Becker, a tattoo artist in New Orleans, was one who anticipated it and immediately began working on designs, such as "Katrina" written on a World War II–era bomb. The customers came, but Becker said most requests had a personal resonance. Many clients were returning residents, rescue workers, and those helping to rebuild. The most popular designs in celebration of New Orleans were fleur-de-lis patterns and the city's abbreviation "NOLA." There was even a symbol modeled after a weather-map depiction of hurricanes.[93]

Creating a design for a tattoo may, because of its permanence, be more serious than finding the right words to describe a hurricane's power, but language helps render the storm memorable. Sam Eifling, a writer from Fort Lauderdale, Florida, surveyed several newspapers in 2004 and discovered a paucity of descriptive terms for hurricanes. Like tornadoes, the tropical cyclones were routinely described as "snapping trees like twigs," "tossing cars like toys," and sounding like "a locomotive." Eifling attributed these clichés to a collision of barometric and deadline pressures, or a reporter's hope of catching an editor's notice with vivid similes, such as "Fat metal light poles crimped at the center, bowing to the ground like the twisty-neck straws in a retro diner."[94]

Although I have proposed that tornadoes are the most "American" of storms because they occur in the heart of the country and have captured our attention in popular culture, hurricanes also have a claim on the national imagination. In the aftermath of every destructive hurricane, promises are made to rebuild in ways that will prevent the wrecking of homes from future storms. Nature can ac-

Hurricane House

WEATHER-VANE DWELLING DESIGNED FOR

Rotating on triple circular tracks, this unusual house turns itself to point its rounded end into the wind, defying even gales of hurricane force. In ordinary weather, it can be made to face in the best direction with reference to sun and breeze

By CARL WARDEN

"Hurricane House," *Popular Science,* October 1939, 68.

complish in a day what might require years of urban renewal. Residents of poor neighborhoods have every reason to fear that businessmen and public officials do not want them to rebuild, if there is any possibility that more expensive housing can attract new, wealthier residents. Architects impart a utopian spirit by designing hurricane-proof homes and offices.

Two designs, sixty-seven years apart, illustrate this aspect of hurricane culture. After the notorious Labor Day hurricane of 1935 and the New England hurricane of 1938, Edwin A. Koch, a New York City architect, designed a streamlined, teardrop-shaped steel-framed house that rotated on a platform, enabling the house to face the strongest winds like a weather vane. With exterior walls of waterproof plywood and sliding glass doors, Koch's fantasy would probably have been demolished by flying debris from a Category 1 hurricane, but the dream of defying nature by technology refuses to die. On the heels of Katrina, the *New York Times* reported that builders and developers in Texas and Florida are building reinforced concrete homes, some on 20-foot stilts. Hundreds of customers are willing to live in houses with smaller windows, little or no landscaping, and concrete roofs to ride out the storms.[95]

A test of Koch's house plans may be in the offing. In 2007, Florida International University received several million dollars from the state of Florida and RenaissanceRe Holdings Ltd., an insurance company, to develop the Wall of Wind (WOW), a prototype two-fan blower that will blast existing and experimental houses with Category 1 to Category 5 hurricane-force winds. "The final phase— a 24-fan Wall of Wind built inside the hangar-like facility—will test houses mounted on a turntable so they can be assaulted by hurricane-force winds, rain, and debris from every direction." Fortunately for Lillian Gish and Dorothy Lamour, their directors lacked access to such a big toy. The most telling details of the story of the Wall of Wind seem to me to be that the research is promoted not so much as a way of saving lives, though that may happen if building codes are improved and enforced, but for the purpose of providing insurance companies with better estimates of future losses. Moreover, the wind tests will be videotaped and used to produce public service announcements for television. The assumption is that "just as the effective visualization of car crashes dramatically changed automobile safety through the introduction of air bags, and the earthquake community improved building safety through shake table experiments, the Wall of Wind will alter the way the construction industry approaches building in hurricane-prone areas."[96] Wall of Wind seems to be a classic example of asking a question (How much wind does it take to blow down a house?) because the tools already exist or can be built, rather than asking a question (How do you convince a person not to build on a hurricane-prone beach?) for which new tools will have to be invented.

A more palatable fruit of hurricane consciousness is the opera *The Second Hur-*

Wall of Wind (WOW), a six-fan hurricane simulator to test single-story structures with category 4 wind-rain simulations. International Hurricane Research Center and Laboratory for Coastal Research, Florida International University. Photo courtesy Dr. Stephen P. Leatherman, director.

ricane (1937) by Aaron Copland. Considered by some to be America's greatest composer, partly because he wrote music based on folk tunes of distinctive regions such as the Appalachians, the prairies, and the Southwest, Copland had long wanted to write an opera and eagerly seized the opportunity provided by Grace Spofford, director of the Henry Street Settlement House in New York City, who asked him to write a piece suitable for production by children in the settlement's music school. Collaborating with Edwin Denby, a Harvard-educated dancer and poet, Copland created a two-act "play opera" about six high school students, four boys and two girls, who volunteer to fly on a rescue mission to help deliver food to stranded survivors of a hurricane. A second storm forces the pilot to leave them on an island in the Mississippi River while he flies off to get help. The students and an eight-year-old African American boy they meet on the island are now forced to stop thinking of their adventure as a "picnic" and learn to work together to survive the second hurricane. They accomplish this only after much squabbling and selfish behavior. Realizing that they are lucky to be alive, they sing the moral of the story in the opera's epilogue:

The newspapers made a story out of it like a lot of others.
That's not what we think of now that it's all over.
We got an idea of what life would be like with everybody pulling together,
If each wasn't trying to get ahead of all the rest—
What it's like when you feel you belong together,
With a sort of love making you feel easy.
We'll remember the feeling even if we six drift apart—
A happy easy feeling, like freedom, real freedom.[97]

The Second Hurricane premiered at the Playhouse on Grand Street in New York City on April 21, 1937, just seven months before the opening of Marc Blitzstein's *The Cradle Will Rock,* a musical about a heroic union organizer and wicked businessmen. One link between these two products of Popular Front theater was the formidable, if relatively unknown, twenty-two-year-old director, Orson Welles. Joseph Cotton, Welles's collaborator in radio and film, was the only adult actor in Copland's show. Following its opening in New York City, *The Second Hurricane* was produced in Alaska, California, Ohio, and Tennessee. Leonard Bernstein directed it at the Peabody Playhouse in Boston in 1942 and later conducted a television version with students from New York City's High School for Music and Arts in 1960. Copland recalled that "the opera's theme of cooperation was particularly appropriate during the war years, so the Treasury Department sponsored a radio broadcast for the sale of U.S. War Bonds." Although playwright Thornton Wilder declared the libretto "awful," containing "all the worst features of the Sunday School, the Boy Scout Movement and those radio serials where Fred aged twelve helps the F.B.I. clean up a nest of counterfeiters," it was clearly the American spirit of volunteerism that appealed to audiences, not left-wing ideals of worker solidarity.

Nor did it matter that Denby had placed his hurricane victims rather far inland, in Missouri. What mattered was the creation by the students of an American version of "communitas," a social system based on ability, service, and respect for others that restores order after the chaos of the storm. The message of the opera was that found in the postscript to every weather disaster—discovering unity in diversity. The Department of Homeland Security might want to sponsor a performance.[98]

Cynicism and sentiment are the poles between which much posthurricane commentary swings. Murder rates, FEMA failures, and contractor fraud are the daily stories datelined New Orleans, while op-ed pages shine a weekly ray of hope with pieces like historian William Chafe's in the December 27, 2006, *Washington Post.* Returning from a weekend as a volunteer in New Orleans, the Duke University professor found a resurgence of the spirit of the civil rights movement of the 1960s. "In the Lower Ninth Ward," Chafe writes, "more than a hundred volunteers gather . . . to help secure—with the community—a foothold toward starting anew. . . .

Most spend their days gutting houses so that returning residents can be eligible for federal rebuilding funds. Tearing down sheetrock infested with toxic mold is dangerous work. . . . No one sees this as a lark. The unpaid staff briefs workers on the hazards they will face, insisting that respirators fit snugly so that no toxins are inhaled." There are elements of truth in both stories, and exaggerations.[99]

The federal government's recent approach to hazardous weather is, seemingly, to criminalize it. Communities are encouraged to organize neighborhood weather watches similar to the Justice Department's "Neighborhood Watch Program," an effort to get citizens to report criminal activity. Report suspicious weather and set up pre- and poststorm plans. Practice the plans. These are the recommendations of the Department of Homeland Security, which is trying to organize "Citizen Corps" programs to create neighborhood emergency plans. When this message is translated into articles in the popular press, the analogy between crime and weather may be presented a little too starkly. "Joining a neighborhood weather watch group," a writer for *Better Homes and Gardens* states, "is just one way you can protect family and friends from the wrath of Mother Nature." Knowing what to do in a weather disaster is one thing; thinking that the natural cycle of weather is wrathful is likely to produce unnecessary anxiety.[100]

A little over a week after Katrina, Chuck Doswell, a lifelong student of tornadoes, set down his thoughts on the meaning of the hurricane. "Severe weather," he begins, "simply is part of the biosphere—all life on this planet is intimately connected to the nonliving processes, among which is the weather." Hazardous weather is inevitable; the lure of life along the coast is powerful; repeating past mistakes will be costly, but "unfortunately, it predictably is what we will wind up doing." Doswell's bleak assessment of what he calls "the futility of the indomitable spirit" is tempered somewhat by thoughts about severe storms he had before Katrina. All destructive storms bring some benefit to the earth. Thunderstorms are accompanied by renewing rains; lightning converts some of the air into nitrogen necessary for plants; strong winds, even tornadoes and hurricanes, play a role in culling old trees and creating spaces for a variety of organisms. Where humans see hurricanes as destructive, other parts of nature may experience the storms as renewal. Besides, Doswell concludes, "Some of us find *beauty* in atmospheric processes, including severe storms. For us, they can be awesome in their grandeur and power, and inspirational in their complexity."[101] Not a comforting message for those promising better protection next time, but a fair assessment of our human relationship with weather.

Everyday Weather and Discomfort

Tornadoes and hurricanes are examples of extreme weather, and suffering from them is, with luck and foresight, no more than a once-in-a-lifetime event. But daily suffering from some element of weather is endemic to every human population. The costs of a few hours of rain to the construction industry, or a brief storm-caused power outage to retail merchants, add up to billions of dollars annually, far more than even the most devastating storms. Accurate forecasts save businesses and individuals money, but even in our largely climate-controlled lives, weather makes us suffer by affecting our health, our plans, and our moods. The history of the discovery of weather's effects on daily life, and of efforts to control them, is a complex story, parts of which have been told in histories of medicine, air-conditioning, and air pollution. These seemingly disparate histories reveal that the threats of everyday weather have stimulated lively debate and inspired some ingenious solutions, but we are still vulnerable. The encounter between the vast number of variables in any given moment of weather and the almost infinite combinations of factors producing an individual's physical and mental state has yielded few indisputable truths.

By 1904, all four of the elements that make up the story of everyday weather in the twenty-first century had emerged: (1) recognition of the dangers of extreme heat and cold that were not infrequent in their seasons; (2) air-conditioning, with its promise of total climate control and its "objective" definition of comfort; (3) the discovery of air pollution and the need for the development of air quality monitoring and forecasting; and (4) the awareness of physiological and psychological effects of weather and climate. A hot spell in July 1901 killed thousands in cities from Chicago to New York; in 1902, Willis Carrier installed his first air-conditioning machinery in a lithography plant in Brooklyn; and in 1904, the American Lung Association was founded, first to find a cure for tuberculosis and later to understand and eliminate other respiratory illnesses.[102]

In the same year, a young professor of education, Edwin Grant Dexter, published *Weather Influences: An Empirical Study of the Mental and Physiological Effects of Definite Meteorological Conditions.* Dexter was a pioneer in what is now known as biometeorology, the study of the influences of weather on living organisms, but his approach was more sociological than medical. Matching Weather Bureau data with police, hospital, and school records, Dexter found correlations among all kinds of human behavior and temperature, humidity, wind, and other meteorological factors. While acknowledging that the problem of attributing any specific behavior, assault, suicide, or bad mood, to a degree of temperature or percentage of humidity was unverifiable, and while fully accepting that sickness and death are due to pathological conditions of the body, Dexter was convinced that weather acted as the final straw, exacerbating conditions that the mind and body were already experiencing. Dexter's work was sufficiently well regarded that Cleveland Abbe wrote a twenty-page introduction, cautioning against climate determinism and racism but endorsing Dexter's attempt to isolate what he called "sensible temperature . . . the sensation produced by atmospheric conditions in general." "Thus we speak of pleasant, exhilarating, bracing weather; depressing, gloomy, and muggy weather," Abbe continued. "Now, these sensations may be produced either by changing the temperature, the moisture, the pressure, the sunshine, or even the food and clothing."[103]

More than a century later, researchers are still pursuing Dexter's dream of establishing definite links between weather and behavior, especially criminal conduct. While some scholars predict that an increase in global temperature will produce thousands of additional murders and assaults in the United States, others remain skeptical. James Rotton of Florida International University, surveying dozens of recent studies of weather and social disorder—from assault to horn honking to riots—concludes that "relationships between temperature and antisocial behavior are not as pervasive as prior reviews suggest. Rather than ask *if* temperature and violence are related, we need research aimed at determining *when* and *why* the two are correlated." Rotton, like Dexter, is careful to separate weather-related behavior from climate determinism, to avoid the taint of racism. Weather's effect on behavior is of concern to employers, teachers, and criminologists; weather and human health is the locus of research by physicians and public health officials. What was labeled "iatrometeorology" in 1885 became "meteorotrophism" in 1938 and continues as "biometeorology" today. The extent to which the atmospheric environment causes or contributes to illness is now the subject of many scientific disciplines.[104]

Two pioneers in the study of weather and health are noteworthy for their contrasting approaches and subsequent influence on contemporary medical debates. Charles-Edward Amory Winslow, professor of public health at Yale University, and William Ferdinand Petersen, professor of pathology and bacteriology at the University of Illinois College of Medicine in Chicago, were energetic proponents of the importance of weather on human well-being. Winslow began as an advocate of fresh air and popularized the image of "The Black Hole of Calcutta" in his public lectures on the dangers of heat and humidity. He joined forces with engineers who were developing air-conditioning and trying to improve on natural weather through climate control. In 1925, he and his associates claimed to "have found an average of about 50 million bacteria per gram in the dust of city streets and between 3 and 5 million in indoor dusts." This is considerably higher than the 10,000 bacteria per gram of airborne sediment found in recent investigations, but Winslow's point made Americans aware of unseen health hazards in the air. Mitchell Prudden made *Dust and Its Dangers* a part of Progressive health reform as early as 1903, while Dr. Charles Reed declared the right to breathe pure air "inalienable" in 1905. By 1917, William Scheppegrell had added hay fever pollen to the list of weather effects that needed to be addressed by meteorologists and physicians. The control of dust and "impurities" as well as temperature and humidity in indoor air became the goal of the American Society of Heating and Ventilation Engineers and the cooling industry.[105]

William Petersen, son of German immigrants to Chicago, began his career with research on immune processes and blood chemistry, but in 1929 he began daily physiological and biochemical measurements on a male volunteer, comparing the observed variations with the weather record. From these observations came the first volume of *The Patient and the Weather*, which included a critique of current medical practice. Petersen felt that physicians were paying insufficient attention to chronic ailments—hay fever, migraines, and arthritis—and to environmental conditions, specifically the weather and seasons. His approach emphasized the dynamics of individuals and their diseases in their physical surroundings; he studied the patient as well as his illness. Barometric pressure was the most important of the daily exogenous factors, he felt, so he began corresponding with Charles Marvin, chief of the U.S. Weather Bureau. Marvin, always conservative, was skeptical of Petersen's theories and methods but supplied him with data on barometric variability for Chicago and Washington, D.C., for the years 1930 through 1932. Petersen was convinced that migraines were associated with barometric change and asked a friend and former colleague, Eric A. Fennel, to do a study in Hawaii, where he was practicing. Fennel treated Petersen's request as a joke, commenting that no one knows anything about migraines, but offering an anecdotal account of his own experience, of which he is clearly dismissive:

> I go to bed with a normal Barometer . . . and the trade winds blowing beautifully. Then at two in the morning I awake for [no] reason at all, save that I know . . . that the Barometer has [fallen sharply] . . . and the wind has begun to blow from Kona. Everybody knows it as Kona weather, and everybody gets ready for a fight. The coffee that morning tasted like hell. On the way to the office everybody makes a right turn signal and turns left.

Currently, doctors are still without a satisfactory explanation of the links between migraines and weather, but Petersen's legacy lives on in the popular press, where the notion of "human weather vanes" has great appeal.[106]

Meanwhile, in 1923, two air-conditioning engineers at the University of Pittsburgh, F. C. Houghten and Constantin Yaglou, established what they termed an "effective temperature," later known as the "comfort zone," of between 61.8°F and 68.8°F, with 64.5°F the optimal temperature of most people indoors. The temperature of the zone would shift higher and lower over the next few years, as the concept was tested and refined to take into consideration other relevant factors such as outside temperature and humidity, the nature of the indoor activities, and personal definitions of comfort. What did not change was the utopian notion of creating a measurable and achievable ideal climate. In 1934, C. Theodore Larson, technical news editor for *Architectural Record*, summed up the converging trends in climate control with two fundamental questions: "Just why should we need air-conditioning? What kind of atmosphere is scientifically the most desirable? The answer is in the delicate balance maintained by the human body in response to atmospheric variables and experienced both as *comfort,* a factor about which much is known, and as *health,* about which much has yet to be learned." Just how delicate is the balance of the body and the atmosphere? This is one of the fundamental questions raised by climate change.

Larson's confidence that comfort was the easy part of climate control reflects the technophilia characteristic of American intellectuals in the years between the "Century of Progress" exposition in Chicago in 1933 and the New York World's Fair in 1939, both of which were air-conditioned and promised "a world freed from geographic and climatic necessity." Larson exclaims: "To the layman, air-conditioning seems little short of magic: making climate to order is as thrilling as flying. But we are only at the beginning of what the scientists and engineers are preparing to do for our comfort and health, perhaps conditioning a whole region, eventually giving us air-conditioned clothes in which each man can suit himself as to temperature." PitZips are on the horizon.[107]

"Climafy" was the winning entry in a contest sponsored by *Forbes* magazine in 1936 to coin a word for air-conditioning. Neither the winner, A. N. Williams, nor his word ever emerged from obscurity, but meteorologists were beginning to understand the challenge to their authority that air-conditioning posed. In the June 1939 *Bulletin of the American Meteorological Society,* Anson D. Marston, an engineer for the Kansas City Power and Light Company, put it bluntly: "The meteorologist and the air-conditioning engineer have one thing in common: both seek to forecast the weather." Marston defines air-conditioning as the complete control of indoor weather, winter and summer. He continues in this vein, observing that the meteorologist does his hard work before his forecast, whereas the air-conditioning engineer does his after determining what the weather should be:

> We must spend hours determining what is required to produce the desired results, and then we must spend more hours or days or weeks in selecting the particular pieces of equipment and arranging them so that they will operate satisfactorily. When we finish we have the comforting (?) knowledge that it is impossible to find any one set of air conditions that will satisfy young and old, robust and weakling, office girl and laborer, all at the same time.[108]

Marston's question mark after the word "comforting" suggests some skepticism about the idea of "comfort zone," but he embraces it and acknowledges the importance of meteorology in providing statistics on the number of "degree-days" for heating or cooling to be expected in a given locality. An engineer with a gas utility company in Baltimore, Eugene D. Milener, noticed in 1915 that home gas consumption was proportional to the difference between the average outside temperature and 64°F. In the mid-1920s the American Gas Association corrected this figure to 65°F and began to publicize the concept of "degree-days," the perplexingly named term for the difference between the mean daily temperature and a given standard. If the day's high is 50 and the low 30, the average is 40. Forty from 65 results in 25 heating degree-days for that date. If an average temperature for a summer day is 80, then it contains 15 cooling degree-days.

By the 1950s the concept of comfort zone was well established in the heating and cooling industries, among architects and city planners, and among savvy consumers of their products. New measures of weather comfort and discomfort were emerging, and by the end of the twentieth century, Americans learned that they lived in an atmospheric environment in which everyday weather posed almost as many challenges as severe weather. First came the "windchill index," the invention of polar explorer Paul Siple, who began conducting experiments in 1939. In 1945 he and an associate announced a simple formula based on the cooling effect of wind on an exposed surface. There were many problems with this "index," not the least of which was that it did not take the warming effect of clothing into account, but it became popular, especially as a scare tactic in boosting TV weather forecasts. The NWS revised the windchill chart in 2001, lessening the effect of wind on temperature and increasing the time for the onset of frostbite. The subjectivity of any such index remains, of course, since we all feel cold or heat differently depending on our acclimatization, body build, diet, age, physical condition, activity level, quality of clothing, and mental attitude. Windchill introduced the possibility of outdoor weather comfort zones, an idea that emerged in June 1959 as the "discomfort index" (DI).[109]

In the late 1950s the U.S. Weather Bureau was suffering from one of its periodic crises of identity and struggling to improve its public image. What better way than to offer a new service, a measure of the combined effects of temperature and humidity to the sweltering masses? Government workers in Washington, D.C., had long enjoyed an informal "misery index" that allowed them to go home if their office temperature reached 95°F with 55 percent humidity. On the assumption, perhaps, that misery loves company, the bureau offered cities a way to boast of their weather or to profit from the suffering of others. The public's response was amused outrage. The *New Yorker,* always in the meteorological vanguard, recalled that it had run a story five years earlier on Osborn Fort Hevener, a weather buff, who had coined the term "humiture" for a summer heat index similar to the degree-days scale. Hevener simply added the temperature to the humidity and divided by two, the quotient being the humiture. Earl C. Thom of the Weather Bureau proposed a more complex process, adding the dry-bulb temperature and the wet-bulb temperature, multiplying by .4 and adding 15. The wet-bulb thermometer is an ordinary dry-bulb thermometer covered with a damp muslin wick. As air flows across the wick (at 600 feet per minute), the evaporative cooling yields a cooler temperature reading than the dry-bulb. Wet-bulb temperature is a measure of humidity. In the resulting index, a DI of 75 was thought to make a majority of people uncomfortable, and a DI of 80 would make almost everyone miserable.

Thom's formula was based on Houghten and Yaglou's "effective temperature" calculations of thirty-five years earlier. Comfort's logical counterpart, discomfort, was a tardy and unwelcome arrival on the weather page, sparking a mock theological debate.[110]

The liberal Catholic weekly *Commonweal* found both good and bad possibilities. "We can look for the emergence, here and there, of the superior individual who claims that even when the Index is in the dreaded 80s, it does not bother him. To know the new figure is high will hardly make us feel any cooler. But when has anyone supplied a better reason for taking it easy?" The Jesuit magazine *America* joined in the wry assessment of how the DI might lead to committing several of the Seven Deadly Sins—pride, envy, sloth, gluttony—"Somewhere beyond D.I. 85 everybody in creation becomes a meteorological beatnik; he wants to do nothing but drape himself over an air-conditioner and sip soothing mint juleps." No reader could miss the admonition against guilty pleasures. The editorial continued:

> The inventor of this gimmick, by trying to link suggestible mental variables to scientific facts, qualifies as an enemy of the human race. In the future millions who never noticed the heat and humidity will suffer guilt feelings if they do not experience the misery they are statistically entitled to. Insecure souls will take the D.I. as a criterion of the response they ought to give the weather in order to be accepted as normal individuals. Let's stamp out the D.I. before we become a nation of thermohumidipaths![111]

With this kind of ridicule the Weather Bureau quickly renamed the discomfort index the "temperature-humidity index," a clumsy substitute that invited a naming contest, with submissions such as thermidity, humi-table, humisery, climature, and Hevener's humiture, which actually caught on for a while. Canadians, avoiding U.S. meteorological hegemony, use the term "humidex." The temperature-humidity index eventually became the "heat index" (HI).[112]

Perhaps the idea of a "comfort zone" or a "heat index" is peculiarly American. Italian journalist Beppe Severgnini thinks so. After a year in Washington, D.C., in the early 1990s, experiencing what he felt were overheated houses in the winter and overcooled offices in the summer, he remarked:

> During the summer, it is not sufficient to communicate infernal temperatures. There's also a comfort index, calculated from the combination of heat and humidity. And winter is not just a question of bitter cold. There is also the windchill factor, a temperature that takes wind-speed, and the consequent sensation of increased cold, into account. The phenomenon is well-known in Alpine valleys but the Americans have classified it and turned it into a science. Knowing the exact quantity of discomfort—being able to

say exactly how badly you feel and why—is the first step toward the goal of every U.S. citizen: to feel good.[113]

The continuous redefinition of the comfort zone is apparently still in progress. On a sultry summer day in June 2005, the *New York Times* sent a reporter armed with "a pair of professional-grade Mannix HDT303K digital thermometers" to compare the temperatures of several Manhattan clothing stores. He discovered a 12°F difference between Bergdorf Goodman's 68.3° and Old Navy's 80.3°. The more expensive the merchandise, the colder the store. The Levi's Store was 76.8°F, while Tiffany's was 70.3°. The reporter found little difference in restaurants; a luncheonette and a fancy bistro were both 68.7°, but a McDonald's was 72°. The lesson from a century of "climate control" is obvious: it may raineth on the just and the unjust, but the rich stay cooler and warmer and safer in all kinds of weather.[114]

What else could a good Weather Service do to protect its customers? Beginning in the 1950s, attention again turned to air pollution—not the billowing smoke stacks but the invisible chemicals emitted into the air from automobiles. Although the word "smog" had been coined by an English physician, Harold Des Voeux, in 1905, it became a symbol of the decay of American cities, a brownish cowl on an atmospheric grim reaper, and a visual reminder of the torrent of hazards—hydrocarbons, nitrogen oxides, ozone, ultraviolet rays, acid rain, global warming, El Niño—cascading over us every day.[115]

The American Meteorological Society recognized an atmospheric pollution problem after the first serious smog episode in Los Angeles in 1943 and the death of twenty citizens from poisoned air in Donora, Pennsylvania, in October 1948. These events also led to the passage of the Federal Air Pollution Control Act of 1955, which provided funds for research. As the public became aware of the severity of the problem, city and state governments began to react with laws on factory and automobile emissions. The federal government followed with the first Clean Air Act in 1963, which began the arduous process of establishing air quality standards. Adding to the urgency, New York City experienced deadly air pollution episodes in 1963 and 1965. Air pollution seemed to hold the public's attention more than any other atmospheric condition of the time, perhaps because it was seen as an urban phenomenon, affecting a majority of American city dwellers. As early as 1955, Helmut Landsberg, then chief of the Climatic Service of the U.S. Weather Bureau, summarized what was known about the influence of cities on weather and climate, including chemical pollution, decreased visibility, higher temperatures, and reduction of wind speeds. These generalizations would be modified over the next half century, but it quickly became evident that large cities make their own weather, and that the expansion of suburbs enlarged the "heat island." Seventeen years after his initial observations on the climate of cities, Landsberg described his "meteorologically utopian city" to an AMS-sponsored conference on urban environment. He envisioned "Metutopia" as a place with more trees, fewer cars, wider streets, and lower buildings. "In the future, if new settlements are really planned rather than scattered around at the whim of speculators, or as results of haphazard sprawl, . . . climatic information can offer some important guidance. Metutopia should certainly not be located in the Southwest or in southern California, nor in central or southern Appalachia."[116]

Utopia is, of course, "nowhere," but the quest for better temperatures and air in cities continues one building and one neighborhood at a time. In the mid-1980s, landscape architect Anne Whiston Spirn lauded the Dutch *woonerf,* a plan for residential streets intended to reduce air pollutants from automobiles by limiting them to certain areas of a neighborhood and planting trees along the pathways shared by cars and pedestrians. She also praised the effort by the city manager of Dayton, Ohio, to develop a comprehensive city plan based on weather and climate studies, one element of which was to control strong downtown winds by tree plantings.[117]

Twenty years later, cities are still struggling to limit urban sprawl and implement climate-suitable, "green" building practices, but the American Lung Association estimates that more than half of Americans live in counties where the air is rated "unhealthful" by Environmental Protection Agency (EPA) standards. Bakersfield, California, provides a case study of the failure to reduce air pollution. From 1981 to 1999 the population of Kern County increased 64 percent, from 342,000 to 563,500, while vehicle miles traveled grew 129 percent, from 7.2 million to 16.4 million per year. In 1999 the city began an eighteen-month review of the problem of air pollution that resulted in a plan, "Bakersfield Vision 2020," to lower automobile emissions. The plan encouraged pedestrian-friendly walkways and shopping centers and discouraged annexations and housing developments that created "islands." A "smart growth" plan, similar to the *woonerf,* reduced block lengths, narrowed streets, increased trees, and increased the number of houses per acre. Three years later, none of these proposals had been implemented, and city and country officials openly admitted that "there was 'no feasible mitigation' that could be imposed to reduce the enormous smog and traffic that will occur if the region grows as expected. . . . growth is more important than air quality." Dystopia is Bakersfield and thousands of other automobile-choked cities.[118]

Air pollution was framed as a moral issue in the 1960s. Morris Neiburger, professor of meteorology at UCLA, writing in the Protestant weekly *Christian Century,* while trying to remain optimistic about the standards set by the Clean Air Act of 1965, struck an apocalyptic note:

Does the atmosphere ever *really* rid itself of all contaminants? We do not know for sure. . . . But whether or not there is already an increase, it is clear that as the amount of pollution put into the atmosphere increases a stage will be reached at which the cleansing processes in the atmosphere are no longer adequate to purify the air before it reaches or returns to sources where it receives additional pollution. . . . Eventually the concentration of toxic substances will reach and exceed lethal concentrations, and life on earth will pass away.[119]

Ten days later the secular political journal the *New Republic* defined air pollution as "aerial garbage" and implied that the free market would solve the problem if the government would simply penalize those found guilty of dumping pollutants into the atmosphere. The impetus for its comment was the campaign by Eastern Airlines pilot William L. Guthrie, who, together with other pilots, sought to call attention to the permanent blanket of smoke blowing across the nation. Part of the problem, Guthrie charged, was the Weather Bureau's reporting of "smoke," the man-made mixture of soot and chemicals, as "haze," a natural composition of salt crystals from oceans and dust from the earth.[120]

Once again, the choice of words and the power to define them created controversy for the Weather Bureau that resulted in more technical distinctions and, perhaps, greater obfuscation. "Haze" is currently defined as "suspension in the atmosphere of extremely small, dry aerosols, which individually are invisible to the naked eye, but collectively give the sky an opalescent appearance; sometimes used in reference to the associated reduction in visibility, and, hence, considered as a lithometeor." A "lithometeor" is "anything in the atmosphere consisting of a visible concentration of mostly solid, dry particles that are lifted from the ground by the wind and/or suspended in the atmosphere." Dust and smoke are also lithometeors. For meteorologists, atmospheric pollution is anything injurious to humans, plants, animals, or property, whether natural or anthropogenic. "Contaminants" are substances that are not natural. The distinctions are important, but they obscure other problems.[121]

From the growing concern about air pollution emerged a new specialty in meteorology, air quality forecasting. Today, air quality forecasting is divided into two types: health alerts and emergency-response predictions. From the 1960s the Weather Bureau worked with municipal air quality agencies to predict pollution in urban areas. In the Clean Air Act of 1970, Congress mandated the achievement, within five years, of national ambient air quality standards necessary for the protection of public health and welfare. The EPA, created by Congress in the same year, was instructed "to publish an initial listing of 'hazardous' air pollutants within ninety days and then, within a year of its listing, to publish final emissions standard regulations." As legal scholar Richard J. Lazarus observes,

The statutory undertaking was enormous. It required strict regulation of twenty to forty thousand major stationary sources of air pollution and millions of cars and trucks being driven by average citizens and reductions of 275 toxic air pollutants, many of which were emitted by industries vital to local economies. The short timescale contemplated by the federal law necessarily precluded resolving the tremendous scientific uncertainty associated with the complex mechanisms of air pollution and its reduction.[122]

Nearly forty years later, many of the goals remain unmet. In 1970, EPA administrators needed answers to questions about the behavior of airborne particles that the Weather Service could not provide without further research. Forecasting pollution levels, like forecasting the weather, means educating health officials and the public about the inherent uncertainty arising from errors in data collection and modeling. A recent assessment of research needs for improved air quality forecasting lists dozens of specific topics about which scientists need more information, and, as the authors observe, "Since the 9/11 attacks, the emphasis on developing urban- and building-scale dispersion modeling capabilities has increased."[123]

Despite the uncertainties, scientists developed a "pollution standards index," which, like the discomfort index, was soon renamed. Each of five pollutants—sulfur dioxide, carbon monoxide, nitrogen dioxide, lead, and ozone—is monitored in all metropolitan areas with populations exceeding 200,000. A community's air is rated on a scale of 0 to 500, divided into six verbal and color descriptors: Good (green), Moderate (yellow), Unhealthy for Sensitive Groups (orange), Unhealthy (red), Very Unhealthy (purple), and Hazardous (black). When a sampling station records concentrations of any one of the five pollutants above the Good or Moderate level (above 100) for more than a few hours, an advisory is issued, warning sensitive groups and children to stay indoors and to avoid "outdoor exertion." Now known by the more positive-sounding name, "air quality index" (AQI), it is a daily feature of newspaper weather pages and radio and television weather reports, but its obvious limitations—inadequate sampling, subjective meaning of "exertion," and the unknown effects of combinations of pollutants—have led some states to amend the federal AQI with more detailed explanations and caveats. The state of Georgia, for example, cautions that "although it is uniform across the country, the AQI cannot be used as the sole method for ranking the relative healthfulness of different cities," but praises the AQI for allowing a television reporter to add details such as "carbon monoxide is usually only a problem during morning or evening rush hours with acceptable air quality expected during the rest of the day." Pollution is literally in the eye of the beholder.[124]

Little wonder, then, that some outside the weather communities of forecasters and pollution samplers found the discussion lacking urgency. In 1976, as the EPA was struggling to set air quality standards and atmospheric chemists were calling attention to the depletion of ozone over the South Pole, a young professor of English, Harold Fromm, living on a small farm about 15 miles south of Chicago, wrote of his experience, "On Being Polluted." Soon after moving to his farm, the hapless professor discovered that the drive to his job in a suburb of Gary, Indiana, was a "banquet of senses." "Every few miles has its own distinctive smell," he writes, "its own ambience: one moves along from the 'baked potato' smell to the 'chives' smell to the smell of burning garbage from the city dumps south of Chicago, to the Sherwin-Williams orgy of sights and smells." The prevailing winds bring smoke from the steel mills of Gary to his home, first causing irritations ("severe burning eyes, curious oppressive headaches, and a general strangeness"), then more troubling symptoms ("dizziness, nausea, tingling pains in the extremities, and a dazed, lethargic aimlessness"). Visits to doctors proved unsatisfactory; pills for burning eyes and antihistamines had unpleasant side effects. He contacts the Lake County Health Department and is told that there is no pollution in the county. He also learns that the filter on the average air-conditioner is incapable of removing most pollen and air pollutants. Adding insult to injury, Fromm soon finds out that other residents of the area are not aware of or bothered by the "smokes and smells." They have either grown accustomed to them or they are smokers. Obtaining a copy of the county health department's pollution report, he learns that its data were collected only at times when southerly winds blew pollution away. His letters to EPA director William Ruckelhaus are answered sympathetically and inform him that U.S. Steel has agreed to a five-year plan to reduce emissions. Fromm concludes that most people are unaware of air pollution because they live in cities in which it is "impossible for them to see the weather." What is needed, he thinks, is "a large scale survey . . . to determine what in fact is the relation between people's day-to-day physical and mental states and the quality of the available air."[125]

More than thirty years later, such knowledge is still fragmentary, and Americans' definition of a clear day depends on how much haze they are accustomed to. In March 2007 a fire that destroyed a railroad trestle in Sacramento, California, raised concerns about poisonous chemicals released into the air by the blaze. Monitoring of the air quality by the EPA and atmospheric scientists at the University of California, Davis, led to a disconcerting conclusion by one of the scientists: "The bad news is, the air was crappy. The good news is, it wasn't a big difference from what we usually see." Practical advice came from a spokeswoman for the Sacramento Metropolitan Air Quality Management District: "If you can smell it, you want to avoid it." The definition of pollution is also in the nose of the inhaler.[126]

Micrometeorology has emerged as an important subfield in the atmospheric sciences. Professor Fromm's frustrations over the government's lack of concern with the atmospheric conditions of a few acres of Indiana farmland may finally be addressed by instruments developed by NCAR and various defense contractors. In 2004, an extensive network of sensors installed around the Pentagon was tested. These instruments, including four-dimensional Doppler radar, Doppler lidar, anemometers, and chemical sensors, provide real-time data on the content and movement of air around the nation's military defense command post. If further tests prove the efficacy of such instruments, we can expect to see them installed in every major city in the near future. We may not need a weatherman to tell us which way the wind is blowing, but we will need high-resolution models to tell us when to close and seal our windows.[127]

Even before September 11, 2001, exposed the crucial importance of understanding the chemical hazards released by fires and explosions in densely populated urban areas, public health officials were becoming aware of the importance of microclimates and street-level wind currents. "Much new air pollution research," the *New York Times* reported,

> is based on the idea that there is no such thing as a regional air supply as it has been traditionally defined, but rather a series of microclimates, differing neighborhood by neighborhood and street by street. Each individual, some research suggests, carries around a personalized cloud of contaminants, depending on his or her environment, lifestyle, and consumer product choices—dubbed the "Pig Pen Effect," in honor of the unbathed character in the "Peanuts" comic strip.

Scientists at Columbia University's Lamont-Doherty Earth Observatory and the Harvard School of Public Health fitted forty high school student volunteers with air-sensor packs to wear for forty-eight hours. The data suggest that air differs significantly from block to block—different, but not necessarily better. As writer Joe Sherman concludes in, *Gasp!,* his engaging meditation on air, there is no place on earth where the air is totally free from contamination or disease. While Congress, the EPA, courts, industry, and public health advocates debate which aerosols to monitor and what the level of contaminants should be, we, the breathing masses, will wheeze while our lungs burn.[128]

Air pollution may be the most significant health factor in everyday weather, but there are others. Seasonal affective disorder (SAD) was identified and named in 1984 by South African psychiatrist Norman Rosenthal, who became lethargic and somewhat depressed after moving to New York City in the winter. Labeling the symptoms "winter blues," Rosenthal and his associates speculated that some people needed more sunlight than they were getting during the winter, whether because of shorter days and overcast skies or because they spent less time out-of-doors. Subsequent research suggests that SAD, as well as carbohydrate-craving obesity and premenstrual syndrome, is "affected by biochemical disturbances in two distinct biological systems. One system involves the hormone melatonin, which affects mood and subjective energy levels; the other involves the neurotransmitter serotonin, which regulates a person's appetite for carbohydrate-rich foods. Both systems are influenced by photoperiodism, the earth's daily dark-light cycle." Psychiatrist Peter S. Mueller treated a woman with cyclic bouts of depression with phototherapy, exposing her to 2,500 lux of full-spectrum light for a few hours daily. In less than a week her symptoms vanished. It is estimated that about 100 people in 100,000 suffer from SAD in northern regions of the United States, whereas only 6 in 100,000 who live in southern states show symptoms. More than twenty years after Rosenthal's initial investigations, commercial light boxes for the treatment of SAD are available, as are "dawn simulators," devices that gradually turn on bedroom lights to fool the body into thinking that it is experiencing sunrise—yet another advance in weather management.[129]

If too little sun is a problem for some, too much is dangerous for others. As the links between sunburn and skin cancers became better understood, new tools for sun management appeared on the market, notably, sun-blocking creams containing dozens of chemicals that prevent ultraviolet radiation from reaching exposed flesh. Developed during the 1940s to protect American troops from sunburn in the tropics, sunscreens were less popular than suntanning lotions until the 1970s. The white stripe of zinc oxide cream on the noses of lifeguards was the first warning sign for tan-seeking Americans of the potential risk of too much sun. As news of an increase in malignant melanoma cancer reached the public in the 1970s, sun protection factor (SPF) numbers became as familiar as the temperature-humidity index and windchill factor. Paradoxically, skin cancer rates continue to rise, leading some to doubt the effectiveness of sun blockers and others to develop the "UV index," to provide additional warning about the level of ultraviolet radiation in specific localities. Created in Canada in 1992 and introduced in the United States in 1994, the UV index is a 15+ point scale of UV radiation reaching the earth as measured in watts per square meter and adjusted for altitude and cloudiness. Like the AQI, it has a color-coded chart, from Green (UV 0–2, low danger to the average person) to Violet (UV 11+, extreme risk of harm from unprotected skin exposure). With weather maps color-coded for temperature and precipitation, AQI and UVI charts, and children's weather drawings, the daily newspaper weather page is now as colorful, crowded, and entertaining as the Sunday comics.[130]

For photophiles and photophobes, the air is rife with medical advice. *New York Times* health writer Jane Brody, reporting on the research of Michael F. Holick of the Boston University School of Medicine, notes that Holick recommends that a fair-skinned person living in the northern half of the country can build up enough vitamin D to last through the winter by exposing a quarter of his body surface (hands, arms, and face) to the sun for five to ten minutes a day between 11 A.M. and 3 P.M. A dark-skinned person may need five to ten times more exposure. Vitamin D, the "sunshine vitamin," is known to reduce the occurrence of osteoporosis, hypertension, diabetes, multiple sclerosis, rheumatoid arthritis, depression, and colon, prostate, and breast cancers. The use of a sunscreen with an SPF of 15 blocks 99 percent of the skin's ability to make vitamin D, according to Holick.[131]

As if the choice between wrinkled skin and a healthy prostate were not hard

"Hot Weather Charity." Water provided for horses only during a New York City heat wave, August 1911. Library of Congress, LC-USZ62-99958.

enough, psychologists have now identified a new category of disability, "severe weather phobia," defined as "an intense, debilitating, unreasonable fear of severe weather," specifically thunderstorms and tornadoes. Anxiety, insomnia, and panic are common symptoms, but the most common symptom, reported by 139 self-identified weather phobics, was obsessive monitoring of TV, radio, and Internet weather reports. Treatment workshops team psychologists with meteorologists to help the phobics sort out real dangers from imagined ones. A Web site, Storm-phobia Phobic Forum, created by a sufferer with the apposite name Zeus Flores, encourages everyone, from astraphobics to pluviophobics, to join in a discussion of their problems.[132]

Three final categories of everyday weather, from which most of us suffer, should be considered—excessive heat and cold, and lightning storms. Depending on the source of the data, all three have been shown to kill more people annually than any other kind of weather, including tornadoes and hurricanes. Record-breaking heat waves in the northern United States in the 1930s alerted meteorologists and physicians to the complex relationship between weather and health. Recognizing that "heatstroke" and "sunstroke" were imprecise terms, a Boston public health official, George C. Shattuck, proposed "heat hyperpyrexia" to describe "cases characterized by high bodily temperatures and hot, dry skin."[133]

Two major figures have emerged in the past thirty years who have defined, analyzed, and improved the management of extreme summer and winter weather: Stanley A. Changnon of the Midwestern Climate Center, Illinois State Water Survey; and Laurence S. Kalkstein, Center for Climate Research, University of Delaware. After the winter of 1977–1978, which was the worst on record in terms of low temperatures and snowfall, Changnon sampled seventy households (195 persons) in Champaign-Urbana and environs to determine how they had been affected. His questionnaire asked only for added costs due to the abnormal cold and snow. Predictably, 94 percent mentioned higher heating bills, but there were also increased expenditures for home repair, landscaping (shrubs killed by frost), vehicle repair and cleaning, warmer clothing, medicines, and "unscheduled use of motels." Changnon estimates, based on an average of $258 per household, that the cost to the region was $8.6 million, and to the state of Illinois, $1.05 billion, not including road and other infrastructure repair. It is also estimated that the state lost millions of dollars in income tax revenue. There were also significant social costs: travel delays, absences from work and school, missed social events and canceled vacations, delayed mail and garbage collection, extra labor for removing snow and housecleaning, additional pet care, injuries and illnesses, worry and anger. Fifty-two people died in the state, six in the sampled area. While sketchy, these findings suggest that even less severe winters add expenses. Since "normal" weather is what we think it should be in a particular season, any deviation has costs.[134]

After the deadly midwestern heat wave of mid-July 1995, which killed more than 500 in Chicago alone, Changnon turned his attention to the cost of extreme summer temperatures. The tragedy in Chicago, which is analyzed in Eric Klinenberg's readable *Heat Wave: A Social Autopsy of Disaster in Chicago,* caused meteorologists and climatologists to rethink the relative deadliness of various kinds of weather. Changnon, using journalistic, government, and business statistics, cautiously proposes that heat waves kill far more Americans annually, 1,000, than winter storms, which claim 130 to 200 lives each year. Other fatalities are caused by lightning, 100 to 156; floods, 100 to 160; tornadoes, 82 to 130; and hurricanes, 38 to 63. Extreme heat and cold are also the most lethal single events; the heat wave of 1980 killed an estimated 10,000 in the United States, while the winter storms of 1983 claimed roughly 500 lives. Heat, some claim, is a greater killer than air pollution, although some estimates of deaths from air pollution in the United States run as high as 70,000 annually. The problem, of course, is what a harried doctor or medical examiner lists as the *primary* cause of death, as geographer Randy S. Cerveny and his associates point out in a comparison of the databases of the Centers for Disease Control and Prevention (CDC) and the National Climatic Data Center (NCDC). Because of the way they categorize weather, collect their statistics, and define cause of death, the CDC's National Center for Health Statistics (NCHS) concluded that more people died of hypothermia (freezing) in the years 1979–1999 (13,970) than died from heat (8,015). The NCDC's storm data, on the other hand, are skewed toward heat-related deaths in the period 1993–2003 by a 3-to-1 margin. The CDC relies mainly on the deceased's death certificate, whereas the NCDC compiles its data from law enforcement agencies, emergency managers, and the media. Ascribing death to the weather is as subjective as defining the comfort zone.[135]

Noting that 73 percent of the heat-related deaths in Chicago in 1995 were of people over sixty-five, many of whom lived alone, Changnon surmises that the lower death tolls in the 1931 and 1936 Chicago heat waves were due to social conditions in which more of the elderly lived with families and were less afraid of crime, allowing them to sleep with windows and doors open or in parks and on beaches. The changing racial composition of the city was also a factor, since African Americans were 1.9 times more likely to die from heat than whites. Poverty, as Klinenberg discusses in detail, is always the primary factor in weather-related deaths, which helps explain the apparent paradox that deaths from cold occur frequently in the southern states while deaths from heat are common in the northern. In the 1995 Chicago heat wave, power consumption went up and there were power failures throughout the city, but "no one claimed that this loss of power and air-conditioning was a cause of heat-related deaths." Highways and railroads were damaged by the heat. The heat caused losses in Wisconsin dairy herds and poultry operations, but it boosted tourism in northern Wisconsin and the sale of air-conditioners throughout the region. Following the usual finger-pointing and recriminations, Chicago's mayor announced a new "heat warning plan" and a "weather center" to disseminate NWS information more quickly than before.[136]

More than twenty years after the NWS introduced its discomfort index, climatologist Laurence Kalkstein attempted a revision based on the relative severity of weather. As he points out, "An absolute discomfort index is place specific in its utility, as an apparent temperature of 40°C might hamper human activities in Minneapolis but have minimal effect in Phoenix." Kalkstein offers a "weather stress index" (WSI) that considers temperature, relative humidity, and wind speed that can be used in both summer and winter. Kalkstein credits the research of R. G. Steadman for developing the first algorithms that quantify variables such as human ventilation rate, surface radiation and convection, and clothing resistance to heat and moisture. Steadman came up with "apparent temperature," "the perceived air temperature for an individual." Using apparent temperatures from 100 evenly spaced first-order weather stations around the country, Kalkstein and his associates developed maps of WSI values, with some surprising results. Because of higher humidity, the most sultry and therefore most uncomfortable region of the country is a large swath of Texas, Oklahoma, and Nebraska, adding stress to those already stressed by living in Tornado Alley. Their relative index is based on the number of days in which the apparent temperature is either higher or lower than the mean apparent temperature of a specific place, date, and time.

Kalkstein and Valimont's WSI maps for winter reveal fewer surprises than their summer maps. The upper Midwest suffers the most from extreme cold and frigid winds, while a stressful day in Southern California or Florida might be 2°C (35.6°F). In short, weather stress varies from place to place, day to day, and hour to hour, and the stress ranges from uncomfortable to deadly. The question remains whether the WSI is measuring something we need.[137]

Lightning is the fifth horseman of the weather apocalypse. Lightning strikes are estimated to kill more than 100 Americans annually, 70–84 percent of them men. Unlike heat and cold, which cull the elderly and infirm, lightning strikes the young and athletic, often golfers trapped by a sudden storm on the links, boaters and fishermen on the water, and ball players on open fields. According to one study, the average age of a lightning victim is thirty-six. Scientific research on lightning, which began on a large scale in 1946 with teams of scientists from the Weather Bureau, the military, and the University of Chicago, expanded in the 1970s in dozens of federal agencies, including NASA, NOAA, the Nuclear Regulatory Commission, and many bureaus within the Departments of Agriculture, Defense, Interior, and Transportation. Lightning studies continue today in the work of NCAR, the National Severe Storms Laboratory, and several universities and private laboratories. The history of lightning research is, of course, closely tied to efforts to understand tornadoes, thunderstorms, flooding, and severe storms in general.[138]

A half century of study has yielded a model of a lightning storm that begins with the cold upper layer of air where particles of partly frozen water and ice crystals collide inside clouds. As Donald MacGorman of NSSL explains, "These collisions generate positive and negative electrical charges the same way that walking across a carpet on a dry day can build up static charges and result in a shock when a person touches a doorknob." As the negative particles descend, they cause the positive charges to jump to the earth to complete the circuit, inducing cloud-to-ground lightning strikes, though most lightning moves between clouds. Scientists still do not know why some storms produce rain, tornadoes, and hail, and others do not. Nor do they fully understand the distribution of positive and negative charges in the atmosphere. In June and July 2000, NSF funded an eight-week Severe Thunderstorm Electrification and Precipitation Study (STEPS) near Goodland, Kansas, and at the Yucca Ridge Field Station near Fort Collins, Colorado. At Yucca Ridge, scientists focused on the phenomena labeled "blue jets," "red sprites," "elves," and "trolls"—lights that can be seen in the atmosphere 40 to 60 miles above a cumulonimbus. As one of the researchers explains,

> Blue jets are extremely energetic fields of charged particles that rise up to 30 miles from the tops of clouds. After they occur, lightning stops for several seconds. Red sprites are striated glowing ribbons that rise high above the jets and reach the ionosphere, lasting 3 to 10 milliseconds. . . . Elves are thin,

expanding doughnuts of light found on the lower edge of the ionosphere. They tend to be above sprites after a lightning pulse. Trolls are propagating waves of energy that appear to come back out of cloud tops and hook up with sprites, but nobody really knows what they are.

Unknown, but not unnamed, like Luke Howard's classification, the process of understanding begins with names, continues with art, and may end in numbers.[139]

Lightning fatalities and property damage, like those caused by other weather hazards, may be reduced by better forecasting, warning, and public education. Meteorologists and several businesses are ready to help. Two investigators from the Center for Science and Technology Policy, Cooperative Institute for Research in the Environmental Sciences, University of Colorado, have expressed concern for the safety of spectators at large outdoor events such as football games and rock concerts. Noting that these events often draw between 60,000 and 110,000 people, who are confined in a relatively small space with little shelter and few exits, the analysts cite the National Fire Protection Association's guide for lightning protection, which lists panic as the number one safety concern. They urge event managers to use computer-based lightning monitoring systems, develop a timeline for decision making about evacuation, and print the plan in programs and placards. "Stormtracker," a system made by Boltek in Welland, Ontario, uses a roof antenna that picks up radio signals emitted by lightning, transmits them to a PC loaded with software that calculates the distance and direction of the strikes, and plots them on the screen. This system costs about $500 and can be enhanced by additional software to classify the types of lightning pulses into cloud-to-cloud and cloud-to-ground. Instruments are available for about $2,000 that warn of impending lightning. Spectrum Electronics advertises a handheld instrument, costing less than $600, that tracks thunderstorms within a radius of 75 miles.[140]

Some weather stress, discomfort, and suffering are inevitable, but a century of research and management by NWS meteorologists and other atmospheric scientists, engineers, physicians, and social scientists has provided a modicum of weatherproofing. Society has paid far less for knowledge of the weather than it has for its ignorance. Weather disasters, like all so-called natural disasters, are also failures of public policy. It is no easier to manage people than it is to manage the weather. Weather is a nagging reminder of both the achievements and the failures of technology; it is a sustainer of life and a constant irritant, both vital and trifling—a hurricane and a cat's-paw.

Conclusion

This book has been about the everyday experience of weather and the ways in which those experiences are perceived, marketed, and managed. I think I have demonstrated that weather has been a constant topic of discussion for more than a century because it is part of a broader conversation about regional and national identity, fear of chaos, and the search for order. Today the discussion has shifted from weather to climate, from the immediate to the more distant, but we should not forget that climate is just weather averages. We don't experience climate; we experience weather.

Mr. Dooley just scratched the surface when he declared that there are two kinds of weather, human weather and Weather Bureau weather. As we have seen, there are as many kinds of human weather as there are humans, and there are now many weather services. Weather is an old wine that is regularly poured into new bottles.

Weather matters—to our environment, our work, our health, our minds—and we show our appreciation of its importance by trying to understand it, control it, and celebrate it. In defining the weather, we define ourselves. The climates and weather systems of the continental United States are geographically and seasonally diverse; the winds that blow across North America are part of global systems that tie American weather to Asia and the Pacific, to Europe, Africa, and the Atlantic. We know this, but in order to comprehend weather, we reduce its vastness to meaningful components—a cold wind scouring a historic site in Montana, a computer model of wind vectors around the Pentagon, a satellite image of Hurricane Katrina.

Two arguments run through this book. Weather is an environmental universal, though its specific effects—heat, cold, humidity, winds—are localized. Wherever humans live, they confront the uncertainties of weather. They learn to balance what they want from weather with what they can get, or they perish. People from all cultures have invented ways they think they can lessen the role of chance and chaos in their weather, but they continue to be surprised, sometimes pleasantly, as when an El Niño brings rain to Florida in the usually dry season, and sometimes unpleasantly, as when hurricanes veer from a predicted path and strike the unprepared. Weather systems are so immense that, at present, we can only approximate their organization in maps and models. They remain, in that sense, chaotic. Chaos challenges us in different ways. A scientist wants to reduce uncertainty; an artist may want to enhance it for aesthetic or psychological effects. Weather stimulates creativity whatever the goal.

Weather as a source of metaphors for the human condition and for specific characteristics of life in the United States has been the second theme. Within the chaos of its ethnic and racial diversity, two features of American culture have always stood out: its emphasis on process (the constant creation and re-creation of ideas and things without regard for tradition or specific future goals) and interchangeability (modular design, whether in manufacturing or elective university courses or the formula of the weather report). Weather lends itself smoothly to these modes of action. Send up a balloon and see what we can find out about the atmosphere. Bring back the data and figure out ways to use them in agriculture, business, medicine, and the arts. Meteorology and pragmatism matured in the same era. Verification of theory by practice, in science or the arts, gives us the weather we know.

The U.S. Weather Bureau came into existence in 1891, chiefly because of scandals in the Army Signal Corps Weather Service and personality clashes, not because Congress or anyone else saw a pressing need for weather forecasting. Losses to shipping, agriculture, and other businesses from storms were costly, so any improvement in weather forecasting was appreciated, but few policy makers and even fewer citizens expected science to unravel the mysteries of the weather.

By 1900 that had changed. Under Willis Moore's aggressive leadership, the U.S. Weather Bureau became a household presence and a valued resource, first though newspapers, then telephones and, by 1921, radio. Five-day forecasts were being made throughout the United States by 1940. Despite low pay, sometimes-incompetent leadership, bureaucratic inertia, and occasional congressional and presidential meddling, the Weather Bureau and its weathermen became a part of the scientific establishment. Ridicule of weather forecasters continued, but the joke was on the skeptics. By the 1930s, transcontinental and transatlantic aviation made an understanding of the atmosphere an absolute necessity. The Weather Bureau struggled to keep up with advances in meteorology in Europe, but it excelled at promoting its work to the American public. In this it was aided by the American Meteorological Society, a professional association with a sense of history and the many influences of weather in American life. Through the media, the Girl Scouts and Boy Scouts, public education, and the arts, the Weather Bureau and the AMS helped Americans become weather-conscious in new ways.

The next step was to see the atmospheric sciences as central to environmental protection. The AMS recognized this from its beginning, but meteorologists were working to understand physical laws and create mathematical models and often neglected the larger ecological and cultural issues raised by weather. By changing the name of the Weather Bureau to the National Weather Service and placing it within the newly created National Oceanic and Atmospheric Administration (NOAA) in 1970, Congress seemed to recognize weather as part of the environment; unfortunately, the reorganization also subordinated the Weather Service to a larger bureaucracy. The possessive form (NOAA's) on the Internet site of "NOAA's National Weather Service" suggests that *its* weather service is just another ship contributing to the nation's, or, more likely, one federal agency's, rule of atmospheric and oceanic waves.

Depending on your perspective and needs, weather talk was trivial, serious, or urgent. City dwellers became less vulnerable to everyday weather; farmers and other outdoor workers remained weather-dependent; for fishermen and aviators, weather could be beneficial or fatal. With commercial aviation, men and women turned their eyes skyward, flew into weather's domain, and gained a new perspective on the earth and clouds from above. World War II and the rapid developments in science, especially computers, after the war, accelerated acceptance of meteorology as a science and weather as intrinsically valuable. More and more facets of daily life seemed weather-dependent, and demands arose for more specialized forecasts. Paradoxically, one of the things learned about weather is that it is chaotic. Chaos theory may hold more promise for humanists than for scientists, as artists and poets seek to show that human behavior is analogous to weather. They do not do this as artists did two centuries ago, with the belief that the heavens declare the glory of God, but with the post-Darwinian idea of chance creation. As William James put it in 1902, in *Varieties of Religious Experience,* "It is impossible, in the present temper of the scientific imagination, to find in the driftings of the cosmic atoms, whether they work on the universal or on the particular scale, anything but a kind of *aimless weather,* doing and undoing, achieving no proper history, and leaving no result."[1]

"Aimless weather," a metaphor for our human condition, is also the place in which we experience life, the place where we live, but from which we must emerge in order to understand and explain to others what that experience is. Weather is a place with both spatial and temporal dimensions. Some of its physical effects can be measured and quantified, but its boundaries can only be drawn artificially. When did we enter the cloud, and when did we leave? The forecaster's maps and the artist's compositions allow us to visualize external and internal weather. Weather, as objective and subjective experience, still requires, in the twenty-first century, the tools of meteorology and metaphysics. Weather could become another depressing topic on the daily news, but we contrive it to be both fearful and fun. Weather is comic. The windblown hat, the rain-soaked lovers, the sunburned tourist are predictable tableaux, partly because the predicaments are predictions ignored. We now expect to be protected from weather, even when we disregard it. The gap between Weather Bureau weather and human weather is not quite what Mr. Dooley thought it was. Human weather is not just Clancy's out-of-door leg; it is the weather of science and of art, which seeks to reveal the tortuous relationship between weather and humans.

This is not to claim a dichotomy of nature and culture. As David Nye writes on the interplay of landscape and technology: "The medium of the air itself is affected, carrying traces of smoke, microscopic particles such as pollen and carbon monoxide, and dust raised by our journeys."[2] Weather, the atmosphere, has been affected by human activities for as long as it has affected those same activities.

If the history of the National Weather Service, the American Meteorological Society, The Weather Channel, and dozens of related organizations teaches us anything, it is that the weather is not just about the state of the atmosphere and its effects on life. It is about people who, for one reason or another, find weather interesting, challenging, and inscrutable. It is about how they study, present, remember, and forget the weather; ultimately, it is about how weather is endured, avoided, marveled, and managed. Today, we have many communities of scientists, popularizers, and consumers of weather. These communities have been shaped by the men and women who created and sustained them—Cleveland Abbe, A. Lawrence Rotch, Willis Moore, Mrs. Ross Woods, Charles F. Brooks, Dorothy Scarborough, Carl-Gustaf Rossby, James Fiddler, Elizabeth Madox Roberts, Francis Reichelderfer, George R. Stewart, Eric Sloane, Robert M. White, Wallace Stevens, George Cressman, Ted Fujita, John Day, Jack Borden, Joel Myers, Valerie Voss, Charles A. Doswell III, Robert Misrach, Dozier Bell, and many others.

My pantheon of weather gods is reserved for those who have changed our perceptions of weather in significant ways, or who have managed and marketed those perceptions creatively. Some contributions are bigger and longer-lasting than others, but all of them shaped our current weather chatter. Surrounding these creators of weather are clouds of lesser-knowns who have labored to collect, classify, and analyze the billions of thermometer, barometer, anemometer, and other instrument readings that remain the sacred texts of weather scientists and hobbyists alike. Weather is nothing if not numbers, sheets of statistics blowing in the wind like prayer flags. As historian Allen Guttmann observed of sports, weather has gone "from ritual to record."[3]

"Record," of course, has a double meaning: (1) the highest achievement in a sport or game; (2) the list of accomplishments or failures over the years, the history of a sport. Among the consequences for sports of an emphasis on records was a loss of spontaneity; nevertheless, players gained satisfaction from competing not only against each other but also against past champions. Something similar happened when weather was subjected to careful record keeping. Storms are matched by temperatures, amounts of precipitation, and wind speeds. Weather records contribute to climate models. What have been lost are the poetry of a weather aphorism and the uniqueness of a single storm.

Knowing the hottest, coldest, wettest, driest, windiest day, week, month, year, or century should make us realize something about weather that we often forget. Numbers cannot contain it because weather is always changing. Sometimes and in some places the changes are almost insignificant, but the ocean of air is restless, as are the schools of humans swimming in its depths. When the two collide, the results are never without consequence. It may be an instant of weather sublime, filled with reverence and awe at the beauty of a storm, or it may be the moment when the hurricane party runs out of booze and the winds and tides are still rising. Weather, we have seen, can evoke simultaneous and conflicting emotions. Weather, we have been told by weatherists (the purveyors of weather information), should be a challenge, but not too much of a challenge. When, in weather disasters, the challenge is too great and people die, we cease defining, and weather defines us.

We are taught to avoid the "pathetic fallacy," but storms *are* pitiless, the sea *is* cruel, clouds *do* weep, because weather only makes sense when we glimpse in it familiar signs of the cosmos of which humans and weather are both a part. Weather's share of this order consists of physical laws we are still striving to understand, while our human lives are governed, as the weatherman in Clint McCown's novel devoutly believes, by "imagination" and "accident," as well as "cause-and-effect." This is a litany that echoes Ishmael's—free will, chance, and necessity—in Herman Melville's *Moby Dick*. Freedom with order is certainly not a uniquely American dream, but it is intensely American. Weather, we find, provides evidence of both order and chance in life. Weather furnishes Americans with metaphors for their yearning to prove that they do not live in James's "aimless weather," that they are achieving a proper history and leaving a worthy result.

Weather, however, remains an aleatory event. Uncertainty shines through every forecast; chance casts shadows in every weather model. Luck is still a cherished American belief or Las Vegas wouldn't be sucking up all the water in eastern Nevada, and the Pequot Indians wouldn't have half a million citations on Google. Americans may have become more risk-averse over the past century, but the weather still reflects the way in which Americans think about nature. It's big, it's mean, and it takes luck to escape the worst consequences of severe weather. Weather teaches us to live with risk.[4]

Poets know, as Stanley Kunitz puts it, that

weather is a form of communication. There is an exchange between the self and the atmosphere that sets the tone for an entire day. The changeability, its overwhelming range of possibilities, exercises a more defined influence on human moods than perhaps anything else. There is no weather or promise of weather that is so alien to our inner responsiveness that we cannot match it with a degree of feeling. That's the curious aspect of it. That's why each morning, the first thing I look at is the sky, and that puts me in tune with the day.[5]

We hold conversations with nature in many languages, translating and interpreting as best we can, in the hope that we can understand one another. Actions, of course, may speak louder than words. Clean air is a kind of poem.

Weather is a source for creative expression and spiritual inspiration. If we do not literally "pray hard to weather," we acknowledge its creed in the way we choose to dress, build our houses, plan our outdoor activities, schedule weddings, decide where to live, and pay insurance premiums. Destructive weather—hurricanes, tornadoes, floods, and droughts—mimics the wrathfulness of gods. Even as nuclear weapons seemed to give humans supernatural power, hurricanes and thunderstorms are recognized to be more powerful still. Without the power and beauty of weather, life would be insipid. On this the believers and nonbelievers can agree.

Americans also look to weather as a kind of frontier, and like the receding wilderness, it is a place the significance of which we discover as it may be degrading. As philosopher Edward Casey expresses it: "The *atmosphere* is more thoroughly pervasive of wilderness than any other factor. . . . But precisely because we are so immersed in it and thus have little or no distance from it, the atmosphere is often the *last* factor to be grasped lucidly, if it is grasped at all. First in the order of experiencing is here last in order of knowing."[6] Weather consciousness may also be a product of our growing awareness of closed frontiers and the degradation of the planet. The atmosphere and the weather it creates are now seen as the last good place, *meteortopia,* where, in a moment of time and sliver of space, we glimpse the promise that life holds.

The secular matins and vespers we call the weather report began to bracket our days a century ago and will continue to do so until some other science promises a better way of mirroring our place in nature. The daily weather forecast is a call to prayer whose message is "heed our authority and the laws of nature, but expect to be surprised, and act for yourself."

We landed in a yak pasture in a snowstorm in an old Soviet Antonov-24 jet. We had been circling for several minutes when the intercom came on and the pilot said: "I think I see an opening in the clouds, this is our last chance." Our stomachs rushed to keep up with our bodies as we dropped through thick clouds, caught a glimpse of trees, felt a slight bump, and taxied to a stop. A smoother landing than many I've experienced in the United States. We were somewhere in the taiga of western Mongolia, almost 400 miles from Ulaanbaatar, but only 15 from our destination, Khövsgöl Nuur. "Hurry and get out!" the charter pilot urged in slightly accented English. "I've got to take off before the wings ice up."

As dozens of bird watchers tumbled through the door murmuring their thanks for a safe landing, my wife and I spotted the jeep that had been sent to pick us up. While the shirt-sleeved birders waited for their bus, flapping their arms in the cold, bewildered by this June weather, we sped through a forest of Siberian larch and spruce on a snow-covered dirt road toward the lake. Through the window of the lurching vehicle we glimpsed a white crane, possibly a demoiselle, staring at us, dazed, through a veil of snowflakes each bigger than a dime. The image of that bird, with its thickly feathered, snow-covered body and improbably long and frail legs, remains one of my most vivid weather memories.

As I think of it now, the unexpected bird in the snow is one more reminder of the astounding paradoxes, the sardonic humor of weather.

Notes

Introduction

1. William Lowell Putnam, *The Worst Weather on Earth: History of the Mount Washington Observatory* (Gorham, NH, and New York: Mount Washington Observatory and the American Alpine Club, 1991).

2. The literature on the cultural and political dimensions of climate change and global warming is vast, but I highly recommend Richard P. Horwitz, "Americans' Problem with Global Warming," *American Studies* 45, no. 1 (Spring 2004): 5–37; David G. Victor, *Climate Change: Debating America's Policy Options* (New York: Council on Foreign Relations, 2004); Spencer R. Weart, *The Discovery of Global Warming* (Cambridge, MA: Harvard University Press, 2003); Michael H. Glantz, *Climate Affairs: A Primer* (Washington, DC: Island Press, 2003); Clark A. Miller and Paul N. Edwards, eds., *Changing the Atmosphere; Expert Knowledge and Environmental Governance* (Cambridge, MA: MIT Press, 2001); William K. Stevens, *The Change in the Weather: People, Weather, and the Science of Climate* (New York: Delta, 1999); James Rodger Fleming, *Historical Perspectives on Climate Change* (New York: Oxford University Press, 1998); Myanna H. Lahsen, "Climate Rhetoric: Construction of Climate Science in the Age of Environmentalism" (Ph.D. diss., Rice University, 1998).

The rise of the risk and hazard management bureaucracy, especially FEMA, needs serious historical study. For a beginning, see Richard A. Posner, *Catastrophe: Risk and Response* (New York: Oxford University Press, 2004); David A. Moss, *When All Else Fails: Government as the Ultimate Risk Manager* (Cambridge, MA: Harvard University Press, 2002); Daniel Sarewitz, Roger A. Pielke Jr. and Radford Byerly Jr., eds., *Prediction: Science, Decision Making, and the Future of Nature* (Washington, DC: Island Press, 2000); Ted Steinberg, *Acts of God: The Unnatural History of Natural Disaster in America,* 2nd ed. (New York: Oxford University Press, 2006); Rutherford H. Platt, *Disasters and Democracy: The Politics of Extreme Natural Events* (Washington, DC: Island Press, 1999); Michele L. Landis, "Fate, Responsibility, and 'Natural' Disaster Relief: Narrating the American Welfare State," *Law & Society Review* 33, no. 2 (1999): 257–318; W. J. Maunder, *The Uncertainty Business: Risks and Opportunities in Weather and Climate* (London: Methuen, 1986).

3. Donald R. Whitnah, *A History of the United States Weather Bureau* (Urbana: University of Illinois Press, 1961); Kristine C. Harper, "Boundaries of Research: Civilian Leadership, Military Funding, and the International Network Surrounding the Development of Numerical Weather Prediction in the United States" (Ph.D. diss., Oregon State University, 2003); Frederik Nebeker, *Calculating the Weather: Meteorology in the 20th Century* (San Diego: Academic Press, 1995); James R. Fleming, *Meteorology in America, 1800–1870* (Baltimore: Johns Hopkins University Press, 1990); James R. Fleming, ed., *Historical Essays on Meteorology, 1919–1995* (Boston: American Meteorological Society, 1996); Mark Monmonier, *Air Apparent: How Meteorologists Learned to Map, Predict, and Dramatize Weather* (Chicago: University of Chicago Press, 1999); William B. Meyer, *Americans and Their Weather* (New York: Oxford University Press, 2000); David Laskin, *Braving the Elements: The Stormy History of American Weather* (New York: Anchor Books, 1996). To this list I would add Vladimir Jankovi, *Reading the Skies: A Cultural History of English Weather,*

1650–1820 (Chicago: University of Chicago Press, 2000), a fascinating story of early modern science; and Sarah Strauss and Ben Orlove, eds., *Weather, Climate, Culture* (New York: Berg, 2003), a collection of essays on what the editors call "the anthropology of weather and climate." Gary Alan Fine, *Authors of the Storm: Meteorologists and the Culture of Prediction* (Chicago: University of Chicago Press, 2007).

Chapter One. Talking about Weather

1. F. P. Dunne, "Mr. Dooley: On the Weather Bureau," *Harper's Weekly* 45 (July 27, 1901): 746, http://app.harpweek.com/viewarticletext.asp?webhitsfile+hw19010727000066%2E. Being a "professor" has always been a dubious distinction in many contexts. Daguerreotypists and stage hypnotists were also titled "professor" in the nineteenth century.

2. This passage from the organic act creating the U.S. Weather Bureau is Section 3 (26 *Statutes at Large* 653) and is reprinted in Donald Whitnah, *A History of the United States Weather Bureau* (Urbana: University of Illinois Press, 1961), 60.

3. For the National Weather Service mission statement, see its Web site, www.weather.gov (accessed February 14, 2006).

4. The early history of meteorology is well told in James R. Fleming, *Meteorology in America, 1800–1870* (Baltimore: Johns Hopkins University Press, 1990). For the more recent period, see Frederik Nebeker, *Calculating the Weather: Meteorology in the 20th Century* (San Diego: Academic Press, 1995); and Mark Monmonier, *Air Apparent: How Meteorologists Learned to Map, Predict, and Dramatize the Weather* (Chicago: University of Chicago Press, 1999). Early Weather Bureau history is chronicled in the pages of the *Monthly Weather Review;* analyzed in Donald Whitnah, *A History of the United States Weather Bureau;* and recounted in a sanitized form in Patrick Hughes, *A Century of Weather Service: A History of the Birth and Growth of the National Weather Service, 1870–1970* (New York: Gordon and Breach, 1970). The only full-length biography of Abbe is Truman Abbe, *Professor Abbe and the Isobars: The Story of Cleveland Abbe, America's First Weatherman* (New York: Vantage Press, 1955). Useful sketches of Abbe's importance may be found in Edmund P. Willis and William H. Hooke, "Cleveland Abbe and the Birth of the National Weather Service, 1870–1891," *Proceedings of the International Commission on History of Meteorology* 1, no. 1 (2004): 48–54; and Edmund P. Willis and William H. Hooke, "Cleveland Abbe and American Meteorology, 1871–1901," *Bulletin of the American Meteorological Society* 87 (2006): 315–326. Some background on meteorological education is available in William A. Koelsch, "From Geo- to Physical Science: Meteorology and the American University, 1919–1945," in *Historical Essays on Meteorology, 1919–1995,* ed. James R. Fleming (Boston: American Meteorological Society, 1996), 511–539; and Koelsch, "Pioneer: The First American Doctorate in Meteorology," *Bulletin of the American Meteorological Society* 62 (1981): 362–367.

5. The Weather Bureau was not unique, of course, in its attempt to create a public constituency from consumers of weather information that would support its work, chiefly by electing legislators friendly to government science. All the new federal agencies at the beginning of the twentieth century—for example, the Forest Service, the Park Service, and the Reclamation Service ("service" was a key word in the Progressive Era, although the Reclamation Service later became the Bureau of Reclamation)—spent a significant percentage of their budgets on public education.

6. J. M. L. Stump, "A Word about Weather," *Meteorologist: Published Monthly in the Interest of the Science of Meteorology* 1, no. 10 (December 1879): 1.

7. Cleveland Abbe, "Lectures on Meteorology," *Monthly Weather Review* 28 (April 1900): 164.

8. "The Brooklyn Museum of Meteorology," *Monthly Weather Review* 28 (April 1900): 163–164. D. T. Maring, "Weather Bureau Kiosks," *Monthly Weather Review* 37 (March 1909): 89–90.

9. Harvey Maitland Watts, "The Weather vs. the Newspapers," *Popular Science Monthly* 58 (February 1901): 387.

10. Harvey Maitland Watts, "The Forecaster and the Newspaper," in *Proceedings of the Second Convention of Weather Bureau Officials,* ed. Willis L. Moore (Washington, DC: Government Printing Office, 1902), 43–53. This article is a condensation of a booklet with the same title published in 1901. Monmonier, *Air Apparent,* 164–167; "Weather in Newspapers," *Bulletin of the American Meteorological Society* 4 (1923): 116. Promotion of the Weather Bureau through the press is part of a larger story of science in the public sphere ably explored by Dorothy Nelkin, *Selling Science: How the Press Covers Science and Technology,* rev. ed. (New York: Freeman, 1995); and Marcel C. LaFollette, *Making Science Our Own: Public Images of Science 1910–1955* (Chicago: University of Chicago Press, 1990).

11. Oliver L. Fassig, "The Weather Bureau," *Bulletin of the American Meteorological Society* 14 (1933): 112. This remark, by the first American Ph.D. in meteorology, was made in the context of a defense of the bureau after sharp criticisms from the American Society of Civil Engineers in 1933.

12. Sarah Strauss, "Weather Wise: Speaking Folklore to Science in Leukerbad," in *Weather, Climate, Culture,* ed. Sarah Strauss and Ben Orlove (New York: Berg, 2003), 43; Willis L. Moore, *Moore's Meteorological Almanac and Weather Guide 1901* (Chicago: Rand McNally, 1900), 32–35, 41–42; Stephen Kern, *The Culture of Time and Space 1880–1918* (Cambridge, MA: Harvard University Press, 1983).

13. Charles F. Brooks and Frances V. Tripp, "Phenology: Responses of Life to the Advance in the Seasons," *Bulletin of the American Meteorological Society* 6 (1925): 47–48; "Phenological Surveys in the U.S.," *Bulletin of the American Meteorological Society* 17 (1936): 55–56; "New Beginning of 'Natural Spring' Reported," *Bulletin of the American Meteorological Society* 87 (2006): 727–728. The University Corporation for Atmospheric Research is helping to organize a national phenology network, Project BudBurst. See http://www.windows.ucar.edu/citizen_science/budburst/.

14. Richard Inwards, *Weather Lore: A Collection of Proverbs, Sayings, and Rules Concerning the Weather* (London: Elliot Stock, 1898), i.

15. H. H. C. Dunwoody, *Weather Proverbs,* Signal Service Notes no. 9 (Washington, DC: Government Printing Office, 1883), 5, 41, 45.

16. Edward B. Garriott, "Weather Folklore and Local Weather Signs," *U.S. Weather Bureau Bulletin,* No. 33 (Washington, DC: Government Printing Office, 1902), 148, 8.

17. Franz Boas, "Introduction," in *The Handbook of North American Indians,* Bulletin 40, pt. 1 (Washington, DC: Smithsonian Institution, 1911), 25–26. The subsequent history of the misunderstanding of Eskimo snow vocabulary is told by Laura Martin, "'Eskimo Words for Snow': A Case History in the Genesis and Decay of an Anthropological Example," *American Anthropologist* 88, no. 2 (June 1986): 418–423. William Churchill, "Weather Words of Polynesia," in *Memoirs of the American Anthropological Association,* vol. 2, pt. 1 (Lancaster, PA: New Era Printing, 1907), 5–98. "Pogonip," *Monthly Weather Review* 22, no. 2 (February 1894): 76. *Pogonip* is a Paiute word for a dense winter fog composed of ice crystals. Captain John P. Finley, "Meteorological Terms Used in the Philippines," *Monthly Weather Review* 25, no. 6 (June 1907): 272. Albert Matthews, "The Term Indian Summer," *Monthly Weather Review* 20, no. 1 (January 1902): 19–28; 20, no. 2 (February 1902): 69–79. For an interesting account of the conversation about "Indian Summer," see Adam Sweeting, *Beneath the Second Sun: A Cultural History of Indian Summer* (Hanover, NH: University Press of New England, 2003).

18. W. J. Humphreys, *Ways of the Weather: A Cultural Survey of Meteorology* (Lancaster, PA:

Jacques Cattell, 1942), 1. The poem first appears in Humphreys's *Weather Proverbs and Paradoxes* (1923; Baltimore: Williams and Wilkins, 1934), 4, and it is possible that he is the original parodist.

19. Marion Dix Mosher, comp., *More Toasts: Jokes, Stories and Quotations* (New York: Wilson, 1922), http://www.gutenberg.org/files/15338/15338-h/15338-h.htm (accessed September 22, 2005).

20. Ambrose Bierce, *The Devil's Dictionary*, 1911, reprinted in *The Devil's Advocate: An Ambrose Bierce Reader,* ed. Brian St. Pierre (San Francisco: Chronicle Books, 1987), 320.

21. "Talking about the Weather," *Independent,* July 17, 1913, 126–127.

22. Ibid., 127. On pluviculture, see Clark Spence, *The Rainmakers: American "Pluviculture" to World War II* (Lincoln: University of Nebraska Press, 1980); Carroll Livingston Riker, *Power and Control of the Gulf Stream* (New York: Baker and Taylor, 1912). Riker's scheme will be discussed in chapter 2.

23. J. Warren Smith, "Speaking of the Weather," in *Yearbook of the Department of Agriculture, 1920* (Washington, DC: Government Printing Office, 1921), 202; Jeffrey K. Lazo and Lauraine G. Chestnut, "Economic Value of Current and Improved Weather Forecasts in the U.S. Household Sector," Stratus Consulting, Inc., November 22, 2002. I thank Dr. Lazo for e-mailing me a copy of his report. Charles F. Brooks, secretary of the American Meteorological Society, claimed that the cost per capita in the United States was only two cents in 1922. Charles F. Brooks, "Reclassification," *Bulletin of the American Meteorological Society* 3 (1922): 164.

24. Emerson Hough, "Does the Weather Bureau Make Good?" *Everybody's Magazine* 20, no. 5 (May 1909): 617; emphasis in original. See also Hough's follow-up, "Our 'Foozling' Weather Bureau," *Everybody's Magazine* 22, no. 4 (April 1910): 568–569.

25. Nebeker, *Calculating the Weather,* 40.

26. "International Meteorological Symbols," *Monthly Weather Review* 26, no. 7 (July 1898): 312.

27. U.S. Department of Agriculture, Weather Bureau, *Instructions for Voluntary Observers,* 2nd ed. (Washington, DC: Government Printing Office, 1901), 26–27.

28. U.S. Department of Agriculture, Weather Bureau, *Instructions for Voluntary Observers* (Washington, DC: Government Printing Office, 1892), 9. The institutional history of the voluntary observers is complex. The Smithsonian volunteer network was utilized by the Signal Corps when it became responsible for meteorological work in 1870, then became part of the state weather services organized in 1883 by Lieutenant H. C. Dunwoody. The control of the state services and voluntary observers did not pass to the Weather Bureau until 1895, when they were merged into the Climate and Crop (later Climatological) Service of the bureau. See Eric R. Miller, "The Evolution of Meteorological Institutions in the United States," *Monthly Weather Review* 59, no. 1 (January 1931): 4.

29. Charles D. Reed, "Criteria of Efficiency for an Unpaid Climatological Service," *Bulletin of the American Meteorological Society* 17 (1936): 113. The debate over the reliability of the Cooperative Observer Network (COOP) and the periodic call for replacing human observations with automated data collection continue, but the need for the human observers is vigorously supported by the American Association of State Climatologists. See Paul G. Knight, "Ensuring the Future of the Nation's Climate Observing Network," *Bulletin of the American Meteorological Society* 88 (2007): 834–836. For criticism of COOP, see Christopher Daly, Wayne P. Gibson, George H. Taylor, Matthew K. Duggett, and Joseph I. Smith, "Observer Bias in Daily Precipitation Measurements at United States Cooperative Network Stations," *Bulletin of the American Meteorological Society* 88 (2007): 899–912; and Roger Pilke Sr. et al., "Documentation of Uncertainties and Biases Associated with Surface Temperature Measurement Sites for Climate Change Assessments," *Bulletin of the American Meteorological Society* 88 (2007): 913–928.

30. The figures for the years before 1900 come from Whitnah, *A History of the United States Weather Bureau,* 21. Later figures are from *Instructions for Voluntary Observers* (1901), 5; J. B.

Kincer, "Our Veteran Cooperative Observers," *Monthly Weather Review* 63, no. 11 (November 1935): 313; Sue Halpern, "The People's Weather Service," *Mother Jones,* November/December 2001, 34–38; the Cooperative Weather Observers Web site, www.coop.nws.noaa.gov; and http://www .weather.gov/mission/shtml (accessed February 14, 2006).

31. Of 76 cooperative observers in Illinois in 1922, 15 were educators (6 college, 9 secondary), and 14 were farmers. There were 6 bankers, 4 post office clerks, 4 jewelers, 4 "executives and clerks," and 4 merchants. Among the other 18 occupations were physicians, engineers, and a librarian. "Professions of Co-operative Observers in Illinois," *Bulletin American Meteorological Society* 3 (December 1922): 172. For pride and frustration, see the correspondence from Edward L. Wells of Portland, Oregon, who compared the cooperative observers to the doughboys of the American Expeditionary Force in World War I: little praised, but essential. *Bulletin of the American Meteorological Society* 3 (1922): 128.

32. Kincer, "Our Veteran Cooperative Observers," 314; *Bulletin of the American Meteorological Society* 19 (1938): 17. Kirkwood's record length of service was eclipsed in 1980 by Edward H. Stoll of Elwood, Nebraska, who became a cooperative observer on October 10, 1905. Carl E. Feather, "Meet Betty Thompson the Weather Lady of Grady," *Goldenseal* 32, no. 4 (Winter 2006): 62–63.

33. Mrs. Ross Woods, "Duties and Experiences of a Cooperative Observer," *Bulletin of the American Meteorological Society* 9 (1928): 5–6.

34. "Weather Bureau Terms Used to Designate Storms," *Monthly Weather Review* 43, no. 8 (August 1915): 395. The article concluded with an attempt to minimize the importance of tornadoes in comparison to cyclones. "*Tornado*—This name is applied to certain storms of well-known characteristics. While they occur in connection with certain cyclonic systems and exhibit great intensity, they are, nevertheless, of extremely local geographic extent and of very short duration" (396). The language reflects the ongoing controversies within the weather communities of the period over the nature and predictability of tornadoes. I will return to this in chapter 5.

35. Eleanor Buynitzky, "The Meaning of the Word 'Fair' in Meteorology," *Monthly Weather Review* 43, no. 12 (December 1915): 613–614; Michael Mogil's "Partly Cloudy vs. Partly Sunny," www.weatherworks.com (accessed March 13, 2006). On the use of "fair" by forecasters in the Chicago area, see Gary Alan Fine, *Authors of the Storm: Meteorologists and the Culture of Prediction* (Chicago: University of Chicago Press, 2007), 159–160.

36. Charles F. Marvin, "Nomenclature of the Unit of Absolute Pressure," *Monthly Weather Review,* 46, no. 2 (February 1918): 73. For definitions of these terms, see the AMS Glossary of Meteorology, http://amsglossary.allenpress.com.

37. Marvin, "Nomenclature of the Unit of Absolute Pressure," 74.

38. "News Items, Notes, and Queries," *Bulletin of the American Meteorological Society* 1 (1920): 13.

39. Nebeker, *Calculating the Weather,* 188–194; Robert Marc Friedman, *Appropriating the Weather: Vilhelm Bjerknes and the Construction of Modern Meteorology* (Ithaca, NY: Cornell University Press, 1989).

40. Kristine C. Harper, "Boundaries of Research: Civilian Leadership, Military Funding, and the International Network Surrounding the Development of Numerical Weather Prediction in the United States" (Ph.D. diss., Oregon State University, 2003), 59–63.

41. Jack C. Thompson, "Weather Prediction at the Local Weather Bureau Office as Concepts from the Bergen School Came to the U.S.," *Bulletin of the American Meteorological Society* 66 (1985): 1251. Harper writes that in the late 1930s only 27 percent of U.S. Weather Bureau professional employees had college degrees, and only five held degrees in meteorology. Harper, "Boundaries of Research," 71.

42. Thompson, "Weather Prediction," 1251.

43. Ibid., 1252.

44. Gustavus A. Weber, *The Weather Bureau: Its History, Activities and Organization* (New York: Appleton, 1922), 9–10, 24–36; Edward L. Wells, "The Weather Bureau and the Homeseekers," in *Yearbook of the U.S. Department of Agriculture, 1904* (Washington, DC: Government Printing Office, 1905), 353–358; Whitnah, *A History of the United States Weather Bureau*, 167–171. For a good brief history of the crop bulletins, see Thomas R. Heddinghaus and Douglas M. Le Comte, "A Century of Monitoring Weather and Crops: *The Weekly Weather and Crop Bulletin*," *Bulletin of the American Meteorological Society* 73 (1992): 180–186. The first snow surveys were done by Dr. James E. Church of the University of Nevada in the winter of 1908–1909 in the Lake Tahoe watershed. The Weather Bureau snow surveyors were forced to use less efficient Marvin snow samplers, and the bureau eventually lost this activity to the Soil Conservation Service. See Bernard Mergen, *Snow in America* (Washington, DC: Smithsonian Institution Press, 1997), 121–157.

45. "The Lobby of the News Building," *Bulletin of the American Meteorological Society* 12 (1931): 141–142.

46. The *Printer's Ink* article is excerpted in the *Bulletin of the American Meteorological Society* 18 (1937): 183–184.

47. "Interview of Joseph Smagorinsky," May 16, 1986, University Corporation for Atmospheric Research/American Meteorological Society, NCAR Library and Archives, Boulder, Colorado, 1. Smagorinsky received B.S., M.S., and Ph.D. degrees from New York University, worked on numerical meteorology at the Princeton Institute for Advanced Study with Jule Charney and John Von Neumann, then headed up the Weather Bureau's work on fluid dynamics until 1982.

48. G. Emmons, "Suggestions for Improved Presentation of Weather Information to the Public," *Bulletin of the American Meteorological Society* 21 (1940): 311–312.

49. Ibid., 313–315.

50. H. Landsberg, "Weather Forecasting Terms," *Bulletin of the American Meteorological Society* 21 (1940): 317–320.

51. Ibid., 317.

52. "The Weather," *Fortune*, April 1940, 146.

53. John Sherrod and Hans Neuberger, "Understanding Forecast Terms—Results of a Survey," *Bulletin of the American Meteorological Society* 39 (1958): 36. The results of a similar survey conducted by a professor of psychology at the University of Toronto suggest that Canadian students had about the same level of understanding of forecast terms as those in the United States despite the admission by 33 percent that they hardly ever read or listened to weather reports. "University of Toronto Poll of Students on Weather Terminology," *Bulletin of the American Meteorological Society* 30 (1949): 61–62.

54. George P. Cressman, "Public Forecasting—Present and Future," in *A Century of Weather Progress* (Boston: American Meteorological Society, 1970), 71–77. Cressman graduated from Penn State and received his Ph.D. from Chicago. He worked as a meteorologist for the air force, then worked for the Joint Numerical Weather Prediction Unit, 1954–1960, before returning to the bureau and becoming its chief from 1965 to 1979, overseeing the administrative transition to NOAA in 1970.

55. Ibid., 75.

56. Donald D. Dunlop, "To Meet the Needs of a World at Play," in *A Century of Weather Progress*, 78–81.

57. George P. Cressman, "Uses of Public Weather Services," *Bulletin of the American Meteorological Society* 52 (1971): 544.

58. C. S. Ramage, "Prognosis for Weather Forecasting," *Bulletin of the American Meteorological Society* 57, no. 1 (January 1976): 4–10; Ramage, "Further Outlook—Hazy," *Bulletin of the American Meteorological Society* 59, no. 1 (January 1978): 18–21.

59. *Climates of the States*, vol. 2 (Detroit, MI: Gale, 1985), 1559–1563.

60. C. S. Ramage, "Forecasting in Meteorology," *Bulletin of the American Meteorological Society* 74 (1993): 1864.

61. Frank Trippett, "The Weather: Everyone's Favorite Topic," *Time,* February 6, 1978, 76. Americans are hardly exceptional in their weather talk. Trevor A. Harley, a psychologist at the University of Dundee, Scotland, writes that "Britain's weather is particularly fascinating because it is so variable and unpredictable throughout the year." He classifies British weather talk into four major categories: (1) future weather, (2) severe weather, (3) places to live having good weather, and (4) nostalgia for white Christmases and other childhood weather experiences. Strauss and Orlove, *Weather, Climate, Culture,* 103.

62. Bob Pifer and H. Michael Mogil, "NWS Hazardous Weather Terminology," *Bulletin of the American Meteorological Society* 59 (1978): 1583.

63. Ibid., 1583–1586.

64. Earl G. Droessler, "The Weather Forecaster Today," *Bulletin of the American Meteorological Society* 61 (1980): 195. The NWA, with about 3,000 members in 2006, is considerably smaller than the AMS, with more than 12,000. Created primarily to represent private sector weather forecasters and applied meteorologists who often feel there is too much emphasis on research and academic meteorology in the AMS, the NWA has struggled to establish itself but now has the *Electronic Journal of Operational Meteorology* to legitimize its activities. See http://www.nwas .org/history/index.php (accessed March 21, 2005).

65. Robert W. Burpee and Leonard W. Snellman, "Editorial," *Weather and Forecasting* 1, no. 1 (June 1985): 3. As of 2006, the AMS published nine journals in addition to the *Bulletin,* which still publishes technical articles but has recently been redesigned to appeal to the general reader in the manner of *Scientific American.* Many meteorologists see weather as fundamental to all life on earth, and AMS meetings increasingly address a wide range of environmental issues.

66. Allan H. Murphy, "What Is a Good Forecast? An Essay on the Nature of Goodness in Weather Forecasting," *Weather and Forecasting* 8 (June 1993): 281.

67. Ibid., 283. Murphy uses the feminine pronoun when referring to the forecaster.

68. "Probability of Precipitation (PoP)," in *Glossary of Weather and Climate,* ed. Ira Geer (Boston: American Meteorological Society, 1996), 178. See also the attempts to clarify this in John D. Cox, *Weather for Dummies: A Reference for the Rest of Us* (New York: Hungry Minds, 2000), 21, in which "measurable amount" is defined as .01 inch or more. Cox then advises, "If there is a 50 percent chance that rain will develop anywhere in the forecast area, and 40 percent of the area is expected to get rain, this translates into a 20 percent chance of rain." Environment Canada's Web site tries to be user-friendly by translating the percentages into expectations, thus a 20 percent chance becomes "No precipitation is expected," while a 40 percent PoP is translated into "An umbrella is recommended," and a 60 percent forecast means the odds favor Mother Nature watering your lawn. www.msc.ec.gc.ca (accessed April 16, 2006).

69. Environment Canada Web site, www.msc.ec.gc.ca.

70. Allan H. Murphy and Robert L. Winkler, "Probability Forecasts: A Survey of National Weather Service Forecasters," *Bulletin of the American Meteorological Society* 55 (1974): 1449–1452.

71. Allan H. Murphy, Sarah Lichtenstein, Baruch Fischhoff, and Robert L. Winkler, "Misinterpretations of Precipitation Probability Forecasts," *Bulletin of the American Meteorological Society* 61 (1980): 695–701.

72. Frederick P. Ostby, "Improved Accuracy in Severe Storm Forecasting by the Severe Local Storms Unit during the Last 25 Years: Then versus Now," *Weather and Forecasting* 14 (August 1999): 538. "The SELS definition of a severe thunderstorm is one that produces hail ¾ in. in diameter or greater, or surface wind gusts of 50 kt or more [sic] which is 58 miles per hour or greater, or thunderstorm winds that cause significant damage." Ostby, "Improved Accuracy,"

526. "Significant" remains undefined. Ostby's wind speed of 50 knots equals 75.4 miles per hour, or hurricane force on the Beaufort scale, which is higher than the speed of 58 miles per hour given in the definition of "severe thunderstorm" in Geer, *Glossary of Weather and* Climate, 199. This is another indication that definitions are loose and shifting, further confusing the public.

73. Chuck Doswell, "Thoughts on the 'NOAA' Weather Radio during Storm Chase '98," http://www.flame.org/~cdoswell/NWR_rant.html (accessed July 10, 2003). Note that Dr. Doswell signs his professional publications with his full name, Charles A. Doswell III, and his personal Web site pieces as "Chuck." Support for Doswell's scorn for NOAA radio comes from emergency managers, who estimate that only 5 to 15 percent of residents in storm-prone areas have the special receivers. See "National Alert System in Disarray," CBS News, June 28, 2004, www.cbsnews.com/stories/2004/06/28/terror/printable626452.shtml (accessed September 13, 2007).

74. Christopher Holder, Ryan Boyles, Peter Robinson, Sethu Raman, and Greg Fishel, "Calculating a Daily Normal Temperature Range That Reflects Daily Temperature Variability," *Bulletin of the American Meteorological Society* 87 (2006): 769–774.

75. "Chuck Doswell, "Misconceptions about What Is 'Normal' for the Atmosphere," http://www.cimms.ou.edu/~doswell/Normals/normal.html (accessed July 10, 2003).

76. Chuck Doswell, "A Jaundiced Look at NOAA and NWS Management," http://www.flame.org/~cdoswell/forecasting/NOAA_management.html.

77. Chuck Doswell, "Meteorology and Users," http://www.cimms.ou.edu/~doswell/Users.html.

78. Roberta Klein and Roger A. Pielke Jr., "Bad Weather? Then Sue the Weatherman! Part I: Legal Liability for Public Sector Forecasts," *Bulletin of the American Meteorological Society* 83 (2002): 1791–1799. Even private forecasters have been protected to some extent; see Robert Klein and Roger A. Pielke Jr., "Bad Weather? Then Sue the Weatherman! Part II: Legal Liability for Private Sector Forecasts," *Bulletin of the American Meteorological Society* 83 (2002): 1801–1807. For a sociological study of meteorologists and forecasters and how they communicate with the public, see Fine, *Authors of the Storm.*

79. Roger A. Pielke Jr., "Weather Policy? What's That?" *WeatherZine* 17 (August 1999), http://sciencepolicy.colorado.edu/zine/archives/1–29/17/editorial.html (accessed July 16, 2005).

80. "NCAR Laboratory Dedication," *Bulletin of the American Meteorological Society* 48 (1967): 442–446; "A Short History of NCAR," http://www.ncar.ucar.edu/ncar/hist.html (accessed May 18, 2003); and brochures available at NCAR. For a full analysis of Pei's design, see Carter Wiseman, *I. M. Pei: A Profile in American Architecture* (New York: Abrams, 1990), 72–91.

81. Roger A. Pielke Jr., "Asking the Right Questions: Atmospheric Sciences Research and Societal Needs," *Bulletin of the American Meteorological Society* 78 (1997): 262. See also Roger A. Pielke Jr., "Societal Aspects of Weather: Report of the Sixth Prospectus Development Team of the U.S. Weather Research Program to NOAA and NSF," *Bulletin of the American Meteorological Society* 78 (1997): 867–876; William H. Hooke and Roger A. Pielke Jr., "Short-Term Weather Prediction: An Orchestra in Need of a Conductor," in *Prediction: Science, Decision Making, and the Future of Nature,* ed. Daniel Sarewitz, Roger Pielke Jr., and Radford Byerly Jr. (Washington, DC: Island Press, 2000), 61–83; Roger A. Pielke Jr. and R. E. Carbone, "Weather Impacts, Forecasts, and Policy: An Integrated Perspective," *Bulletin of the American Meteorological Society* 83 (2002): 393–403; "Statement of Dr. Roger A. Pielke, Jr., to the Committee on Environment and Public Works of the United States Senate," March 13, 2002. I thank Dr. Pielke for providing a copy of his testimony.

82. James Gleick, *Chaos: Making a New Science* (New York: Penguin Books, 1987), 8.

83. Ibid., 22.

84. Ibid., 322. Lorenz presented his paper in 1972, not 1979.

85. Edward Lorenz, *The Essence of Chaos* (Seattle: University of Washington Press, 1993), 15.

"A Sound of Thunder" was published in hardcover in Ray Bradbury, *The Golden Apples of the Sun* (Garden City, NY: Doubleday, 1953).

86. George R. Stewart, *Storm* (New York: Random House, 1941), 44.

87. Lorenz, *Essence of Chaos,* 15.

88. Edward Lorenz, "Some Reflections on the Arrival of Numerical Weather Prediction," *Annual Review of Earth and Planetary Sciences* (Palo Alto, CA: Annual Reviews, 2004). Jack C. Thompson, in recalling his early career in the Weather Bureau offices in San Francisco, implies that he was the model for Stewart's nameless protagonist, "The Junior Meteorologist." If this is true, then there is a possibility that Stewart heard the sneezing Chinese theory from Major Edward Bowie, meteorologist-in-charge, San Francisco Weather Bureau. See Thompson, "Weather Prediction," 1252.

89. Charles F. Brooks, "Inversion!" *Bulletin of the American Meteorological Society* 22 (1941): 141.

90. Ibid.

91. William S. Franklin, "A Much Needed Change of Emphasis in Meteorological Research," *Monthly Weather Review* 46 (October 1918): 451.

92. William S. Franklin, *A Calendar of Leading Experiments* (South Bethlehem, PA: Franklin, McNutt, and Charles, 1918), 164. This is a paraphrase of an observation by the British philosopher John Ruskin.

93. *The National Cyclopædia of American Biography,* vol. 22 (New York: James T. White, 1932), 140.

94. A. A. Tsonis and J. B. Elsner, "Chaos, Strange Attractors, and Weather," *Bulletin of the American Meteorological Society* 70 (1989): 14.

95. *The Compact Edition of the Oxford English Dictionary* (New York: Oxford University Press, 1971), 273; *Webster's New International Dictionary of the English Language,* 2nd ed. (Springfield, MA: Merriam, 1936), 450.

96. Napier Shaw, *Manual of Meteorology,* vol. 1 (New York: Cambridge University Press, 1926), 123.

97. Jeffrey Rosenfeld, "The Butterfly That Roared," *Scientific American Presents the Weather,* Spring 2000, 24.

98. Brad Leithauser, *Darlington's Fall* (New York: Knopf, 2002), 242.

99. Peter L. Bernstein, *Against the Gods: The Remarkable Story of Risk* (New York: Wiley, 1998), 202, 332.

100. Susan Allen Toth, *Leaning into the Wind: A Memoir of Midwest Weather* (Minneapolis: University of Minnesota Press, 2003), 117.

101. Gaston Bachelard, *Air and Dreams: An Essay on the Imagination of Movement* (Dallas: Dallas Institute of Humanities and Culture, 1988), 196. This is a translation of *L'Air et les songes, essai sue l'imagination du movement* (Paris, 1943). Toby Miller, "Tomorrow Will Be . . . Risky and Disciplined," in *Reality Squared: Televisual Discourse on the Real,* ed. James Friedman (New Brunswick, NJ: Rutgers University Press, 2002), 215; Lucian Boia, *The Weather in the Imagination* (London: Reaktion Books, 2005), 184. A brief overview of weather in some of the world's religions is provided by Serinity Young, "Religion and Weather," in *Encyclopedia of Climate and Weather,* vol. 2, ed. Stephen Schneider (New York: Oxford University Press, 1996), 639–643.

102. *Webster's New International Dictionary of the English Language,* 2nd ed. (Springfield, MA: G. and C. Merriam, 1936), 27; *Black's Law Dictionary,* pocket edition (Berkeley, CA: West Group, 1996), 13; Ted Steinberg, *Acts of God: The Unnatural History of Natural Disaster in America,* 2nd ed. (New York: Oxford University Press, 2006).

103. Russell R. Dynes and Daniel Yutzy, "The Religious Interpretation of Disaster," *Topic* 10 (Fall 1965): 34–48.

104. Leland W. Mann, "The Packers Take Up Theology," *Christian Century,* June 20, 1934, 833–

834. Earlier in the decade a prominent liberal minister wrote: "No imaginable connection exists . . . between man's inward spiritual state and a rainstorm. . . . Evidently this still needs to be said in this benighted and uncivilized country. The crude, obsolete supernaturalism which prays for rain is a standing reproach to our religion." Dr. Harry Emerson Fosdick, "Does Prayer Change the Weather?" *Christian Century,* September 10, 1930, 1084, quoted in William B. Meyer, *Americans and Their Weather* (New York: Oxford University Press, 2000), 169.

105. According to various media reports, Michael Marcavage of Repent America blamed gays for the hurricane, while Louis Farrakhan of the Nation of Islam saw Katrina as punishment for the U.S. invasion of Iraq. Douglas Brinkley quotes New Orleans mayor Ray Nagin as saying, "God is mad at America." *The Great Deluge: Hurricane Katrina, New Orleans, and the Mississippi Gulf Coast* (New York: William Morrow, 2006), 204.

106. "I admit I have prayed because droughts, the lack of rain, drives me batty what with having come from farm families way back when, and owning a farm at present, an ancient prayer to be sure." Jim Harrison, *Off to the Side: A Memoir* (Boston: Atlantic Monthly Press, 2002), 121–122. I am equally sure that baseball fans pray hard to stop rain.

107. Lance Morrow, "The Religion of Big Weather," *Time,* January 22, 1996, 72.

108. Barbara C. Farhar, "A Societal Assessment of the Proposed Sierra Nevada Snowpack Augmentation Project" (typescript of report for Bureau of Reclamation, U.S. Department of the Interior, 1977), 106; Barbara C. Farhar and Julia Mewes, *Social Acceptance of Weather Modification: The Emergent South Dakota Controversy,* monograph no. 23 (Boulder, CO: Institute of Behavioral Science, Program on Environment, Technology, and Man, 1976), 50; J. Eugene Haas, "Sociological Aspects of Weather Modification," in *Weather and Climate Modification,* ed. W. N. Hess (New York: Wiley, 1974), 800.

109. Brian McCammack, "Hot Damned America: Evangelicalism and the Climate Change Policy Debate," *American Quarterly* 59, no. 3 (September 2007): 645–668.

110. Christopher H. Achen and Larry M. Bartels, "Blind Retrospection: Electoral Responses to Droughts, Flu, and Shark Attacks" (unpublished paper, January 2004), 1. I thank Professors Achen and Bartles for sending me a copy of their work in progress. Their paper was mentioned in the *New Yorker,* among other magazines.

111. Ibid., 34.

112. Ibid., 25.

113. Ibid., 26–27.

114. Ibid., 29. As early as 1920, the economist Alvin H. Hansen noted a correlation between rainfall and politics in Minnesota, noting that "the lower the precipitation the larger is the nonpartisan vote," but he attributed this to a social philosophy arising from climatic conditions. Hansen, "The Effect of Rain on Politics," *Independent,* August 28, 1920, 251. Likewise, a study of the Populist Party vote in the 1890 election in Nebraska attributed the party's victory to drought but concluded, "To suggest that the farmer held the politician responsible for the shortage of rainfall would be an unwarranted exaggeration for the thoughtlessness of the voters." John D. Barnhart, "Rainfall and the Populist Party in Nebraska," *American Political Science Review* 19, no. 3 (August 1925): 540.

115. "Reasons for Not Voting, by Sex, Age, Race and Hispanic Origin, November 2000," U.S. Census Bureau, Internet release date, February 27, 2002. See also "Vote or Stay Dry?" *Bulletin of the American Meteorological Society* 86 (2005): 10. The Democrats were quick to use Katrina to their advantage; see Adam Nagourney, "Democrats Looking to Use New Orleans as G.O.P Used 9/11," *New York Times,* April 22, 2006, A11.

116. "Original Melted Snowman," Ace Enterprises, DeLand, Florida, 1982, patent pending.

117. Frederick Chambers, Brian Page, and Clyde Zaidins, "Atmosphere, Weather, and Baseball: How Much Further Do Baseballs Really Fly at Denver's Coors Field?" *Professional Geographer* 55,

no. 4 (2003): 491–504. A study of the effect of humidity on baseballs concluded that keeping the balls at 90°F and 40 percent humidity might actually make the ball livelier and give hitters an advantage. David Kagan and David Atkinson, "The Coefficient of Restitution of Baseballs as a Function of Relative Humidity," *Physics Teacher* 42 (September 2004): 89–92. In 2006, batters complained that Coors Field had become a pitchers' park; see Lee Jenkins, "At Coors Field, It's Not the Hits, It's the Humidor," *New York Times,* August 17, 2006, C15, C17.

118. For a brief description of "Smog," see Jean Belch, comp., *Contemporary Games,* vol. 1 (Detroit: Gale, 1973), 410; and "Daniel Zwerdling, "On a Dirty Day You Can Play Forever," *New Republic,* August 29, 1970, 11. "The Weather" is described in an advertisement in *Weatherwise,* April 1983, 86. On "Eastern Front," see David Myers, "Chris Crawford and Computer Game Aesthetics," *Journal of Popular Culture* 24, no. 2 (Fall 1990): 17–32. For "Second Front," see Ed Dille, "Second Front: Germany Turns East," *Video Games and Computer Entertainment,* March 1991, 114, 116; Fantasy Weather League, http://weathergeek.net/LEAGUE.html (accessed May 11, 2006).

119. Charles F. Talman, "Weather Toys," *Mentor* 14 (August 1926): 59–60.

120. S. E. Schlosser, "Drought Buster," American Folklore Web site, http://www.american-folklore.net/folktales/ne.html (accessed November 10, 2006). For additional Feboldson and other weather lore, see B. A. Botkin, *A Treasury of American Folklore: Stories, Ballads, and Traditions of the People* (New York: Crown, 1944), 227–230, 279–281.

121. Joanne Palmer Bollenbacher, "'But I Don't See Anything!' Weathering That First Sonogram," *Parents,* November 1998, 278.

122. *New Yorker,* March 19, 2001, 12; April 7, 2003, 52; April 14, 2003, 59.

123. All cartoons from the Library of Congress Prints and Photographs Online Catalog. Opper's Zephyr Trust may have been inspired by a satiric piece in the reform journal the *Arena* in 1899, which reported on a landowner in Kansas who sold plots for the building of Olathe, then charged its citizens for breathing his air, the rights to which he had retained. This well-crafted article makes some telling points about the legality of split estates, science, and the courts. Bolton Hall, "The Lords of the Air," *Arena* 21, no. 3 (March 1899): 293–295. Cartoons about the environmental issues of the twenty-first century continue the reform tradition.

124. *Mark Twain's Speeches,* ed. Albert Bigelow Paine (New York: Harper, 1929), 53–57.

125. Mark Twain, *The American Claimant* (New York: Charles L. Webster, 1892), 644. M. Quad was the pseudonym of Charles Bertrand Lewis (1842–1924). Twain also quotes a passage of purple weather prose from Ernest William Hornung, author of the popular Raffles books.

126. My version is a synthesis of several versions I have heard and seen on the Internet. I thank Jack Borden for first telling me this joke.

127. James E. Cook, *Dry Humor: Tales of Arizona Weather* (Baldwin Park, CA: Gem Guides Book, 1992); e-mail from Sarah Brown to author, October 17, 2004; e-mail from Andrew Mergen to author, March 11, 2004.

128. See www.theonion.com/content/note/27461 (accessed June 2, 2004); "Moojita Scale," http://operations.mesonet.org/~nebain/weather/humor/moojita.htm (accessed February 20, 2008).

129. *National Lampoon,* February 1984, 50–51; obituary for Don Knotts, *Washington Post,* February 26, 2006, C8.

130. William Ferrel, *A Popular Treatise on the Winds* (New York: Wiley, 1889); Douglas Archibald, *The Story of the Earth's Atmosphere* (New York: D. Appleton, 1897); Mark Harrington, *About the Weather* (New York: D. Appleton, 1899); Moore, *Moore's Meteorological Almanac;* Edwin C. Martin, *Our Own Weather: A Simple Account of Its Curious Forms, Its Wide Travels, and Its Notable Effects* (New York: Harper and Brothers, 1913); Alexander McAdie, *Man and Weather* (Cambridge, MA: Harvard University Press, 1926); Willis L. Moore, *The New Air World: The Sci-*

ence of *Meteorology Simplified* (Boston: Little, Brown, 1922); Charles F. Brooks, *Why the Weather?* (1924; revised and enlarged edition, New York: Harcourt, Brace, 1935).

131. Stuart Mackenzie, "Don't Apologize for Talking about the Weather!" *American Magazine* 101, no. 2 (February 1926): 44–46, 198–200; Mackenzie, "When Winter Comes," *American Magazine* 101, no. 3 (March 1926): 38–40, 161–166; Mackenzie, "How Many Tons of Rain Will Fall on Your Town?" *American Magazine* 101, no. 4 (April 1926): 38–39, 199–204; Mackenzie, "Wind! A Star Performer in the Drama of American Life," *American Magazine* 101, no. 5 (May 1926): 40–41, 201–206; Mackenzie, "Our Tornadoes Are the Fiercest of All Storms," *American Magazine* 101, no. 6 (June 1926): 38–40, 80–88. The series is interesting for a number of reasons. The topics of the articles are obviously timed to the changing seasons, from winter to tornado. Coordination with advertising is evident in the March issue, where an ad for "Southern California, Year 'Round Vacation Land Supreme," appears with the article on winter. Finally, Mackenzie credits Charles Talman for all the information in the series. Charles Fitzhugh Talman, *The Realm of the Air: A Book about Weather* (Indianapolis: Bobbs-Merrill, 1931); W. J. Humphreys, *Weather Rambles* (Baltimore: Williams and Wilkins, 1937); Humphreys, *Ways of the Weather*.

132. Hans H. Neuberger and F. Briscoe Stephens, *Weather and Man* (New York: Prentice-Hall, 1948); Irving P. Krick and Roscoe Fleming, *Sun, Sea, and Sky: Weather in Our World and in Our Lives* (Philadelphia: Lippincott, 1954); Eric Sloane, *Eric Sloane's Weather Book* (New York: Little, Brown, 1952); David Ludlum, *Early American Hurricanes, 1492–1870* (Boston: American Meteorological Society, 1963); Paul E. Lehr, R. Will Burnett, and Herbert S. Zinn, *Weather: Air Masses, Clouds, Rainfall, Weather Maps, Climate* (New York: Golden Press, 1957, 1975, 1987); Philip D. Thompson, Robert O'Brien, and the editors of Time-Life Books, *Weather* (Alexandria, VA: Time-Life Books, 1965, 1968, 1973, 1975, 1980).

133. Cox, *Weather for Dummies;* Mel Goldstein, *The Complete Idiot's Guide to the Weather,* 2nd ed. (Indianapolis: Alpha, 2002); Robert Henson, *The Rough Guide to Weather* (New York: Penguin Putnam, 2002); Jack Williams, *USA Today Weather Almanac* (New York: Vintage, 1995). The American Meteorological Society annual report for 2005 mentions that Williams has been hired to write the AMS's own popular weather book to be used in outreach programs "and other educational efforts to promote scientific literacy."

134. Ford A. Carpenter, *The Climate and Weather of San Diego California* (San Diego: San Diego Chamber of Commerce, 1913); Ben Gelber, *The Pennsylvania Weather Book* (New Brunswick, NJ: Rutgers University Press, 2002); Jon Nese and Glenn Schwartz, *The Philadelphia Area Weather Book* (Philadelphia: Temple University Press, 2002). University presses seem to see a market for local weather books; see Harry P. Bailey, *Weather of Southern California* (Berkeley and Los Angeles: University of California Press, 1966); George Bomar, *Texas Weather* (1983; 2nd ed., Austin: University of Texas Press, 1995); Val Eichenlaub, *Weather and Climate of the Great Lakes Region* (South Bend, IN: University of Notre Dame Press, 1979); Harold Gilliam, *Weather of the San Francisco Bay Region* (Berkeley and Los Angeles: University of California Press, 1962); Jerry Hill, *Kentucky Weather* (Lexington: University Press of Kentucky, 2005); Thomas W. Schmidlin and Jeanne Appelhaus Schmidlin, *Thunder in the Heartland: A Chronicle of Outstanding Weather Events in Ohio* (Kent, OH: Kent State University Press, 1998); George H. Taylor and Raymond H. Hatton, *The Oregon Weather Book: A State of Extremes* (Corvallis: Oregon State University Press, 1999); Morton D. Winsberg, *Florida Weather,* 2nd ed. (Gainesville: University Press of Florida, 2003); Gregory A. Zielinski and Berry D. Klein, *New England Weather, New England Climate* (Hanover, NH: University Press of New England, 2003).

135. For the total employment, see Sky-Fire Web site, http://sky-fire.tv/index.cgi/careers.html (accessed June 9, 2006). For bloggers, see Rebecca Dana, "Forecast: Plenty of Activity on the Weather Blog Front," *Washington Post,* August 15, 2004, D1–D2; Nolan Doesken, "Let It Rain," *Weatherwise* 60, no. 4 (July/August 2007): 51–55.

136. Maria Aspan, "Everybody Talks about the Weather; All of a Sudden, It's Controversial," *New York Times,* June 4, 2007, C1, C4. See also "Frequently Asked Questions of Cumul.us," http://www.cumul.us/help/.

137. Quoted in W. C. Devereaux, "A Meteorological Service of the Future," *Bulletin of the American Meteorological Society* 29 (1939): 215.

138. Quoted in ibid., 218.

Chapter Two. Managing Weather

1. For recent controversies and the articulation of opposition in rural New York, Nantucket, and West Virginia, see Stanley Fish, "Blowin' in the Wind," http://fish.blogs.nytimes.com/2007/08/26/blowin-in-the-wind/; Glenn Collins, "Windmills on Their Minds," *New York Times,* August 28, 2002, A16; Karen Lee Ziner, "Offshore Harvest of Wind Is Proposed for Cape Cod," *New York Times,* April 16, 2002, D3; Felicity Barringer, "Debate over Wind Power Creates Environmental Rift," *New York Times,* June 6, 2006, A14; Anthony L. Rogers, James F. Manwell, and Sally Wright, "Wind Turbine Acoustic Noise," Renewable Energy Research Lab, University if Massachusetts, Amherst, January 2006, 18, www.ceere.org/rerl/publications/whitepapers/WindTurbineNoise Issues.pdf (accessed July 7, 2006).

2. Amanda S. Adams and D. W. Keith, "Climate Impacts of Increased Wind Energy," *Bulletin of the American Meteorological Society* 88 (2007): 307–309.

3. More than 200 letters to the PSC are available on the Protect Pendleton Web site, http://www.protectpendleton.com (accessed August 18, 2007).

4. *Wind Energy Resource Atlas,* http://rredc.nrel.gov/wind/pubs/atlas/chap3.html#east (accessed August 15, 2007); Public Service Commission of West Virginia, Case No. 05-1740-E-CS, http://www.psc.state.wv.us/webdocket/default.htm (accessed August 18, 2007).

5. Committee on Environmental Impacts of Wind-Energy Projects, *Environmental Impacts of Wind-Energy Projects* (Washington, DC: National Academies Press, 2007), http://books.nap.edu/catalog.php?record_id=11935 (accessed August 18, 2007).

6. For good, but dated, fog maps and a discussion of the changing terminology of thick and dense fog, see Arnold Court and Richard D. Gerston, "Fog Frequency in the United States," *Geographical Review* 56, no. 4 (October 1966): 543–550.

7. Peter Barnes, *Who Owns the Sky? Our Common Assets and the Future of Capitalism* (Washington, DC: Island Press, 2001), 14, 16–17. Breathing 15,000 times a day yields a rate of 10.4 breaths a minute, a resting rate. By other estimates we breathe about 19,000 times a day. See Joe Sherman, *Gasp! The Swift and Terrible Beauty of Air* (Washington, DC: Shoemaker and Hoard, 2004), 3.

8. Barnes, *Who Owns the Sky?* 88.

9. Karen Ordahl Kupperman, "The Puzzle of the American Climate in the Early Colonial Period," *American Historical Review* 87 (1982): 1262–1289; Kupperman, "Fear of Hot Climates in the Anglo-American Colonial Experience," *William and Mary Quarterly* 41, no. 2 (1984): 213–240; Joyce E. Chaplin, "Climate and Southern Pessimism: The Natural History of an Idea, 1500–1800," in *The South as an American Problem,* ed. Larry J. Griffin and Don H. Doyle (Athens: University of Georgia Press, 1995), 57–82; Conevery Bolton Valenčius, *The Health of the Country: How American Settlers Understood Themselves and Their Land* (New York: Basic Books, 2002).

10. Bushrod W. James, *American Resorts; with Notes upon Their Climate* (Philadelphia: F. A. Davis, 1889); Willis L. Moore, *Climate: Its Physical Basis and Controlling Factors,* Weather Bureau Bulletin no. 34 (Washington, DC: Government Printing Office, 1904), 14; Ellsworth Huntington, *Civilization and Climate,* reprint of the 3rd rev. ed. (Hamden, CT: Archon Books, 1971), 220–221 (originally published New Haven, CT: Yale University Press, 1915; 3rd rev. ed., 1925).

11. Frances V. Tripp, "Heliotherapy," *Bulletin of the American Meteorological Society* 6 (1925): 33–34; Ellsworth Huntington and C. F. Brooks, "Is Colorado Sunshine Injurious?" *Bulletin of the American Meteorological Society* 6 (1925): 121–122.

12. Charles F. Brooks, "Local Climates of Worcester, Mass., as a Factor in City Zoning," *Bulletin of the American Meteorological Society* 4 (1923): 83–85.

13. "Nice sells the sun in winter, as Manchester sells cotton fabric." V. Conrad, "Problems of Health-Resort Climatology," *Bulletin of the American Meteorological Society* 27 (1946): 152–154.

14. Cleveland Amory, "Turnabout in Arizona," *Saturday Evening Post,* January 22, 1944, 72. See also Victoria Clark, *How Arizona Sold Its Sunshine: The Historical Hotels of Arizona* (Sedona, AZ: Blue Gourd, 2004).

15. Donald A. Beattie, "The Early Years of Federal Solar Energy Programs," in *History and Overview of Solar Heat Technologies,* ed. Donald A. Beattie (Cambridge, MA: MIT Press, 1997), 23–33; Arthur Inkersley, "Sunshine as Power," *Sunset,* April 1903, reprinted in *The Early Sunset Magazine, 1898–1928,* ed. Paul C. Johnson (San Francisco: California Historical Society, 1973), 32–33.

16. Verner E. Suomi, "Atmospheric Research for the Nation's Energy Program," *Bulletin of the American Meteorological Society* 56 (1975): 1060, 1063; emphasis in original.

17. Ibid., 1063.

18. Peter L. Palmer, "Modeling Crop Water Use in the Pacific Northwest: The USBR AgriMet Program" (paper presented at the Western Snow Conference, South Lake Tahoe, California, April 20, 1999), 1, 3; William J. Broad, "A Web of Sensors, Taking Earth's Pulse," *New York Times,* May 10, 2005, D1, D4. The increasingly complicated relation of science and society is illustrated in the conflict that arises when researchers want to put sensors in wilderness areas where federal law forbids man-made structures and equipment. "To conform to the Wilderness Act, an artificial pine tree is being developed at the University of Nevada, Reno. This 'tree' camouflages the data collection and telemetry equipment, thus allowing it to be installed in sensitive areas." John A. Kleppe and William J. Norris, "Fallen Leaf Lake: A Microcosm of Lake Tahoe," *Proceedings of the Western Snow Conference,* April 19–22, 1999, 30.

19. Suomi, "Atmospheric Research for the Nation's Energy Program," 1068.

20. Beattie, "The Buildup Years: 1974–1977," 45–108; Frederick H. Morse, "The Growth Years: 1977–1980," 109–137; Morse, "The Contraction and Redirection Years: 1981–1988 (and Beyond)," 139–180, all in *History and Overview of Solar Heat Technologies.* A Btu (British thermal unit) is a unit of energy with several definitions, but it is used in the United States to describe the heat value of fuels.

21. Optimism is expressed in Marilyn Berlin Snell, "Power Lunch," *Sierra* 67, no. 4 (July/August 2002): 28–39.

22. The history of wind power is well told in Robert Righter, *Wind Energy in America: A History* (Norman: University of Oklahoma Press, 1996). Wind energy growth since the 1970s is covered in Paul Gipe, *Wind Energy Comes of Age* (New York: Wiley, 1995). Peter Asmus, a journalist, provides a lively and more recent account in *Reaping the Wind: How Mechanical Wizards, Visionaries, and Profiteers Helped to Shape Our Energy Future* (Washington, DC: Island Press, 2001). Estimates of the federal investment in solar and wind power come from Morse, "The Contraction and Redirection Years," 143, and Righter, *Wind Energy in America,* 180. Coal-fueled plants provide a little more than 50 percent of American electricity; nuclear plants 20 percent; natural gas 18 percent; hydropower 7 percent; oil 4 percent; and solar and wind about 1 percent. For a recent survey of nuclear energy, see Jon Gertner, "Atomic Balm?" *New York Times Magazine,* July 16, 2006, 36–47, 56, 62, 64.

23. E. Wendell Hewson, "Generation of Power from the Wind," *Bulletin of the American Meteorological Society* 56 (1975): 660–675. The Venturi effect, a local increase in wind as it blows through a narrow mountain pass, is named for the Italian physicist G. B. Venturi. Ira W. Geer, ed.,

Glossary of Weather and Climate (Boston: American Meteorological Society, 1996), 246; Matthew L. Wald, "It's Free, Plentiful and Fickle," *New York Times,* December 28, 2006, C4.

24. Robert L. Thayer and Carla M. Freeman, "Altamont: Public Perceptions of a Wind Energy Landscape," *Landscape and Urban Planning* 14 (1987): 379–398; Martin J. Pasqualetti, "Morality, Space, and the Power of Wind-Energy Landscapes," *Geographical Review* 90, no. 3 (July 2000): 381–394. See also *Wind Power in View: Energy Landscapes in a Crowded World,* ed. Martin J. Pasqualetti, Paul Gipe, and Robert W. Righter (San Diego: Academic Press, 2002).

25. Righter, *Wind Energy in America,* 276, 334.

26. *Air France Magazine* 67 (November 2002): 27.

27. "Wind & Weather," www.windandweather.com.

28. Francis Rolt-Wheeler, *The Boy with the U.S. Weather Men* (Boston: Lothrop, Lee and Shepard, 1917). According to *Contemporary American Authors,* vols. 89–92, 436, Rolt-Wheeler was born in London in 1876 and died in 1960.

29. Rolt-Wheeler, *The Boy with the U.S. Weather Men.*

30. For a study of boys' aviation series books and the characteristics of their heroes, see Fred Erisman, *Boys' Books, Boys' Dreams and the Mystique of Flight* (Fort Worth: Texas Christian University Press, 2006).

31. Rolt-Wheeler, *The Boy with the U.S. Weather Men,* 48.

32. Ibid., 126.

33. A. Lawrence Rotch, *Sounding the Ocean of Air* (New York: E. and J. B. Young, 1900), 124. These altitudes were achieved using a single kite. Using ten linked kites, Eddy set a record of 23,385 feet in 1910. See Bob White, "Diamonds in the Sky: The Contributions of William Abner Eddy to Kiting," *Kitelife Magazine,* November 1999, online at http://best-breezes-squarespace .com/william-abner-eddy/ (accessed February 7, 2008).

34. "The American Meteorological Society," *Bulletin of the American Meteorological Society* 1 (1920): 1.

35. For the history of consumption in general, see Charles E. McGovern, *Sold American: Consumption and Citizenship, 1890–1945* (Chapel Hill: University of North Carolina Press, 2006).

36. *Weather,* Merit Badge Series (New York: Boy Scouts of America, 1928). For the early history of the BSA, see David I. Macleod, *Building Character in the American Boy: The Boy Scouts, YMCA, and Their Forerunners, 1870–1920* (Madison: University of Wisconsin Press, 1983). The reality of scouting in more recent times is well told in Jay Mechling, *On My Honor: Boy Scouts and the Making of American Youth* (Chicago: University of Chicago Press, 2001).

37. *Weather* (New York: Boy Scouts of America, 1928).

38. *Weather,* Merit Badge Series No. 3816 (New Brunswick, NJ: Boy Scouts of America National Council, 1943).

39. *Weather,* Merit Badge Series (Irving, TX: Boy Scouts of America, 2002 printing of 1999 revision).

40. I am very grateful to Ms. Yevgeniya Gribov, archivist, National Historic Preservation Center, Girl Scouts of the USA, for the relevant pages from the Girl Scout handbooks.

41. C. F. Talman, "Public Information," *Bulletin of the American Meteorological Society* 1 (1920): 17; Humphreys's response to Dr. Henry E. Kock, "The Value of Meteorology in Science Teaching in Secondary Schools," *Bulletin of the American Meteorological Society* 5 (1924): 43; David R. Smith, I. W. Geer, R. S. Weinbeck, P. R. Chaston, and J. T. Snow, "AMS Project ATMOSPHERE– University of Oklahoma 1993 Workshop for Atmospheric Education Resource Agents," *Bulletin of the American Meteorological Society* 75 (1994): 95–100; Lin H. Chambers et al., "The CERES S'COOL Project," *Bulletin of the American Meteorological Society* 84 (2003): 759–765; see also the programs of the annual meeting of the AMS, from 2003 to the present.

42. Chambers et al., "The CERES S'COOL Project," 759–765. Also see http://asd-www.larc .nasa.gov/SCOOL/schedule.html (accessed May 15, 2006).

43. See http://asd-www.larc.nasa.gov/SCOOL/groundtruth.html (accessed May 5, 2006).

44. Chambers et al., "The CERES S'COOL Project," 759–765; NOAA's Weather Education site, http://www.weather.gov/om/edures.shtml (accessed May 18, 2006).

45. "Popularize Meteorology: Why Not?" *Bulletin of the American Meteorological Society* 9 (1928): 212–213.

46. Robert Allen Ward, "A National Center of Meteorology," *Bulletin of the American Meteorological Society* 14 (1933): 260.

47. Oliver L. Fassig, "The Weather Bureau," *Bulletin of the American Meteorological Society* 14 (1933): 111–113; C.F.B. [Charles Brooks], "The U.S. Weather Bureau," *Bulletin of the American Meteorological Society* 14 (1933): 135–142.

48. Norman A. Phillips, "Carl-Gustaf Rossby: His Times, Personality, and Actions," *Bulletin of the American Meteorological Society* 79 (1998): 1097–1112; John D. Cox, *Storm Watchers: The Turbulent History of Weather Prediction from Franklin to El Niño* (New York: Wiley, 2002), 179–188; Kristine C. Harper, "Boundaries of Research: Civilian Leadership, Military Funding, and the International Network Surrounding the Development of Numerical Weather Prediction in the United States" (Ph.D. diss., Oregon State University, 2003), 28–132.

49. Harper, "Boundaries of Research," 55–65.

50. Daniel J. Kevles, *The Physicists: The History of a Scientific Community in America* (New York: Knopf, 1978), 88–89, 252–266; "For Improving U.S. Weather Service," *Bulletin of the American Meteorological Society* 14 (1933): 273–281.

51. Harper, "Boundaries of Research," 62–72; Patrick Hughes, "Francis W. Reichelderfer, Part II: Architect of Modern Meteorological Services," *Weatherwise*, August 1981, 148–157.

52. The history of the Weather Bureau/Service since 1950 has not been written. Donald R. Whitnah's *A History of the United States Weather Bureau* (Urbana: University of Illinois Press, 1961) treats the decade of the 1950s in fewer than thirty pages. Patrick Hughes, *A Century of Weather Service 1870–1970* (New York: Gordon and Breach, 1970), adds another twenty, but neither author analyzes the important changes taking place in these years. The "NOAA Legacy Time Line—1807–2000," http://www.history.noaa.gov/, is, needless to say, highly selective. The pages of the *Bulletin of the American Meteorological Society* provide the best record. For example, Neal M. Dorst, "The National Hurricane Research Project: 50 Years of Research, Rough Rides, and Name Changes," *Bulletin of the American Meteorological Society* 88 (2007): 1566–1588. For the changes in American environmentalism, see Samuel P. Hays, *Beauty, Health, and Permanence: Environmental Politics in the United States, 1955–1985* (New York: Cambridge University Press, 1987).

53. George T. Mazuzan, "Up, Up, and Away: The Reinvigoration of Meteorology in the United States: 1958 to 1962," *Bulletin of the American Meteorological Society* 69 (1988): 1152–1153.

54. Ibid., 1157–1161; *Preliminary Plans for a National Institute for Atmospheric Research*, prepared for the National Science Foundation by the University Committee on Atmospheric Research, February 1959; and the UCAR Web site, www.ucar.edu.

55. "Educational Program of the American Meteorological Society," *Bulletin of the American Meteorological Society* 41 (1960): 496–501; "Educational Activities of the AMS," *Bulletin of the American Meteorological Society* 65 (1984): 881–883. I will discuss *Unchained Goddess* in the following chapter.

56. "Policy of the American Meteorological Society with Respect to Unwarranted Statements and Claims in the Field of Meteorology," *Bulletin of the American Meteorological Society* 44 (1963): 133.

57. Robert M. White, "The Organization of the Environmental Sciences in the Federal Government," *Bulletin of the American Meteorological Society* 45 (1964): 317, 320. More than forty years later,

White emphasized the role of the Commission on Marine Sciences, Engineering and Resources, chaired by Julius Stratton of MIT and the Ford Foundation, whose report, *Our Nation and the Sea* (1969), prompted Senators Warren Magnusson and Fritz Hollings to take the lead in creating NOAA. White got most of what he wanted, but not independent agency status. Buried in the Department of Commerce, NOAA still struggles with bureaucratic inertia. Robert M. White, "The Making of NOAA, 1963–2005," *History of Meteorology* 3 (2006): 55–63. White's lecture was given December 1, 2005, at the National Air and Space Museum and is available on the Web site of the International Commission on History of Meteorology, www.meteohistory.org (accessed January 2, 2007).

58. Harper, "Boundaries of Research," chap. 8.

59. AMS/UCAR, interview with George Cressman, August 24, 1992, 36–37, UCAR Archives, Boulder, Colorado.

60. Cressman, interview, 40–41. Cressman appreciated NWS employees, but it was during his administration that the workers organized, creating the National Weather Service Employees Organization, which represents NOAA employees in disputes with management. See the organization's Web site, www.nwseo.org.

61. Cressman, interview, 49. The idea for the WWW is attributed to Dr. Harry Wexler, director of research at the Weather Bureau, who had also been an early advocate of weather satellites.

62. Clark A. Miller, "Scientific Internationalism and American Foreign Policy: The Case of Meteorology, 1947–1958," in *Changing the Atmosphere: Expert Knowledge and Environmental Governance,* ed. Clark A. Miller and Paul N. Edwards (Cambridge, MA: MIT Press, 2001), 168–214; WMO World Weather Watch, http://www.wmo.ch/pages/prog/www/index_en.html.

63. Shelia Jasanoff, "Image and Imagination: The Formation of Global Environmentalism," in Miller and Edwards, *Changing the Atmosphere,* 312–324.

64. National Weather Service Modernization Committee, *A Vision for the National Weather Service: Road Map for the Future* (Washington, DC: National Academy Press, 1999), 1–6.

65. U.S. Congress, Subcommittee on Space of the Committee on Science, Space, and Technology, U.S. House of Representatives, 103rd Congress, September 14, 1993, *NOAA's Response to Weather Hazards—Has Nature Gone Mad* (Washington, DC: Government Printing Office, 1993).

66. Elbert W. Friday Jr., "The Modernization and Associated Restructuring of the National Weather Service: An Overview," *Bulletin of the American Meteorological Society* 75 (1994): 43–52; Mark Monmonier, *Air Apparent: How Meteorologists Learned to Map, Predict, and Dramatize Weather* (Chicago: University of Chicago Press, 1999), 137–152.

67. Friday, "Modernization and Associated Restructuring," 45–49; a map of AWOS/ASOS sites in the United States may be found at http://www.faa.gov/airports_airtraffic/weather/asos/ (accessed May 30, 2006).

68. "Automated Surface Observations," http://www.nws.noaa.gov/asos/obs.htm, which is part of the National Weather Service's Automated Surface Observing System, http://www.nws .noaa.gov/asos/index.html; Friday, "Modernization and Associated Restructuring," 45; Leslie Maitland, "U.S. Will Overhaul Weather Service," *New York Times,* June 11, 1989; Aiguo Dai, Thomas R. Karl, Bomin Sun, and Kevin E. Trenberth, "Recent Trends in Cloudiness over the United States," *Bulletin of the American Meteorological Society* 87 (2006): 604.

69. Harry R. Glahn and David P. Ruth, "The New Digital Forecast Database of the National Weather Service," *Bulletin of the American Meteorological Society* 84 (2003): 195–201; Jonathan D. W. Kahl and Kevin A. Horwitz, "Daily Low or Overnight Low? Confusion in Icon-Based Forecasts," *Bulletin of the American Meteorological Society* 84 (2003): 155–156; Robert T. Ryan, "Digital Forecasts: Communication, Public Understanding, and Decision Making," *Bulletin of the American Meteorological Society* 84 (2003): 1002.

70. "Cutback Is Planned by Weather Service," *New York Times,* September 10, 1990, B10; Noam Cohen, "Weather Service Faces Tough Test," *New York Times,* January 13, 1991, 16; Randolph E.

Schmid, "National Weather Service Nervously Awaits Forecast on Its Future," *Washington Post*, October 14, 1997, A15; National Weather Service biography of David L. Johnson, http://www.weather.gov/com/files/johnsonbio.pdf; NOAA Office of Program Planning and Integration, "Mary M. Glackin," http://www.noaa.gov/glackin.html (accessed August 15, 2007).

71. Roger A. Pielke Jr. and William H. Hooke, "Short-Term Weather Prediction: An Orchestra in Need of a Conductor," in *Prediction: Science, Decision Making, and the Future of Nature*, ed. Daniel Sarewitz, Roger A. Pielke Jr., and Radford Byerly Jr. (Washington, DC: Island Press, 2000), 61–83.

72. Clifford Mass, "The Uncoordinated Giant: Why U.S. Weather Research and Prediction Are Not Achieving Their Potential," *Bulletin of the American Meteorological Society* 87 (2006): 573–584.

73. Clifford F. Mass et al., "Regional Environmental Prediction over the Pacific Northwest," *Bulletin of the American Meteorological Society* 84 (2003): 1353–1366.

74. Mass, "Uncoordinated Giant," 574.

75. Charles F. Brooks, "Enter: The Consulting Meteorologist," *Bulletin of the American Meteorological Society* 1 (1920): 45. Two months earlier, the *Bulletin* had listed the daily work of the New York City Weather Bureau office, which included answering 1,500 telephone calls, some of which sought the services of a forensic meteorologist, doing a dozen newspaper interviews, giving lectures at schools and colleges, and providing climatological data. All these functions could be shared with private meteorologists.

76. Brooks, "Enter," 47. In a prophetic aside, Brooks notes the need of these businesses to be able to predict the extremes of weather and suggests "that for about $100,000 a connection between the water surface temperatures of the ocean and the weather over the oceans and adjacent lands far into, if not across, the continents could be established and seasonal weather forecasts instituted on a basis not of probabilities, . . . but of cause" (46).

77. Lawrence Drake, "Implications of the Current Widespread Interest in Applied Meteorology," *Bulletin of the American Meteorological Society* 28 (1947): 307–308.

78. Eugene Van Cleef, "Weather, Climate and Advertising," *Bulletin of the American Meteorological Society* 2, no. 2 (February 1921): 17–18.

79. "Miscellaneous Notes," *Bulletin of the American Meteorological Society* 2 (1921): 52.

80. David B. Spiegler, "A History of Private Sector Meteorology," in *Historical Essays on Meteorology, 1919–1995*, ed. James R. Fleming (Boston: American Meteorological Society, 1996), 417–441. For a brief update, see David B. Spiegler, "The Private Sector in Meteorology: An Update," *Bulletin of the American Meteorological Society* 88 (2007): 1272–1275.

81. Charles C. Bates, "Industrial Meteorology and the American Meteorological Society—A Historical Overview," *Bulletin of the American Meteorological Society* 52 (1976): 1322–1323. For the importance of Kenneth Spengler, an MIT-trained army air force weather officer in World War II who served as AMS executive secretary for forty-two years, see H. Taba, "The *Bulletin* Interviews: Dr. Kenneth Spengler," *Bulletin of the American Meteorological Society* 68 (1987): 1549–1555.

82. Bates, "Industrial Meteorology," 1324. For the "Six-Point Program," see "Report of the Executive Committee," *Bulletin of the American Meteorological Society* 30 (1949): 140–141.

83. Chairman George, in a commentary on the committee's work, claims that the title was selected after a tour of the country brought the realization that "the operations of so many enterprises were directly affected" by weather. Joseph J. George, "On 'Weather Is the Nation's Business,'" *Bulletin of the American Meteorological Society* 35 (1954): 44. George also remarks, "To me, the outstanding deficiency of private meteorology, and I could easily leave out the word 'private,' is a lack of salesmanship. This is partly due to the fact that the profession is very young. . . . It is also partly due to the incompatibility of being both a salesman and a meteorologist. . . .

I submit that this means more than selling a case of soap. It means selling ideas; it means showing a client what your firm can do for him, it means convincing a superior that your idea for research is a good one and should be implemented" (46).

84. "Establishment of a Board for Certified Consulting Meteorologists," *Bulletin of the American Meteorological Society* 38 (1957): 555.

85. Spiegler, "A History of Private Sector Meteorology," 425–426; "About Our Corporate Members: The Travelers Research Center, Inc.," *Bulletin of the American Meteorological Society* 45 (1964): 608–609; John A. Russo Jr., "The Economic Impact of Weather on the Construction Industry of the United States," *Bulletin of the American Meteorological Society* 47 (1966): 967–972; Robert Henson, *Television Weathercasting: A History* (Jefferson, NC: McFarland, 1990), 126.

86. Henry J. Cox, "Report," *Monthly Weather Review* 28 (March 1900): 115. Cox reported that he had testified on weather conditions in thirty-three cases in the past year and consulted on many other property damage suits that were settled out of court.

87. Fowler Spencer Duckworth, "The Meteorologist as Expert Witness," *Bulletin of the American Meteorological Society* 42 (1961): 447.

88. Myrna Silverman, "Now Meteorologists Are Solving Crimes," *Science Digest*, February 1974, 42–46. A forensic meteorologist helped to acquit Ethel Kennedy's nephew William Kennedy Smith when he was accused of rape in Palm Beach, Florida, in 1991. Sean Potter, "The Practice of Forensic Meteorology," *Weatherwise* 57, no. 3 (May/June 2004): 29–33.

89. Although confident that a well-trained meteorologist can overcome the problems associated with reconstructing microscale weather, Fred Ward and Peter Leavitt of Weather Services, Inc., raise similar doubts. See "The Limitations of Official Weather Records in Legal Proceedings," *Bulletin of the American Meteorological Society* 52 (1971): 1190–1191. Among the unsolved problems of forensic meteorology is that of determining the ambient temperature while a body is in situ. A study in Australia to test the accuracy of retrospective weather data correction using linear regression between stations and sites found that "mean predicted body in situ temperatures for sites differed significantly between correlation periods. . . . practitioners should be cautious in making correlations if weather patterns during correlation differ greatly from those while the body was in situ." M. S. Archer, "The Effect of Time after Body Discovery on the Accuracy of Retrospective Weather Station Ambient Temperature Corrections in Forensic Entomology," *Journal of Forensic Science* 49, no. 3 (May 2004): abstract. Because insect and parasite activity is often determined by atmospheric conditions, forensic meteorologists and entomologists need to cooperate closely.

90. "Weather Is the Nation's Business," Report of the Department of Commerce Advisory Committee on Weather Services, GPO, 1953; Bates, "Industrial Meteorology," 1324–1326; and www.ncim.org.

91. Joseph H. Golden, "On the Role of the Private Sector in Disseminating Hurricane Forecasts and Warnings," *Bulletin of the American Meteorological Society* 65 (1984): 972.

92. Ibid., 975.

93. Ibid., 977–979.

94. Fred Guterl, "The Nerds of Weather," *Newsweek,* September 30, 2002, 48–49. On weather as an amenity, see William B. Meyer, *Americans and Their Weather* (New York: Oxford University Press, 2000), 140, 214.

95. I visited AccuWeather headquarters and interviewed Mr. Bastardi by telephone on September 26, 2002. I am grateful to Joe Bastardi and Jesse Ferrell for their time and helpfulness. See also Jeffrey Rosenfeld, "Do We Need the National Weather Service?" *Scientific American Presents Weather,* Spring 2000, 28–31.

96. See http://www.weatherbug.com/aws/aboutus.asp. For an example of NWS's competing services, see Steve Timko, "Web Site Zeroes in on Local Forecasts," *Reno Gazette-Journal,* Febru-

ary 16, 2003, 1C, 3C; and Andrew C. Revkin, "Weather Forecasters Look Ahead, Far Ahead," *New York Times,* November 13, 2001, D1. On The Weather Channel, see Leonard Ray Teel, "The Weather Channel," *Weatherwise* 35, no. 4 (August 1982): 157–163; and Frank Batten, with Jeffrey L. Cruikshank, *The Weather Channel: The Improbable Rise of a Media Phenomenon* (Boston: Harvard Business School Press, 2002).

97. Ellen McCarthy, "Weather Junkies Keep This Bug in Business," *Washington Post,* June 22, 2004, E1, E6; "WeatherBug and Send Word Now Launch Smart Notification Weather Service to Deliver Nation's First Location-Based, Two-Way Mobile Weather Alerts and Notification Based on Live WeatherBug Network Data," Business Wire, June 12, 2006, http://home.businesswire.com.

98. Weather Communications Group, http://www.met.psu.edu/dept/WxComm/; "Report of the U.S. Weather Research Program Workshop on the Weather Research Needs of the Private Sector," September 27, 2001 (unpublished paper given to me by Roger A. Pielke Jr.); Committee on Partnerships in Weather and Climate Services, National Research Council, *Fair Weather: Effective Partnerships in Weather and Climate Services* (Washington, DC: National Academies Press, 2003), 29, 144–145; on Vaisala, see http://www.vaisala.com.

99. Committee on Partnerships in Weather and Climate Services, National Research Council, *Fair Weather: Effective Partnerships in Weather and Climate Services;* H. Michael Mogil, "How the WeatherWorks," http://www.weatherworks.com; Keith Heidorn, The Weather Doctor, http://www.islandnet.com/~see/weather/doctor.htm.

100. Roger Pielke Jr., "Public-Private Provision of Weather and Climate Services: Defining the Policy Problem," appendix B, *Fair Weather,* 132.

101. Sean Potter, "Name That Storm," *Weatherwise,* 58, no. 1 (January/February 2005): 26–28; for Internet gambling, see www.betonweather.com, started in 2003; and Nikki Waller, "Gamblers Put Bucks on Big Storms," *MiamiHerald.com,* May 29, 2006.

102. See, for example, Frederick L. Hoffman, *Windstorm and Tornado Insurance,* 3rd ed. (New York: Spectator, 1902); A. H. Palmer, "Weather Insurance," *Bulletin of the American Meteorological Society* 3 (1922): 67–70; John R. Weeks, "Basis of Rain Insurance Rates," *Bulletin of the American Meteorological Society* 4 (1923): 81–82; Charles F. Brooks, "Unsatisfactory Rain Insurance Policies," *Bulletin of the American Meteorological Society* 4 (1923): 82–83; Dr. Andrew H. Palmer, "Recent Developments in Weather Insurance," *Bulletin of the American Meteorological Society* 6 (1925): 65–73.

103. Peter L. Bernstein, *Against the Gods: The Remarkable Story of Risk* (New York: Wiley, 1996), 304; James Surowiecki, "The Financial Page: What Weather Costs," *New Yorker,* July 23, 2001, 29.

104. *Annual Report of the Department of Agriculture for the Year Ended June 30, 1914,* 51, and *Annual Report of the Department of Agriculture for the Year Ended June 30, 1915,* 66, http://www.earlyradiohistory.us/agric.htm.

105. *Annual Report of the Department of Agriculture for the year ended June 30, 1926,* 56. Rideout is mentioned in Roy Leep, "The American Meteorological Society and the Development of Broadcast Meteorology," in *Historical Essays on Meteorology, 1919–1995,* 482.

106. Not to be confused with Jimmy Fidler, Hollywood gossip columnist of the same era.

107. James C. Fidler, "Broadcasting the Weather," *Bulletin of the American Meteorological Society* 18 (1937): 172–175.

108. James C. Fidler, "Popularizing the Weather Broadcast," *Bulletin of the American Meteorological Society* 19 (1938): 310–317. On the place of radio in Muncie in this period, see Robert S. Lynd and Helen Merrell Lynd, *Middletown in Transition: A Study in Cultural Conflicts* (New York: Harcourt Brace and World, 1937). For the impact of radio in the 1930s, with the Welles broadcast as a case study, see Hadley Cantril, *The Invasion from Mars* (Princeton, NJ: Princeton University Press, 1940).

109. Henson, *Television Weathercasting,* 130–131.

110. John D. Thomas, "For the 'Weather Engaged,' Manna from Heaven," *New York Times,* April

28, 2002, 10. A possible model for combining meteorological information with other environmental news is the "Earthweek" column in the *San Francisco Chronicle* and more than eighty other newspapers, created by meteorologist Steve Newman in 1987, which provides a roundup of weather, health, earthquake, volcanic, and other environmental events. Although there is no effort to link any of the phenomena, the global perspective helps the reader place American environmental hazards in perspective.

111. Henson, *Television Weathercasting*, 58; Tom McNichol, "Whose Weather Is It, Anyway?" *Washington Post Magazine*, February 24, 1991, 20, 22–23, 32–33.

112. K. H. Jehn, "The Challenge of Television Weather Programs," *Bulletin of the American Meteorological Society* 37 (1956): 351–353; Leep, "The American Meteorological Society and the Development of Broadcast Meteorology," 488; Henson, *Television Weathercasting*, 7; James C. Fidler, "Weather via Television," *Bulletin of the American Meteorological Society* 29 (1948): 329–331; Francis K. Davis, "The Professional Meteorologist in Radio," *Bulletin of the American Meteorological Society* 30 (1949): 86–89.

113. Henson, *Television Weathercasting*, 20–22, 52–54; K. H. Jehn, "Radio and Television Weathercasting—The Seal of Approval Program after Five Years," *Bulletin of the American Meteorological Society* 45 (1964): 489–491; "Seal of Approval Program for Radio and Television Weathercasting," *Bulletin of the American Meteorological Society* 65 (1984): 872–873. In 1984, provisions were added to suspend or revoke the seal; see Werner A. Baum, "The Radio and Television Seal of Approval Program of the Society: Augmentation of Policies and Procedures," *Bulletin of the American Meteorological Society* 65 (1984): 874–875.

114. Robert M. White, "Broadcasting the Weather," *Bulletin of the American Meteorological Society* 47 (1966): 21–24; Henson, *Television Weathercasting*, 56–57.

115. Johan Huizinga, *Homo Ludens: A Study of the Play Element in Culture* (1944; Boston: Beacon Press, 1955), 21. For the refinement and extension of Huizinga's ideas on the centrality of play in dealing with risk and uncertainty, see Brian Sutton-Smith, *The Ambiguity of Play* (Cambridge, MA: Harvard University Press, 1997).

116. Henson, *Television Weathercasting*, 79–91. Henson places his discussion of "weather girls" in the context of gender and racial discrimination in the media and provides good background to these issues. David Laskin, *Braving the Elements: The Stormy History of American Weather* (New York: Anchor Doubleday, 1996), 179–181; "Television and Radio Seals of Approval Granted by AMS," *Bulletin of the American Meteorological Society* 45 (1964): 492–493; "Television Seals of Approval Granted in 2005," *Bulletin of the American Meteorological Society* 87 (2006): 826–827; Margaret A. LeMone and Patricia L. Waukau, "Women in Meteorology," *Bulletin of the American Meteorological Society* 63 (1982): 1266–1278; Robert Henson, "Show and Tell," *Weatherwise* 46 (1993): 14; on Thurman, see "The Monitor Tribute Pages, http://www.monitorbeacon.com/miss monitor.html (accessed June 26, 2006); on Karna Small, see "Modern Television," http://www .moderntv.com/modtvweb/media/van2.htm (accessed January 26, 2005); and for "Bobbie," see Evan Morgan, "Bobbie the Weather Girl—AFVN-TV Saigon," http://illyria.com/women/ vnweather.html (accessed July 31, 2002).

117. Telephone interview with Mrs. Karna Small Bodman, August 2, 2006. My sincere thanks to Mrs. Bodman for her time and information. The backward writing gimmick worked, since I was told about Karna Small by Ann Shumard, curator of photography, National Portrait Gallery, Smithsonian Institution, who remembered it after more than thirty years and looked her up on the Internet. I thank Ms. Shumard for the suggestion. Karna Small Bodman went on to become deputy press secretary and later senior director and spokesman in the National Security Council in the Reagan administration. She has recently published her second novel.

118. Jonathan Yardley, "Which Way the Wind Blows," *Washington Post*, June 17, 2002, C2. In his book on storm chasing, *Big Weather* (New York: Henry Holt, 2005), 196, Mark Svenvold tries

to make a humorous connection between female weathercasters and porn stars, citing www
.kapturedforyou.com, a Web site that provides pictures of female TWC personalities. The female
weathercaster as sex object is exploited by the makers of the movie *To Die For* (1995), in which a
teenage boy becomes so obsessed with a beautiful TV weathercaster that he murders her husband.

119. Leonard Ray Teel, "The Weather Channel," *Weatherwise* 35, no. 4 (August 1982): 158, 160.

120. John Seabrook, "Selling the Weather," *New Yorker,* April 3, 2000, 44–53; Batten, with
Cruikshank, *The Weather Channel.*

121. D. T. Max, "Ominous Clouds," *New York Times Magazine,* March 3, 2002, 19–20. For Landmark's style, see http://landmarkcom.com/about/culture.php (accessed March 3, 2007). I thank
Read Mercer Schuchardt for calling my attention to the connection between Landmark and EST.

122. Batten, *The Weather Channel,* 155.

123. Ibid., 71–75, 165, 216–221, 224.

124. David Browne, "The Barometric Pressure Is Just the Beginning," *New York Times*, March
11, 2007, Arts & Leisure, 29.

125. Sam McManis, "Taking the Temperature," *Sacramento Bee,* February 27, 2007, E1–E2.

126. Paul Farhi, "Blanketing the D.C. Area with Snow Coverage," *Washington Post,* January
26, 2000, C1; "Summary of Questionnaire Results—1980 Weathercasters Conference, Denver,
Colo.," *Bulletin of the American Meteorological Society* 62 (1981): 341. The AMS has held an annual
conference on radio and television weathercasting since 1971, separate from its own annual
meeting.

127. Interview with George Hirschmann in the studios of VHSV, Harrisonburg, Virginia, January 9, 2003. I thank Mr. Hirschmann for his time and assistance.

128. Marita Sturken, "Desiring the Weather: El Niño, the Media, and California Identity," *Public
Culture* 13, no. 2 (2001): 165, 172.

129. Early accounts in nonscientific journals are Howard T. Orville, "Weather Made to Order?"
Collier's, May 28, 1954, 25–20; and John Lear, "Shepherding the Wind," *Saturday Review,* April 2,
1966, 49–52. Lear's article drew letters from two important actors in cloud seeding, James E.
Howell and Bernard Vonnegut, *Saturday Review,* May 7, 1966, 43. A recent popular account is
Daniel Pendick, "Cloud Dancers," *Scientific American Presents Weather* 11, no. 1 (Spring 2000):
64–69. For a lengthy technical account, see William R. Cotton and Roger A. Pielke, *Human Impacts on Weather and Climate* (New York: Cambridge University Press, 1995). The best scholarly
accounts, based on extensive primary research, are James R. Fleming, "Fixing the Weather and
Climate: Military and Civilian Schemes for Cloud Seeding and Climate Engineering," in *The Technological Fix: How People Use Technology to Create and Solve Problems,* ed. Lisa Rosner (New York:
Routledge, 2004), 175–200; Fleming, "The Pathological History of Weather and Climate Modification: Three Cycles of Promise and Hype," *Historical Studies in the Physical and Biological Sciences* 37, no. 1 (2006): 3–25; and Fleming, "Distorted Support: Pathologies of Weather Warfare,"
in *Science in Uniform: Science, Technology, and American Military Institutions, from the Revolu-
tionary War to the Present,* ed. Barton C. Hacker and Margaret Vining (Lanham, MD: Scarecrow
Press, in press). Fleming's more popularly written essay "The Climate Engineers," *Wilson Quarterly* 32 (Spring 2007): 46–60, contains additional details. An analytical, policy-oriented study
may be found in Chunglin Kwa, "The Rise and Fall of Weather Modification: Changes in American Attitudes toward Technology," in *Changing the Atmosphere: Expert Knowledge and Environmental Governance,* ed. Clark A. Miller and Paul N. Edwards (Cambridge, MA: MIT Press, 2001),
135–164. Theodore Steinberg deals with some episodes in cloud seeding history in two of his
books; see "Cloud Busting in Fulton County," in *Slide Mountain, or The Folly of Owning Nature*
(Berkeley: University of California Press, 1995), 106–195, and "The Neurotic Life of Weather Control," in *Acts of God: The Unnatural History of Natural Disaster in America,* 2nd ed. (New York:

Oxford University Press, 2006), 127–147. For an account that focuses on snowpack augmentation, see Bernard Mergen, *Snow in America* (Washington, DC: Smithsonian Institution Press, 1997), 190–198. Earlier accounts by advocates of weather modification include Vincent J. Schaefer, "Artificially Induced Precipitation and Its Potentialities," in *Man's Role in Changing the Face of the Earth,* ed. William L. Thomas (Chicago: University of Chicago Press, 1956), 607–618; Vincent J. Schaefer, "The Early History of Weather Modification," *Bulletin of the American Meteorological Society* 49 (1968): 337–342. A good overview of the first twenty-five years is Horace R. Byers, "History of Weather Modification," in *Weather and Cloud Modification*, ed. W. N. Hess (New York: Wiley, 1974), 3–44.

130. W. Findeisen, "Colloidal Meteorological Processes in the Formation of Atmospheric Precipitation," *Meteorology* 2, no. 55 (1938): 121–133, quoted in Schaefer, "Artificially Induced Precipitation," 608; Fleming, "Fixing the Weather and Climate," 177.

131. Fleming, "Fixing the Weather and Climate," 175–181. Fleming notes that "in 1965 the U.S. military employed 14,300 meteorologists and the Weather Bureau, 4,500 . . . the total number of employees in the Soviet Hydrometeorological Service was in excess of 70,000" (185). See also Arthur H. Westing, ed., *Environmental Warfare: A Technical, Legal, and Policy Appraisal* (London: Taylor and Francis, 1984), 3–12; and Joe Chew, *Storms above the Desert: Atmospheric Research in New Mexico, 1935–1985* (Albuquerque: University of New Mexico Press, 1987), 33.

132. Henry Fountain, "Wallace E. Howell, 84, Dies; Famed Rainmaker in Drought," *New York Times,* July 6, 1999, A15; E. B. White, "Rainmakers," in *Poems and Sketches of E. B. White* (New York: Harper and Row, 1981), 29.

133. Jack C. Oppenheimer and W. Henry Lambright, "Technology Assessment and Weather Modification," *Southern California Law Review* 45, no. 570 (1972): 585–587. For more on Howell's role in the ill-fated Fulton County, Pennsylvania, rain enhancement experiment and a thoughtful discussion of the issues, see Steinberg, *Acts of God,* 109–134.

134. Irving P. Krick and Roscoe Fleming, *Sun, Sea, and Sky: Weather in Our World and in Our Lives* (Philadelphia: Lippincott, 1954), 142, 141, 232, 238, 233.

135. Fleming, "Fixing the Weather and Climate," 175–176, 183.

136. *Final Report of the Advisory Committee on Weather Control* (Washington, DC: Government Printing Office, 1957), vol. 1, 6; vol. 2, 201.

137. Kwa, "Rise and Fall of Weather Modification," 140–161; House Committee on Science and Aeronautics, *Weather Modification* (Washington, DC: Government Printing Office, 1959), 10–12; Archie M. Kahan, "The Place of Government Programs in Weather Modification," *Bulletin of the American Meteorological Society* 49 (1968): 242–246; Werner A. Baum, "Congressional Action on Weather Modification," *Bulletin of the American Meteorological Society* 49 (1968): 234–237; Mason T. Charak and Mary T. DiGiulian, "A Review of Federal Legislation on Weather Modification," *Bulletin of the American Meteorological Society* 55 (1974): 755–758; Fleming, "Fixing the Weather and Climate," 186–188; Deborah Shapley, "Weather Warfare: Pentagon Concedes 7-Year Vietnam Effort," *Science* 184 (July 7, 1974): 1059–1061; R. Cecil Gentry, "Project Stormfury," *Bulletin of the American Meteorological Society* 50 (1969): 404–409; H. E. Willoughby, D. P. Jorgensen, R. A. Black, and S. L. Rosenthal, "Project Stormfury: A Scientific Chronicle 1962–1983," *Bulletin of the American Meteorological Society* 66 (1985): 505–514. A readable account of Project Stormfury by one of its directors may be found in Bob Sheets and Jack Williams, *Hurricane Watch: Forecasting the Deadliest Storms on Earth* (New York: Vintage, 2001), 165–177.

138. Kwa, "Rise and Fall of Weather Modification," 152–163; W. R. Derrick Sewell, "Emerging Problems in the Management of Atmospheric Resources: The Role of Social Science Research," *Bulletin of the American Meteorological Society* 49 (1968): 326–342; W. R. Derrick Sewell, ed., "Human Dimensions of Weather Modifications" (Research Paper no. 105, Department of Geography, University of Chicago, Chicago, 1995); Howard J. Taubenfeld, ed., *Weather Modification*

and the Law (Dobbs Ferry, NY: Oceana Publications, 1968); J. Eugene Haas, "Social Aspects of Weather Modification," *Bulletin of the American Meteorological Society* 54 (1973): 647–657; Haas, "Sociological Aspects of Weather Modification," in Hess, *Weather and Climate Modification*, 787–811; Leo Weisbecker, *The Impacts of Snow Enhancement: Technology Assessment of Winter Orographic Snowpack Augmentation in the Upper Colorado River Basin* (Norman: University of Oklahoma Press, 1974); Barbara C. Farhar, "A Societal Assessment of the Proposed Sierra Snowpack Augmentation Project," unpublished report to Bureau of Reclamation, 1976; Cotton and Pielke, *Human Impacts on Weather and Climate*, 57–60. For an update of Taubenfeld, see Ronald B. Standler, "Weather Modification Law in the U.S.A.," 2002, http://www.rbs2.com/weather.htm.

139. Quoted in Richard Moreno, "Cloud Rustlers," *Nevada Magazine,* March/April 1991, 10. The song was inspired by Dick Haman, a ranch manager who, in 1947, tried to file for water rights to all the clouds above his 12,300-acre spread near Topaz Lake on the California border south of Carson City. The claim was made in jest, but it raised the same issue, who owns the clouds, that was present in later cloud-seeding suits.

140. Richard A. Kerr, "Cloud Seeding: One Success in 35 Years," *Science* 217, no. 6 (August 1982): 519–521; Ronald List, "Weather Modification—A Scenario for the Future," *Bulletin of the American Meteorological Society* 85 (2004): 59.

141. Michael Garstang et al., "Weather Modification: Finding Common Ground," *Bulletin of the American Meteorological Society* 86 (2005): 640; Andrew Freedman, "Back in Business? The Government Takes Another Look at Weather Modification," *Weatherwise* 59, no. 4 (July/August 2006): 24–27; "Weather Modification Research and Technology Transfer Authorization Act of 2005," http://www.theorator.com/bills109/s517.html (accessed May 13, 2005).

142. Mindy Sink, "Can Rain Be Bought? Experts Seed Clouds and Seek Answers," *New York Times,* October 14, 2003, D2; Henry Fountain, "The Science of Rain-Making Is Still Patchy," *New York Times,* October 19, 2003, D12. A similar conclusion from California is Carrie Peyton Dahlberg, "Cloud Seeding Validity Is Still Up in the Air," *Sacramento Bee,* October 20, 2003, A1. On the use of bezoars, enterroliths from the stomachs of ruminants, and black horses in rainmaking by Mongols, see Ádám Molnár, *Weather Magic in Inner Asia* (Bloomington: Indiana University Research Institute for Inner Asian Studies, 1994), 124–142.

143. Edward C. Greiner, "Department of Water Resources Activities in Weather Modification," *Proceedings of the Western Snow Conference,* 1968, 77–83; "Utah Division of Water Resources," http://water.utah.gov/cloudseeding/History/Default.asp (accessed May 13, 2005); Harold Klieforth, "On the Potential of Weather Modification for Enhanced Water Supply in Nevada," unpublished report to the Atmospheric Sciences Center of the Desert Research Institute, University of Nevada System, 1969; "Nevada State Cloud Seeding Program: A Synopsis 2002," http://cloudseeding.dri.edu/Program/Programs.html (accessed July 9, 2006); Jeff Hull, "The Modern Rain Dance," *New York Times Magazine,* July 2, 2006, 16–17; Colonel Tamzy J. House et al., "Weather as a Force Multiplier: Owning the Weather in 2025," http://www.au.af.mil/au/2025/volume 3/chap15/v3c15-1.htm.

144. House et al., "Weather as a Force Multiplier," 1–5.

145. Peter Schwartz and Doug Randall, "An Abrupt Climate Change Scenario and Its Implications for United States National Security," unpublished paper prepared for the Department of Defense, October 2003, http://www.environmentaldefense.org/documents/3566_Abrupt ClimateChange.pdf. In April 2007, eleven retired generals released a sixty-three-page report, "The National Security Implications of Global Climate Change," and in May several Democratic congressmen sought to add an amendment to the bill funding the intelligence agencies that would require an interagency study of global climate change on national defense. Juliet Epstine, "Military Sharpens Focus on Climate Change," *Washington Post,* April 15, 2007, A6; Mark

Mazzetti, "Bill Proposes Climate Study Focused on U.S. Defense," *New York Times*, May 4, 2007, A20; William J. Broad, "How to Cool a Planet (Maybe)," *New York Times*, June 27, 2006, D1, D4; Paul J. Crutzen, "Albedo Enhancement by Stratospheric Sulfur Injections: A Contribution to Resolve a Policy Dilemma? An Editorial Essay," *Climatic Change* 77, nos. 3–4 (August 2006): 211–220; Ralph J. Cicerone, "Geoengineering: Encouraging Research and Overseeing Implementation," *Climatic Change* 77, nos. 3–4 (August 2006): 221–226; "Radical Plan Presented by Nobel Winner," *Bulletin of the American Meteorological Society* 87 (2006): 1303–1305.

146. Lucian Boia, *The Weather in the Imagination* (London: Reaktion Books, 2005), 115–119.

147. Carroll Livingston Riker, *Power and Control of the Gulf Stream: How It Regulates the Climates, Heat and Light of the World* (New York: Baker and Taylor, 1912), 9, 72–73.

148. Ibid.; "A Plan to Revolutionize the Climate of Europe and North America," *Current Opinion*, June 1913, 478–479; Wallace S. Broecker, "The Biggest Chill," *Natural History* 96, no. 10 (October 1987): 74–80; Curt Suplee, "Threat of Localized Cooling Flows from Global Warming," *New York Times*, December 1, 1997, A3.

149. John M. Barry, *Rising Tide: The Great Mississippi Flood of 1927 and How It Changed America* (New York: Simon and Schuster, 1997), 407. Barry does not mention Riker or analyze the Jadwin plan. Carroll Livingston Riker, *Control and Utilization of the Mississippi and the Drainage of Its Valley* (Washington, DC: n.p., 1929), 42. The plans of Riker and the Corps of Engineers illustrate many of the misconceptions of flood control discussed in Roger A. Pielke Jr., "Nine Fallacies of Floods," *Climatic Change* 42 (1999): 413–438.

150. "Climate Control," *Bulletin of the American Meteorological Society* 25 (1944): 391–392.

151. Penelope Green, "Not Enough Snow for You? Talk to Your Father," *New York Times*, February 15, 2007, D1, D6.

Chapter Three. Seeing Weather

1. W. E. Knowles Middleton, *Invention of the Meteorological Instruments* (Baltimore: Johns Hopkins University Press, 1969), 13.

2. Ibid., 182–214.

3. Stephen Kern, *The Culture of Time and Space, 1880–1918* (Cambridge, MA: Harvard University Press, 1983).

4. Tom D. Crouch, *The Eagle Aloft: Two Centuries of the Balloon in America* (Washington, DC: Smithsonian Institution Press, 1983), 595.

5. "Abbott Lawrence Rotch," *Dictionary of American Biography*, ed. Dumas Malone (New York: Scribner's, 1943), vol. 16, 183–184. See also Crouch, *The Eagle Aloft*, 594.

6. Alexander McAdie, "The Discovery of the Stratosphere," *Bulletin of the American Meteorological Society* 15 (1934): 175. McAdie dates the "discovery" to March 1902.

7. Douglas Archibald, *The Story of the Earth's Atmosphere* (New York: D. Appleton, 1897), 15; A. Lawrence Rotch, *Sounding the Ocean of Air* (New York: E. and J. B. Young, 1900), 174.

8. Examples of the use of the ocean of air image, more than fifty years after Rotch, include David Blumenstock, *The Ocean of Air* (New Brunswick, NJ: Rutgers University Press, 1959); and James G. Edinger, *Watching for the Wind: The Seen and Unseen Influences on Local Weather* (Garden City, NY: Anchor Books, 1967), 23, 97.

9. Edwin C. Martin, *Our Own Weather: A Simple Account of Its Curious Forms, Its Wide Travels, and Its Notable Effects* (New York: Harper and Brothers, 1913), 1, emphasis in original.

10. Ibid., 9–10. His global perspective is noteworthy because the USWB maps of the time never extend across the Canadian or Mexican borders; anything outside the United States is *pelago d'aria incognita*. The remainder of Martin's book is more prosaic, offering standard explanations

of air movements, storms, winds, seasons, clouds, precipitation, and weather superstitions. According to the *Dictionary of North American Authors*, Martin was a lawyer and publisher who was born in Hamilton, Ohio, and died in Watchung, New Jersey.

11. Alexander McAdie, *A Cloud Atlas* (Chicago: Rand McNally, 1923), 8.

12. Alexander McAdie, *Man and Weather* (Cambridge, MA: Harvard University Press, 1926), 37.

13. Ibid., 40.

14. George R. Stewart, *Storm* (New York: Random House, 1941), 19.

15. Guy Murchie, *Song of the Sky* (Boston: Houghton Mifflin, 1954), 94. For biography and obituary, see *Contemporary American Authors*, vol. 27, 337, and vol. 159, 299–300. Curiously, *Song of the Sky* is subtitled *An Exploration of the Ocean of Air* in *Contemporary Authors*, but the subtitle does not appear in the printed volume.

16. Murchie, *Song of the Sky*, 94.

17. Gordon W. Wares, "Terminology of Atmospheric Shells," *Bulletin of the American Meteorological Society* 34 (1953): 221.

18. Leonard Engle, "Mystery of the Air We Explore," *New York Times Magazine*, April 15, 1956, 21; Frederick K. Lutgens and Edward J. Tarbuck, *The Atmosphere: An Introduction to Meteorology* (Upper Saddle River, NJ: Prentice-Hall, 2001), 20; "Jetstream—An Online School for Weather," http://www.srh.weather.gov/srh/jetstream/atmos/layers.htm (accessed March 7, 2007).

19. Louise B. Young, *Earth's Aura* (New York: Knopf, 1977), xi, 18. On the mountain wave project, see Harold Klieforth, "Meteorological Aspects of the Sierra Wave," *Organisation Scientifique et Technique Internationale du Vol a Voile* 7 (March 1957): 1–8; and Vanda Grubiši and John M. Lewis, "Sierra Wave Project Revisited," *Bulletin of the American Meteorological Society* 858 (2004): 1127–1142.

20. See Alan Burdick, "Here Comes the Sun," *New York Times Magazine*, February 4, 2001, 9–10; Michael J. Carlowicz and Ramon E. Lopes, *Storms from the Sun: The Emerging Science of Space Weather* (Washington, DC: Joseph Henry Press, 2002); and the University Corporation for Atmospheric Research Web site, http://www.windows.ucar.edu/spaceweather/basic_facts.html.

21. Lewis Thomas, *Lives of a Cell* (New York: Viking, 1974), 170–174.

22. For a history of the discovery of air and its importance for life, see Joe Sherman, *Gasp! The Swift and Terrible Beauty of Air* (Washington, DC: Shoemaker and Hoard, 2004).

23. John Cox, *Weather for Dummies* (New York: Hungry Minds, 2000), 37.

24. Samuel Hays, *Beauty, Health, and Permanence: Environmental Politics in America, 1955–1985* (New York: Cambridge University Press, 1987), 97–98. I am very grateful for Borden's cooperation and assistance. I have benefited enormously from his work.

25. Conversations with Jack Borden, August 6–7, 2002; Joanna J. Berkman, "For Spacious Skies," *Amicus Journal* 5 (Spring 1983): 32–38.

26. Ervin Zube and Charles Law, "Perceptions of the Sky in Five Metropolitan Areas," *Urban Ecology* 8 (1984): 199–208.

27. Leonard J. Duhl, "Health and the Inner and Outer Sky," *Journal of Humanistic Psychology* 26, no. 3 (Summer 1986): 46–61.

28. Charles Roth, "Education for Life in the Sky," *Environmentalist* 1, no. 4 (1980): 293–297.

29. Owen E. Thompson, "Summary of the AMS Workshop on Meteorology as a Unifying Educational Strategy for Improving Pre-college Science, Mathematics, and Technology Education 21–23 September 1984, Crystal City, Virginia," *Bulletin of the American Meteorological Society* 67 (1986): 426–428.

30. Elizabeth Levitan Spaid, "For These Children, School Is Looking Up," *Christian Science Monitor*, April 18, 1994, 14.

31. Matthew L. Wald, "Seeing the Sky: Not a Limit but Infinity," *New York Times*, June 21, 1987, 1.

32. Jerome Washington was born in Trenton, New Jersey, in 1939, attended Columbia Univer-

sity, served in the army in Vietnam, and was active in the civil rights and peace movements in the 1960s. In 1973 he was convicted of murder and served sixteen years in prison. After his release he organized writing workshops for the elderly and for "at-risk" teenagers. He published two short-story collections, *A Bright Spot in the Yard* (Trumansburg, NY: Crossing Press, 1981), and *Iron House: Stories from the Yard* (Fort Bragg, CA: QED Press, 1994). Biographical information from Jerome Washington collection 1979–1988, Special Collections of the Lloyd Sealy Library, John Jay College of Criminal Justice, New York, http://www.lib.jjay.cuny.edu/info/speccoll/Jerome%20Washington%20Papers.pdf (accessed August 20, 2007).

33. Scott Allen, "Inmates Find 'Freedom' in the Sky," *Boston Globe,* December 5, 1994; and John Larrabee, "Students Look to the Skies for Inspiration," *USA Today,* August 7, 1996, 4D.

34. See www.cloudman.com.

35. Interview with John Day, February 18–19, 2003. My deepest gratitude to John and his wife, Mary, for their kindness and hospitality when I visited them in Oregon. In late 2007, Day was at work on a revised edition of the *Field Guide to the Atmosphere* with Jay Pasachoff.

36. Gaston Bachelard, *Air and Dreams: An Essay on the Imagination of Movement* (1943; Dallas: Dallas Institute of Humanities and Culture, 1988), 172.

37. Edward S. Casey, *Getting Back into Place: Toward a Renewed Understanding of the Place-World* (Bloomington: Indiana University Press, 1993); Casey, *The Fate of Place: A Philosophical History* (Berkeley: University of California Press, 1997); Robert B. Bechtel and Arza Churchman, eds., *Handbook of Environmental Psychology* (New York: Wiley, 2002); Keith H. Basso, *Wisdom Sits in Places: Landscape and Language among the Western Apache* (Albuquerque: University of New Mexico Press, 2001), 121.

38. Edinger, *Watching for the Wind,* 33.

39. "10 Reasons for Looking Up!" http://www.cloudman.com/look/look_up.htm; John A. Day, *The Book of Clouds* (New York: Barnes and Noble, 2003), 8.

40. Rudolf Arnheim, *Visual Thinking* (Berkeley: University of California Press, 1971), 136–138; Percy Bysshe Shelley, "The Cloud" (1820).

41. Richard Hamblyn, *The Invention of Clouds: How an Amateur Meteorologist Forged the Language of the Skies* (New York: Farrar, Straus and Giroux, 2001); John Ruskin, *Modern Painters,* 2 vols. (New York: Wiley and Putnam, 1847–1848); Henry D. Thoreau, *Journal,* vol. 5, 1852–1853, ed. Patrick F. O'Connell (Princeton, NJ: Princeton University Press, 1997), 216.

42. Middleton, *Invention of the Meteorological Instruments,* 269. Now museum objects, nephoscopes were used by the Weather Bureau through the 1920s. Benjamin C. Kadel, *Instructions for Erecting and Using Weather Bureau Nephoscope 1919 Pattern* (Washington, DC: Government Printing Office, 1920). A nephoscope of about this vintage may be seen in the exhibit area of the Mount Washington Observatory in New Hampshire. Early ceilometers used beams of light reflected from a cloud base; more recent models use lasers. A good introduction to all aspects of the study of clouds is Roger Clausse and Léopold Facy, *The Clouds,* translated from French (New York: Evergreen Profile Books of Grove Press, 1961).

43. Andrew C. Revkin, "NASA's Goals Delete Mention of Home Planet," *New York Times,* July 22, 2006, A1, A10; Graeme L. Stephens et al., "The Cloudsat Mission and the A-Train: A New Dimension of Space-Based Observations of Clouds and Precipitation," *Bulletin of the American Meteorological Association* 83 (2002): 1778; *Calipso* site, http://www.nasa.gov/mission_pages/calipso/main/index.html.

44. John C. Van Dyke, *The Desert: Further Studies in Natural Appearances* (New York: Scribner's, 1901; reprint, Baltimore: Johns Hopkins University Press, 1999), 78–79, 97–98.

45. Ibid., 96, 101–104. Van Dyke was probably thinking of altostratus clouds that occur as much as 9 miles above the earth, rather than the simple stratus often produced by the lifting of a fog layer.

46. Frank H. Bigelow, "The Formation and Motions of Clouds," *Popular Science Monthly* 60 (April 1902): 495–502. Although cloud classification uses Latin words, they are by convention used only in the singular; thus more than one stratus cloud is stratus, not strati. See Todd S. Glickman, ed., *Glossary of Meteorology,* 2nd ed. (Boston: American Meteorological Society, 2000), at http://amsglossary.allenpress.com/glossary/search?id+cloud-classification1.

47. Charles F. Brooks, "Cloud Nomenclature," *Monthly Weather Review,* September 1920, 513–514.

48. Ibid., 514–515.

49. Ibid., 518–519; Linton Weeks, "When There Are Gray Skies," *Washington Post,* August 28, 2003, C1, C7.

50. Weather Bureau, *Manual of Cloud Forms and Codes for States of the Sky,* Circular S, 2nd ed. (Washington, DC: Government Printing Office, January 1949), 1; World Meteorological Organization, *International Cloud Atlas,* rev. ed. (Geneva: World Meteorological Organization, 1987); Glickman, *Glossary of Meteorology,* 145–146; Day, *Book of Clouds,* 138–141, 172–173.

51. "Clouds, Clouds, Clouds," NCAR Web site, http://www.ucar.edu/news/features/; David Randall, Marat Khairoutdinov, Akia Arakawa, and Wojciech Grabowski, "Breaking the Cloud Parameterization Deadlock," *Bulletin of the American Meteorological Society* 84 (2003): 1549; James T. Bunting, "Nephanalysis," in *The Encyclopedia of Climatology,* ed. John E. Oliver and Rhodes W. Fairbridge (New York: Van Nostrand Reinhold, 1987), 607.

52. McAdie, *A Cloud Atlas,* 3.

53. Ibid., 15–16.

54. Matthew Luckiesh, *The Book of the Sky: Journeys in Cloudland on the Wings of Experience and Knowledge,* revised and enlarged (New York: Dutton, 1933), 6, 13, 15, 16.

55. Ibid., 117, 120.

56. Murchie, *Song of the Sky,* 178.

57. Ibid., 187; Leo Marx, *The Machine in the Garden: Technology and the Pastoral Ideal in America* (New York: Oxford University Press, 1964); Annette Kolodny, *The Lay of the Land: Metaphor as Experience and History in American Life and Letters* (Chapel Hill: University of North Carolina Press, 1975).

58. Alfred H. Thiessen, "Measuring Visibility," *Monthly Weather Review,* June 1919, 401–402; A. H. Palmer, "Visibility," *Bulletin of the American Meteorological Society* 2 (1921): 1; W. R. Gregg, "Visibility," *Bulletin of the American Meteorological Society* 2 (1921): 75; R. C. Jacobson, "A Vertical Visibility Scale," *Bulletin of the American Meteorological Society* 16 (1935): 239–240; "visibility," *Glossary of Weather and Climate,* ed. Ira W. Geer (Boston: American Meteorological Society, 1996), 247.

59. Middleton, *Invention of the Meteorological Instruments,* 214–221; Louvan E. Wood, "The Development of a New Wind-Measuring System," *Bulletin of the American Meteorological Society* 26 (1945): 361–370.

60. Scott Huler, *Defining the Wind: The Beaufort Scale and How a 19th Century Admiral Turned Science into Poetry* (New York: Crown, 2004); David Longshore, *Hurricanes, Typhoons, and Cyclones* (New York: Checkmark Books, 2000), 279–280.

61. A. C. Spectorsky, ed., *The Book of the Sky* (New York: Appleton-Century-Crofts, 1956); Edinger, *Watching for the Wind;* Lyall Watson, *Heaven's Breath: A Natural History of the Wind* (New York: William Morrow, 1984), 300–301; Jan DeBlieu, *Wind: How the Flow of Air Has Shaped Life, Myth, and the Land* (Boston: Houghton Mifflin, 1998); Marq de Villiers, *Windswept: The Story of Wind and Weather* (New York: Walker, 2006). Some of these writers are collected in Dave Thurlow and C. Ralph Adler, eds., *Soul of the Sky: Exploring the Human Side of Weather* (North Conway, NH: Mount Washington Observatory, 1999).

62. Gavin Pretor-Phinny, *The Cloudspotter's Guide: The Science, History, and Culture of Clouds* (New York: Perigee Book, 2006); Billie Alonzo, "Faces in the Clouds," pamphlet in author's possession. I am not sure that it was actually Billie Alonzo who accosted me that day, but that's the name on the pamphlet. On pareidolia, see Robert Todd Carroll, *The Skeptic's Dictionary,* at http://skepdic.com/pareidol.html; and Elizabeth Svoboda, "Faces, Faces Everywhere," *New York Times,* February 13, 2007, D1. On *yukigata,* see Yutaka Yamada, "On the Topographical Origin of Some Remaining Snow Patterns, 'Yukigata,'" *Proceedings of the International Snow Science Workshop 1996* (Banff, Alberta: ISSW, 1996), 94–99; Yasoichi Endo and Kaoru Izumi, "Remaining Snow Patterns in Mountains, Yukigata, as Seen in Spring," in *Snow Engineering: Recent Advances,* ed. Masanoriizumi Izumi, Tsutomu Nakamura, and Ronald L. Sack (Rotterdam: Balkema, 1997), 549–552; Yasuaki Noguchi et al., "Activities of International Yukigata Research Group," *Proceedings of the International Snow Science Workshop 1998* (Sunriver, OR: ISSW, 1998), 318–323. I thank Ed Adams of Montana State University for calling *yukigata* to my attention.

63. Howard Gardner, *Artful Scribbles: The Significance of Children's Drawings* (New York: Basic Books, 1980), 99. Most of my collection of "kids' weather eye" drawings comes from the *Bakersfield Californian,* but I have seen them in other papers. NOAA's National Weather Service Weather Forecast Office in Jackson, Kentucky, offers a selection of almost 200 children's drawings of weather from Highland Turner Elementary School on its Web site, http://www.crh.noaa.gov/jkl/?n=kid (accessed August 10, 2007).

64. Sarah Greenough, "Of Charming Glens, Graceful Glades, and Frowning Cliffs," in *Photography in Nineteenth-Century America,* ed. Martha Sandweiss (Fort Worth, TX: Amon Carter Museum; New York: Harry N. Abrams, 1991), 275.

65. Hertz quoted in Geraldine Wojno Kiefer, "Alfred Stieglitz and *The Steerage:* An Empirio-Critical Correlation," *Word & Image* 7 (January–March 1991): 58–60; Charles Chetham, introduction to *In Pursuit of Clouds: Images and Metaphors,* by Ralph Steiner (Albuquerque: University of New Mexico Press, 1985), 11; Bachelard, *Air and Dreams,* 185.

66. Beverly W. Brannan, "Discovering Theodor Horydczak's Washington," *Quarterly Journal of the Library of Congress* 36, no. 1 (Winter 1979): 38–67. The index to Horydczak's photos and the photos themselves are accessible through the search engine at the Library of Congress American Memory Web site, http://memory.loc.gov/ammem/index.html.

67. See, for example, Hans Kaden, "The Sky in Your Landscape," *American Photographer* 46 (October 1952): 19–26; Joseph Foldes, "You Can Make Attractive Cloudscapes," *Popular Photography,* November 1955, 54; Henry Lansford, "Shooting the Sky," *Weatherwise* 35, no. 2 (April 1982): 73–81; John Field, *Colorado Skies* (Englewood, CO: Westcliffe, 1998); Steiner, *In Pursuit of Clouds;* Raymond Bial, *From the Heart of the Country: Photographs of the Midwestern Sky* (Champaign, IL: Sagamore, 1991); Wyman Meinzer, *Texas Sky* (Austin: University of Texas Press, 1998); Meinzer, *Between Heaven and Texas,* introduction by Sarah Bird, poems selected by Naomi Shihab Nye (Austin: University of Texas Press, 2006); Richard Misrach, *The Sky Book: Photographs of Richard Misrach,* text by Rebecca Solnit (Santa Fe, NM: Arena Editions, 2000).

68. Steiner, *In Pursuit of Clouds,* 11.

69. Ibid., 12, 16, 27, 30, 47, 92–93.

70. Rebecca Solnit, "Excavating the Sky," introduction to Misrach, *The Sky Book,* 23.

71. Richard Misrach, *Golden Gate* (New York: Aperture Foundation, 2005).

72. K. McGuffie and A. Henderson-Sellers, "Almost a Century of 'Imaging' Clouds over the Whole Sky," *Bulletin of the American Meteorological Society* 70 (1989): 1243–1253. Observer bias can be corrected by using a mirror marked with grid lines to objectively determine the cloud amount. An okta is a measurement of cloud cover. One okta is one-eighth of the sky.

73. C. C. Clark, "The Use of Motion Pictures in Illustrating Meteorological Problems," *Bulletin*

of the American Meteorological Society 6 (1925): 90–92; L. F. Richardson, "The Supply of Energy from and to Atmospheric Eddies," *Proceedings of the Royal Society London. Series A* 97 (1920): 354–373.

74. Delbert M. Little, "Clouds in Motion," *Bulletin of the American Meteorological Society* 15 (1934): 221–222; Edinger, *Watching for the Wind,* 75; Mogil interview, March 10, 2005.

75. Finis Dunaway, *Natural Visions: The Power of Images in American Environmental Reform* (Chicago: University of Chicago Press, 2005), 176; Ansel Adams to Jack Borden et al., January 20, 1981, Borden Papers.

76. Lillian Gish, *The Movies, Mr. Griffin, and Me* (Englewood Cliffs, NJ: Prentice-Hall, 1969), 232–233.

77. *American Film Institute Catalog: Feature Films of 1921–1930* (Berkeley: University of California Press, 1971), 906; Dorothy Scarborough, *The Wind* (New York: Harper and Brothers, 1925; reprint, Austin: University of Texas Press, 1979), 34.

78. Scarborough, *The Wind,* 171.

79. Ibid., 175.

80. Ibid., 197, 334, 336–337.

81. *The Hurricane, American Film Institute Catalog: Feature Films 1931–1940* (Berkeley: University of California Press, 1973), 977–978; Bill Levy, *John Ford: A Bio-Bibliography* (Westport, CT: Greenwood Press, 1998), 133–134.

82. Tom Shales, "'Category 6': CBS Opens a Storm Window," *Washington Post,* November 14, 2004, N1, N4.

83. Keay Davidson, *Twister: The Science of Tornadoes and the Making of an Adventure Movie* (New York: Pocket Books, 1996).

84. *Twister: The Dark Side of Nature,* Warner Bros. and Universal Pictures, 1996.

85. Eric Adler, "Meteorologists Say 'Twister' Is Exciting, but Distorts Reality," http://www.cimms.ou.edu/~stumpf/twistint.html (accessed June 22, 2005); Davidson, *Twister,* 150; Howard B. Bluestein, "A History of Severe-Storm-Intercept Field Programs," *Weather and Forecasting* 14, no. 4 (August 1999): 558–577.

86. Rachel Shortt, "Twister Movie Put NSSL on the Map," http://www.nssl.noaa.gov/stories/twister.html (accessed August 8, 2006); Twister Museum, http://www.twistercountry.com.

87. Scott MacDonald, *The Garden in the Machine: A Field Guide to Independent Films about Place* (Berkeley: University of California Press, 2001), 127; David Nye, *American Technological Sublime* (Cambridge, MA: MIT Press, 1999).

88. *The Day after Tomorrow,* 20th Century Fox, 2004; Andrew C. Revkin, "NASA Curbs Comments on Ice Age Disaster Movie," *New York Times,* April 25, 2004, 14; Revkin, "When Manhattan Freezes Over," *New York Times,* May 23, 2004, 19, 22.

89. Revkin, "When Manhattan Freezes Over," 22; Greenpeace, "Big Screen vs Big Oil," http://www.greenpeace.org/international/news/the-day-after-tomorrow (accessed August 8, 2006).

90. Mark Svenvold, *Big Weather: Chasing Tornadoes in the Heart of America* (New York: Henry Holt, 2005), 97, 99.

91. On children's stories as narratives of self in dramatic and dangerous situations, see Brian Sutton-Smith, *The Folkstories of Children* (Philadelphia: University of Pennsylvania Press, 1991).

92. *The Weather Man,* Paramount Pictures, 2005.

93. *Groundhog Day,* Columbia Pictures, Special Edition DVD, 1993; Ryan Gilbey, *Groundhog Day* (London: British Film Institute, 2004); Suzanne M. Daughton, "The Spiritual Power of Repetitive Form: Steps toward Transcendence in *Groundhog Day,*" *Critical Studies in Mass Communication* 13, no. 2 (1996): 138–154; NOAA "Special Climate Report," http://www.ncdc.noaa.gov/oa/climate/extremes/2006/groundhog.html (accessed August 11, 2006). On the invention of

tradition in the late nineteenth and early twentieth century, see David Glassberg, *American Historical Pageantry: The Uses of Tradition in the Early Twentieth Century* (Chapel Hill: University of North Carolina Press, 1990); and Ellen Litwicki, *America's Public Holidays: 1865–1920* (Washington, DC: Smithsonian Institution Press, 2000).

94. Jonah Goldberg, "A Movie for All Time," *National Review,* February 14, 2005, 35–37; Miguel de Unamuno, *Tragic Sense of Life* (New York: Dover, 1954), 268.

95. *Unchained Goddess,* 1958, Bell Science Series, available on DVD from Rhino Home Video, ©1991; Shamus Culhane, *Talking Animals and Other People* (New York: St. Martin's Press, 1986), 342–364. Some of the other films in the series were *Mr. Sun, Hemo the Magnificent* (human biology and physiology), and *The Strange Case of Cosmic Rays.*

96. MacDonald, *The Garden in the Machine,* 12.

97. Ibid., 140. Tornadoes, toilets, and drains in the Northern Hemisphere usually rotate counterclockwise, but they can rotate clockwise as well. Gravity has nothing to do with the rotation.

98. Ibid., 146n8, 407; Jesse Lerner, "Storm Squatting at El Reno," http://www.othercinema.com/otherzine/otherzine3/pweather.html (accessed August 15, 2006).

99. *Weather Diary 1* and *Weather Diary 4* rented from Video Data Bank, School of the Art Institute of Chicago, 112 S. Michigan Ave., Chicago, IL 60603.

100. Gregg Mitman, *Reel Nature: America's Romance with Wildlife on Film* (Cambridge, MA: Harvard University Press, 1999), 208.

101. Hans Neuberger, "Climate in Art," *Weather* 25, no. 2 (February 1970): 40–56. NOAA has a Web site of historic art, mostly nineteenth-century book engravings of cloud types and water spouts and photographs of clouds, storms, and Weather Bureau instruments and personnel, http://www.history.noaa.gov/art/nws_gallery/.

102. Stanley David Gedzelman, "Cloud Classification before Luke Howard," *Bulletin of the American Meteorological Society* 70 (1989): 381–395; Neuberger, "Climate in Art," 50; Kenneth Clark, *Landscape into Art* (1949; reprint, Boston: Beacon Press, 1961), 102. I viewed the temporary exhibit of Monterey artists at the Crocker Museum in Sacramento, April 1, 2006. Gedzelman's comprehensively researched and beautifully illustrated book *The Soul of All Scenery: A History of the Sky in Art* covers the sky in world art from antiquity to the present. It is available on his Web site, http://www.sci.ccny.cuny.edu/~stan. Gedzelman's book combines meteorological knowledge with a good eye for details in the paintings; it is a superb synthesis of art and science.

103. Barbara Novak, *Nature and Culture: American Landscape Painting 1825–1875* (New York: Oxford University Press, 1995), 78, 100.

104. Hubert Damisch, *A Theory of /Cloud/ Toward a History of Painting* (1972; Stanford, CA: Stanford University Press, 2002), 15.

105. Joy Waldron Murphy, "P. A. Nisbet: Yearning for Place," *Southwest Art,* August 1987, 55. A year earlier Nisbet wrote a piece for another magazine on his painting techniques; P. A. Nisbet, "Painting the World Aloft," *Artist's Magazine* 3, no. 5 (May 1986): 52–59. The Cloud Appreciation Society maintains Web pages for its members' artwork. Some of the paintings are quite good. See www.cloudappreciationsociety.org.

106. Biographical information from an unpublished manuscript by Delsie Hoyt Eaton, whose parents were friends of the Sibers, and from Eric Pender, "Painting Up a Storm," *Weatherwise* 50, no. 3 (June/July 1997): 12–17. I thank Ms. Eaton for sharing her material with me and Jack Borden for calling my attention to Siber.

107. Pinder, "Painting Up a Storm," 15–16.

108. Ibid.

109. Sloane created his public persona in more than fifty books, several of them autobiographical. Surprisingly, there is no published scholarly biography. Eric Sloane, *Eric Sloane's I Remember America* (New York: Funk and Wagnalls, 1971); Sloane, *Recollections in Black and White* (New

York: Walker, 1974); Sloane, *For Spacious Skies: A Meteorological Sketchbook of American Weather* (New York: Crowell, 1978), 27–38; Sloane, *Return to Taos: A Twice-Told Tale* (New York: Hastings House, 1982); Sloane, *Eighty: An American Souvenir* (New York: Dodd, Mead, 1985). See also James William Mauch, *Aware: A Retrospective of the Life and Work of Eric Sloane* (Laurys Station, PA: Garrigues House, 2000).

110. Eric Sloane, *Look at the Sky . . . and Tell the Weather* (New York: Duell Sloan and Pearce, 1961), 65, 87.

111. Edward S. Casey, *Earth-Mapping: Artists Reshaping Landscape* (Minneapolis: University of Minnesota Press, 2005), xx–xxii. Casey does not deal with maps as art, though there have always been cartographers whose work was considered art. Even the development of satellite and digital mapping has not replaced the skilled mapmaker-artist such as Hal Sheldon and Tibor Tóth, whose natural color relief maps are highly prized.

112. Dan Stillman, "The Art of Science," *Weatherwise* 59, no. 3 (May/June 2006): 52–53; Graeme Stephens, "The Useful Pursuit of Shadows," *American Scientist* 91 (September–October 2003): 442–449.

113. Stephens, "The Useful Pursuit of Shadows," 442, 449; L. C. W. Bonacina, "Landscape Meteorology and Its Reflection in Art and Literature," *Quarterly Journal of the Royal Meteorological Society* 65, no. 282 (October 1939): 485.

114. Jacobshagen's painting hangs in the Nelson-Atkins Museum of Art in Kansas City, Missouri, while Gornik's *Gyre* is in a private collection. Reproductions of both, and more than a hundred other land/skyscapes, may be found in Joni L. Kinsey, *Plain Pictures: Images of the American Prairie* (Washington, DC: Smithsonian Institution Press, 1996). Some of Gornik's paintings, drawings, and prints can be seen on her Web site, http://www.aprilgornik.com/ .

115. Carr's paintings are in the Vancouver Art Gallery. O'Keeffe's *Light Coming on the Plains II* is in the Amon Carter Museum, Fort Worth, Texas, and is reproduced in Kinsey, *Plain Pictures,* 112.

116. "Walter De Maria *The Lightning Field,*" http://www.lightningfield.org; The Collector's Guide: The Lightning Field, http://www.collectorsguide.com/ab/abfa06.html; John Beardsley, "Art and Authoritarianism: Walter De Maria's *Lightning Field,*" *October* 16 (Spring 1981): 25–38; Beardsley, *Earthworks and Beyond: Contemporary Art in the Landscape,* 3rd ed. (New York: Cross Rivers Press, 1998), 59–63; Cornelia Dean, "Drawn to the Lighting," *New York Times,* September 21, 2003, Travel, 9–10. Beardsley provides the information that the Dia Foundation is largely supported by Phillipa Pellizzi, heiress to the De Menil oil fortune. According to "Handbook of Texas Online," her father, John De Menil, was a French-born lawyer who married Dominique Schlumberger, whose family made a fortune in oil prospecting. In 1942 he and his family settled in Houston, Texas, where he built a museum to house his collection of modern art and supported the civil rights movement and the Democratic Party, http://www.tsha.utexas.edu/handbook/online/articles/MM/fmeny.html.

117. Walter De Maria, *The Lightning Field,* quoted in Beardsley, *Earthworks and Beyond,* 205. A summary of De Maria's philosophy may be found in an interview conducted by Paul Cummings in New York City, October 4, 1972, just before he began *The Lightning Field.* The interview is in the Archives of American Art, Smithsonian Institution, Washington, DC.

118. Michael Kimmelman, "Inside a Lifelong Dream of Desert Light," *New York Times,* April 8, 2001, sec. 2, pp. 1, 34; David Hay, "Using the Sky to Discover an Inner Light," *New York Times,* April 8, 2001, Art, 1. On the enhancement of the falls by aerialists, see Jeremy Elwell Adamson, *Niagara: Two Centuries of Changing Attitudes, 1697–1901* (Washington, DC: Corcoran Gallery of Art, 1985), 107; on the aesthetics of the via ferrata, see Stephen Regenold, "Hit the Heights, but Take the Stairs," *New York Times,* September 15, 2006, "Escapes," 1, 5.

119. James Turell, *Air Mass* (London: South Bank Centre, 1993); Mario Diacono, "Iconographia

Coelestis," in *Mapping Spaces: A Topological Survey of the Work by James Turrell* (New York: Peter Blum Editions, 1987), 48; Craig Adcock, "Light, Space, Time: The Visual Parameters of Roden Crater," in *Occluded Front James Turrell* (Los Angeles: Lapis Press, 1985), 128. See also Craig Adcock, *James Turrell: The Art of Light and Space* (Berkeley: University of California Press, 1990), and Almine Rech, *Recontres 9: James Turrell* (Paris: Almine Rech Editions Images Modernes, 2005).

120. "Exhibitions and Cultural Programs: Exploring the Intersections of Art and Science," National Academies, January–August 2004; Dozier Bell's Web site, http://www.dozierbell.com/trace_home.html.

121. Exhibitions, Art Institute of Chicago, "Focus: Iñigo Manglano-Ovalle," http://www.artic.edu/aic/exhibitions/inigo.html.

122. Deborah Solomon, "Ars Brevis," *New York Times Magazine,* February 11, 2001, 32; Roberta Smith, "Forecast: Cloudy, with a 50% Chance of Sculpture," *New York Times,* April 25, 2006, B5.

123. Randy Kennedy, "A Most Public Artist Polishes a New York Image," *New York Times,* August 20, 2006, Arts & Leisure, 1, 25; Benjamin Genocchio, "Sky Mirror," *New York Times,* October 13, 2006, B22. A measure of the acceptance of "Cloud Gate" as a way of viewing clouds may be seen in the awarding of an honorable mention to Talbot Merloti for his photograph of altocumulus reflected in the sculpture by the judges of *Weatherwise* magazine's 2007 contest. *Weatherwise* 60, no. 5 (September/October 2007): 40.

124. Claire Dederer, "Looking for Inspiration in the Melting Ice," *New York Times,* September 23, 2007, 35. A catalog for the exhibit is available from the museum, http://www.bmoca.org/artist.php?id=74 .

125. For more information on the Sphere, see http://sos.noaa.gov/about/. There are traveling versions of SOS and a permanent display at the Global Systems Division (formerly Forecast Systems Laboratory) of NOAA's Earth System Research Laboratory (ESRL) in Boulder, Colorado. Susan Stewart, *On Longing: Narratives of the Miniature, the Gigantic, the Souvenir, the Collection* (Durham, NC: Duke University Press, 1993), 48.

126. The term "Blue Marble," apparently coined by one of the *Apollo 17* crew, Eugene Cernan, Ronald Evans, and Jack Schmitt, on December 7, 1972, when the first photographs of the whole earth from space were taken, has become a slogan for NASA's Earth Observatory program, which includes producing detailed images of Earth's surface using a year's worth of photographs, from the *Terra* and *Aqua* satellites, in a montage. While acknowledging the limitations of its "Blue Marble: Next Generation" photos, NASA, like The Weather Channel, promotes a brand name and image, http://earthobservatory.nasa.gov/Newsroom/BlueMarble/. NASA archivist Mike Gentry "speculated that 'The Blue Marble' is the most widely distributed image in human history."

127. W. E. Knowles Middleton, *Catalog of Meteorological Instruments in the Museum of History and Technology* (Washington, DC: Smithsonian Institution Press, 1969). On the controversy over changes to the National Museum of Natural History's exhibit on climate change in the Arctic, see Brett Zongker, "Smithsonian Accused of Altering Exhibit," May 21, 2007, www.washingtonpost.com/wp-dyn/content/article/2007/05/21/AR2007052100860.html. For a glimpse of the exhibit at the Koshland Museum, see http://www.koshland-science-museum.org/exhibitgcc/index.jsp (accessed September 23, 2006).

128. David H. Shayt, "Artifacts of Disaster: Creating the Smithsonian's Katrina Collection," *Technology and Culture* 47 (April 2006): 358; additional information in an e-mail from David Shayt, October 30, 2007.

129. Jean Spraker, "'Come to the Carnival in Old St. Paul': Souvenirs from a Civic Ritual Interpreted," in *Prospects: An Annual of American Cultural Studies* 11 (New York: Cambridge University Press, 1987), 233–246; "'When the Weatherball Is Red, Warmer Weather Is Ahead . . . ,'" *Member*

News, Minnesota Historical Society, January/February 2002, 1–2; Peg Meier, "Weathering the Storm," *Star Tribune*, January 20, 2002, E1, E10; Minnesota Historical Society, http://www.mnhs .org/places/historycenter/exhibits/weather/tour.html (accessed January 18, 2005); Johnson County Museums, http://www.jocomuseum.org/ (accessed February 2, 2006). Material on the Minnesota Historical Society was provided by my George Washington University colleague Professor Christine Meloni. I gratefully acknowledge her help.

130. Weather Discovery Center, Mount Washington Observatory, www.mountwashington.org; Children's Museum of Houston, "Kicks Up a Storm," www.cmhouston.org.

131. Weather Research Center, John C. Freeman Weather Museum, http://www.wxresearch .org/. "The Dvorak method is based upon the analysis of cloud patterns . . . from both polar orbiting and geostationary satellites." See also Todd S. Glickman, *Glossary of Meteorology*, 2nd ed. (Boston: AMS, 2000), 244, which cites Vern Dvorak's 1975 and 1984 papers on tropical cyclone intensity analysis from satellite imagery. A full review and explanation of what is now called "the Dvorak tropical cyclone (TC) intensity estimation technique" appeared in *Bulletin of the American Meteorological Society* in September 2006; see Christopher Velden et al., "The Dvorak Tropical Cyclone Intensity Estimation Technique: A Satellite-Based Method That Has Endured for over 30 Years," *Bulletin of the American Meteorological Society* 87 (2006): 1195–1210.

132. "Wall-to-Wall Nature: Tour Guide," Montréal Biodôme brochure; Web site, http://www2 .ville.montreal.qc.ca/biodome/site/site.php?langue=en.

133. Interview with Roger Brickner, August 9, 2002. I thank Mr. Brickner for his tour and hospitality. See also Mary Reed, "Museum of Dreams," *Weatherwise* 49, no. 3 (June/July 1996): 33.

134. Trudi Hahn, "Obituary: Wendell Mordy, President of the Science Museum of Minnesota from 1977 to 1985, Dies at 82," *Star-Tribune*, July 18, 2002, http://www.startribune.com/466/ story/46680.html. "Remarks of Professor Wendell A. Mordy at the Dedication of the Charles and Henriette Fleischmann Atmospherium—Planetarium," November 14, 1963, AC 454, folder 1, Fleischmann Planetarium Records, Special Collections Department/University Archives, University of Nevada, Reno. For the technical context for the atmospherium, see Jordan D. Marché, *Theaters of Space and Time: American Planetaria, 1930–1970* (New Brunswick, NJ: Rutgers University Press, 2005), 143–146.

135. "The Fleischmann Atmospherium-Planetarium of the Desert Research Institute, University of Nevada," *Bulletin of the American Meteorological Society* 45 (1964): 394; "Remarks of Professor Mordy"; brochure of the Fleischmann Atmospherium and Planetarium, 1963; typescript of press release to save the Fleischmann Atmospherium, 3 pages, undated; "Fleischmann Atmospherium/Planetarium Fact Sheet," 2 pages, undated, AC 454, folders 8 and 13, Fleischmann Planetarium Records, Special Collections Department/University Archives, University of Nevada, Reno; quotations by permission.

Chapter Four. Transcribing Weather

1. I consulted several dozen anthologies of American poetry—U.S., Canadian, and Mexican to provide a truly "North American" perspective—and *The Reader's Guide to Periodical Literature* from 1900 to 2005, searching for weather poems. While there seem to be at least eight anthologies of poetry about the night sky and astronomy and several on the poetry of flight, the Library of Congress catalog lists none in English for "weather poetry." (There are anthologies in Japanese, Chinese, Bulgarian, Greek, and Spanish, and four collections for children.) Things may be changing, however, with the publication of *Between Heaven and Texas*, a volume of photographs

by Wyman Meinzer and poetry about the sky selected by Naomi Shihab Nye (Austin: University of Texas Press, 2006).

2. John Neubauer, "Science and Poetry," in *The New Princeton Encyclopedia of Poetry and Poetics,* ed. Alex Preminger and T. F. Brogan (Princeton, NJ: Princeton University Press, 1993), 1120–1127.

3. Eduardo Cadava, "Literature and the Weather," *Encyclopedia of Climate and Weather,* vol. 1, ed. Stephen H. Schneider (New York: Oxford University Press, 1996), 471. Cadava's study of Ralph Waldo Emerson's use of weather metaphors in discussing society and politics, *Emerson and the Climates of History* (Stanford, CA: Stanford University Press, 1997), is an elegant example of literary scholarship informed by science. Arden Reed's *Romantic Weather: The Climates of Coleridge and Baudelaire* (Hanover, NH: University Press of New England, 1983), is a pioneering effort in this same area. Schneider's *Encyclopedia* also contains groundbreaking essays by Marsha Ackermann titled "Drama, Dance, and Weather" and "Music and Weather," from which I have learned a great deal.

4. Stephen Jay Gould, *The Hedgehog, the Fox, and the Magister's Pox: Mending the Gap between Science and the Humanities* (New York: Harmony Books, 2003), 167. The relation between science and poetry is elegantly explored in Gary Paul Nabhan, *Cross-Pollinations: The Marriage of Science and Poetry* (Minneapolis, MN: Milkweed Editions, 2004).

5. Walt Whitman, *Leaves of Grass and Selected Prose,* ed. John Kouwenhoven (New York: Modern Library), 407.

6. *Webster's International Dictionary of the English Language,* 2nd ed. (Springfield, MA: G. & C. Merriam, 1936), 2078; *The Compact Edition of the Oxford English Dictionary,* vol. 1 (New York: Oxford University Press, 1971), 135.

7. Harriet Monroe, "Winds of Texas," *Poetry* 31, no. 6 (March 1928): 301–302. Monroe's poem was published in the same year as the movie *The Wind.* She may have been inspired by Scarborough's novel and/or advanced notice of the film, which was released November 23.

8. Carl Sandburg, "Wind Song," in *Modern American Poetry, Modern British Poetry,* ed. Louis Untermeyer (New York: Harcourt, Brace, 1950), 221.

9. May Swenson, "Weather," in *Nature: Poems Old and New* (Boston: Houghton Mifflin, 1994), 66. I thank my daughter, Alexandra Mergen, for calling my attention to Swenson's and several other poems used in this chapter.

10. Charles Wright, "Weather Report," in *POETRY: An Introductory Anthology,* ed. Hazard Adams (Boston: Little, Brown, 1968), 375.

11. David McCord, "Weather Words," in *Imagination's Other Place: Poems of Science and Mathematics,* ed. Helen Plotz (New York: Crowell, 1955), 51.

12. Albert Goldbarth, "A Theory of Wind," in *The Morrow Anthology of Younger American Poets,* ed. Dave Smith and David Bottoms (New York: Quill, 1985), 245–246.

13. Robinson Jeffers, "Distant Rainfall," in *The Selected Poetry of Robinson Jeffers,* ed. Tim Hunt (Stanford, CA: Stanford University Press, 2001), 501.

14. Hildegarde Flanner, "The Rain," *Poetry* 30, no. 6 (September 1927): 300–302.

15. Dorothy Livesay, "Green Rain," in *The New Oxford Book of Canadian Verse in English,* ed. Margaret Atwood (New York: Oxford University Press, 1982), 134.

16. Robert Creeley, "The Rain," in *Contemporary American Poetry,* 4th ed., ed. A. Poulin Jr. (Boston: Houghton Mifflin, 1985), 90–91.

17. Loren Eiseley, "It Is the Rain That Tells You," in *The Innocent Assassins* (New York: Scribner's, 1973), 85.

18. Jay Parini, "Mizzle," in *Poems for a Small Planet: Contemporary American Nature Poetry,* ed. Robert Pack and Jay Parini (Hanover, NH: University Press of New England, 1993), 173.

19. Linda Hussa, "Give Us Rain!" in *Cowboy Poetry Matters: From Abilene to the Mainstream,* ed. Robert McDowell (Ashland, OR: Story Line Press, 2000), 85–86.

20. Elizabeth Madox Roberts, "The Sky," in *Modern American Poetry,* 291.

21. Leonard Nathan, "The Sky Hung Idiot Blue," *Poetry* 87, no. 4 (January 1956): 223.

22. Naomi Shihab Nye, "El Paso Sky," in *Between Heaven and Texas* (Austin: University of Texas Press, 2006), 96.

23. Wallace Stevens, "A Clear Day and No Memories," in *Wallace Stevens: Collected Poetry and Prose,* ed. Frank Kermode and Joan Richardson (New York: Library of America, 1997), 475.

24. Amy Lowell, "Night Clouds," in *Modern American Poetry,* 165.

25. Robinson Jeffers, "Clouds at Evening," in *Selected Poetry,* 139. This poem appears with slight differences—"curdled" instead of "curded," for example—in Louis Untermeyer's *Modern American Poetry,* 381.

26. Richard Hugo, "The Clouds of Uig," in *Making Certain It Goes On: The Collected Poems of Richard Hugo* (New York: Norton, 1991), 383–384.

27. Kenneth Koch, "A Big Clown-Face-Shaped Cloud," in *Possible World* (New York: Knopf, 2002), 57. The New York City transit program "Poetry in Motion" put Koch's poem in subway cars and buses throughout the city.

28. Gaston Bachelard, *Air and Dreams: An Essay on the Imagination of Movement* (1943; Dallas: Dallas Institute of Humanities and Culture, 1988), 185–196.

29. Billy Collins, *Sailing Alone around the Room: New and Selected Poems* (New York: Random House, 2001), 32.

30. Andrew C. Revkin, "The Ununited States, When It Comes to the Weather," *New York Times,* January 7, 2007, sec. 4, p. 4. The two scientists quoted in Revkin's article, Michael MacCracken and Michael Mann, are both outstanding climatologists, and I certainly agree with MacCracken that "climate is a mental construct," but he and Mann seem to be responding to a question about why the federal government has been slower than California to respond to concerns about global warming. This may have caused them to miss the essential unity of Americans on the meaning of their weather. "Variables of state," in *Glossary of Weather and Climate,* ed. Ira W. Geer (Boston: American Meteorological Society, 1996), 246.

31. For an extended discussion of snow poetry, see Bernard Mergen, *Snow in America* (Washington, DC: Smithsonian Institution Press, 1997), 225–235; William Matthews, "Spring Snow," in *Rising and Falling* (Boston: Little, Brown, 1979), 3.

32. Jun Fujita, "Projections," *Poetry* 32 (July 1928): 202. Jun Fujita is not, as far as I can tell, related to Tetsuya "Ted" Fujita (1920–1998), the internationally renowned tornado scientist, although both made their careers in Chicago and lived there in the decade 1953–1963.

33. May Swenson, "The Fluffy Stuff," in *Nature,* 65–66.

34. Peter Viereck, "A Walk on Snow," in *Terror and Decorum* (New York: Scribner's, 1948), 52–53.

35. Collins, *Sailing Alone around the Room,* 141.

36. Ibid.

37. Radcliffe Squires, "Storm in the Desert," in *Waiting in the Bone* (Omaha: University of Nebraska Abattoir Editions, 1973), 76–77.

38. Ibid., 77.

39. Stephen [*sic*] Hazo, "Rains," *Samizdat #6,* Fall 2000, http://www.samizdateditions.com/issue6/hazo1.html.

40. Samuel Hazo, "The World That Lightning Makes," in *And Not Surrender: American Poets on Lebanon* (Washington, DC: Arab American Cultural Foundation, 1982). Hazo was Pennsylvania state poet until the post was abolished by Governor Ed Rendell in 2004. I found this poem in the excellent anthology of Arab American poetry *Grape Leaves,* ed. Gregory Orfalea and Sharif Elmusa (Salt Lake City: University of Utah Press, 1988), 123–124.

41. A. R. Ammons, "Standing Light Up," in *Poems for a Small Planet,* 9.

42. Charles Martin, "Reflections on a Dry Spell," in *Poems for a Small Planet,* 144.

43. Robert Hayden, "Ice Storm," in *American Journal,* 2nd ed. (New York: Liveright, 1982), 39.

44. Woody Guthrie, "Dust Storm Disaster," in *The Woody Guthrie Song Book,* ed. Harold Leventhal and Marjorie Guthrie (New York: Grosset and Dunlop, 1976); "The Great Dust Storm," NOAA Photo Library, http://www.photolib.noaa.gov/nws/woddie.html.

45. Donald Worster, *Dust Bowl: The Southern Plains in the 1930s* (New York: Oxford University Press, 1979), 164–180. There are several Internet sites for Hurricane Katrina poetry, for example, a fund-raising site sponsored by Mississippi State University, http://www.msstate.edu/dept/IH/KatrinaSubmissions.html, and an inspirational site, "Poetry in a Cup," http://poetryinacup.org/hurricanekatrina/tribute.html. Some of these poems do express anger and point the finger of blame.

46. Karl Shapiro, "Tornado Warning," in *Selected Poems* (New York: Random House, 1968), 318.

47. Carol Muske, "An Octave above Thunder," in *An Octave above Thunder* (New York: Penguin, 1997), 7.

48. Ibid., 9.

49. Robert Hedin, "Tornado," in *Tornadoes* (Memphis, TN: Ion Books, 1990), 14. Most of the fourteen poems in this slender volume make reference in some way to storms. I have seen slightly different versions of this poem.

50. James Merrill, "Waterspout," in *Collected Poems,* ed. J. D. McClatchy and Stephen Yenser (New York: Knopf, 2001), 502.

51. Hart Crane, "The Hurricane," in *The Complete Poems of Hart Crane* (Garden City, NY: Doubleday Anchor Books, 1958), 128–129. For a beautifully illustrated treatment of hurricanes in science, history, literature, and art, see Kerry Emanuel, *Divine Wind: The History and Science of Hurricanes* (New York: Oxford University Press, 2005).

52. Victor Hernández Cruz, "Problems with Hurricanes," in *Maraca: New and Selected Poems, 1965–2000* (Minneapolis, MN: Coffee House Press, 2001), 148. Cruz's longer poem in the same volume, "Atmospheric Phenomenon: The Art of Hurricanes," contrasts the styles of Hurricanes Hugo, Iris, and Marilyn.

53. Alfonso Reyes, "Monterrey Sun," in *An Anthology of Mexican Poetry,* ed. Octavio Paz (Bloomington: Indiana University Press, 1958), 194–196.

54. Stevens, "The News and the Weather," in *Collected Poetry and Prose,* 237–238; Mark Van Doren, "This Amber Sunstream," in *A Winter Diary and Other Poems* (New York: Macmillan, 1935), 113.

55. David Colosi, "Sun with Issues," in *From Totems to Hip-Hop: A Multicultural Anthology of Poetry across the Americas, 1900–2002,* ed. Ishmael Reed (New York: Thunder's Mouth Press, 2003), 8–11.

56. Ted Kooser, "Weather Central," in *Weather Central* (Pittsburgh: University of Pittsburgh Press, 1994), 87.

57. George W. Mindling, "Soliloquy of the Weather Man," "Weather Man Poems," NOAA, http://www.history.noaa.gov/art/weatherpoems.html (accessed August 4, 2004).

58. Joy Whiteside, "Weather Woman," *Breeze* 3, no. 5 (June 10, 1945): 4, NOAA, http://www.history.noaa.gov//art/breeze.html (accessed July 14, 2004).

59. George W. Mindling, "The Raymette and the Future," NOAA, Weather Man Poems.

60. Archibald MacLeish, "Weather," in *Modern American Poetry,* ed. Louis Untermeyer (New York: Harcourt, Brace, 1950), 475.

61. Hyatt H. Waggoner, *American Poets: From the Puritans to the Present* (Baton Rouge: Louisiana State University Press, 1968), 486; MacLeish, "Memorial Rain," in *Modern American Poetry,* 473–474.

62. Reed Whittemore, "The Weather This Spring," *Poetry* 89, no. 2 (November 1956): 107–108.

63. Howard Nemerov, "The Weather of the World," in *The Collected Poems of Howard Nemerov* (Chicago: University of Chicago Press, 1977), 483–484.

64. Richard Hugo, "Bear Paw," in *Making Certain It Goes On,* 215–216.

65. Mark Strand, *The Weather of Words: Poetic Invention* (New York: Knopf, 2001), 139–142.

66. Wallace Stevens, "A Collect of Philosophy," in *Collected Poetry and Prose,* 858, 866.

67. Charles Berger, *Forms of Farewell: The Late Poetry of Wallace Stevens* (Madison: University of Wisconsin Press, 1985). Berger is one of the few scholars who reads Stevens as a political poet engaged in his own time. "I would argue," Berger writes, referring to an earlier poem, "Esthétique du Mal," "what triggered the finding of the auroras by Stevens was not so much a text as an event: the dropping of the atomic bomb, the epitome of all the great explosions prefigured in the volcano's trembling." While I think the reference to nuclear annihilation in "Auroras" is obvious, I believe Berger and other critics undervalue the purely meteorological aspects of the poem.

68. See http://science.nasa.gov/headlines/y2002/23sep_auroraseason.htm (accessed November 9, 2006); Robert H. Eather, *Majestic Lights: The Aurora in Science, History, and the Arts* (Washington, DC: American Geophysical Union), 1980.

69. Wallace Stevens, "The Auroras of Autumn," in *Collected Poetry and Prose,* 355.

70. Ibid., 363.

71. Gyorgyi Voros, *Notations of the Wild: Ecology in the Poetry of Wallace Stevens* (Iowa City: University of Iowa Press, 1997), 6, 12.

72. Ibid., 16–17. Other literary critics who have glimpsed a piece of Stevens's environmental ethic are George S. Lensing, *Wallace Stevens and the Seasons* (Baton Rouge: Louisiana State University Press, 2001); and Justin Quinn, *Gathered beneath the Storm: Wallace Stevens, Nature, and Community* (Dublin: University College Dublin Press, 2002). I cannot fully agree with Quinn's narrowing of Stevens epistemology when he writes: "The 'reality' that Stevens finds in landscape and nature is not some solid basis on which psychologies and polities may be safely built, but rather the place where all such systems are recognised as simulacra and fictions" (78). This returns us to Napier Shaw's comment about scientific theories and fairy tales, but I think Stevens saw the atmosphere as more than simulacra. Like his "Snow Man," it is "the nothing that is." After it is named and numbered, nature becomes "the nothing that is not there."

73. Elizabeth Madox Roberts, *Jingling in the Wind* (New York: Viking, 1928), 256.

74. Ibid., 26.

75. Ibid., 27–39.

76. Ibid., 183–184.

77. Ibid., 213–215.

78. Allan Nevins, "Jingling in the Wind," in *Elizabeth Madox Roberts: An Appraisal by J. Donald Adams* (New York: Viking Press, 1958), 26. Nevins's review originally appeared in the *Saturday Review of Literature.*

79. George R. Stewart, *Storm* (New York: Random House, 1941), xi–xiii.

80. According to geographer Randy Cerveny, the Weather Bureau began naming storms in the North Atlantic in 1950, using the alphabetical form "Able," "Baker," etc. The bureau used women's names from 1954 to 1979. Cerveny also mentions that, as early as 1887, Clement L. Wragge, director of the Australian Central Weather Bureau, named every storm that appeared on the Australasian weather map, often using the names of politicians he disliked. Randy Cerveny, "Weather Talk: What's in a Name?" *Weatherwise* 56, no. 5 (September/October 2003): 46–47. See also Hanna Rosin, "From: Agusta Wynd: Re: Hurricane Names," *Washington Post,* August 9, 2004, C1–C2, who calls attention to the fact that the National Hurricane Center receives hundreds of letters suggesting names for hurricanes or protesting names already selected by the WMO. The most outrageous protest came in 2001 from Abraham Foxman of the Jewish Anti-

Defamation League who forced the National Hurricane Center to change "Israel" to "Ivo." Rosin also identifies a Florida feminist, Roxcy Bolton, who badgered the Hurricane Center to stop using women's names exclusively. See also Charles Fitzhugh Talman, "The Vocabulary of Weather and Climate," *Bulletin of the American Meteorological Society* 6 (1925): 10–11. An article in the *Bulletin of the American Meteorological Society,* January 1948, 35, quotes an editorial in the *Honolulu Advertiser* of October 13, 1947, that military meteorologists began the practice of bestowing names on typhoons during the war. Stewart may have inspired the Frederick Loewe and Alan J. Lerner song "They Call the Wind Maria," from their 1951 musical *Paint Your Wagon,* but I have no evidence to support this supposition.

81. Stewart, *Storm,* 48–49.

82. Ibid., 62–63, 244.

83. Ibid., 289–290.

84. Gore Vidal, *Williwaw* (New York: Dutton, 1946); *The Compact Edition of the Oxford English Dictionary,* vol. 2 (New York: Oxford University Press, 1971), 3781. The *Glossary of Arctic and Subarctic Terms,* compiled by the Research Studies Institute, of the Arctic, Desert, Tropic Information Center, Air University, Maxwell Air Force Base, Alabama, in September 1953, p. 88, claims "williwaw" for Alaska and describes it at some length. Marjory Stoneman Douglas, *Hurricane* (New York: Rinehart, 1958), 7.

85. Vidal, *Williwaw* 6–7, 63, 66, 53.

86. Paul Quarrington, *Storm Chasers* (New York: St. Martin's, 2005), 32.

87. Ibid., 30.

88. Ibid., 236.

89. Ibid., 135–136.

90. Rick Moody, *The Ice Storm* (New York: Warner Books, 1994), 135. Moody grew up in suburban Connecticut, graduated from Brown and Columbia Universities, and has published eight books. Two of his novels have been made into movies, *Garden State* and *The Ice Storm.*

91. Ibid., 32–33, 41.

92. Ibid., 95.

93. Ibid., 175, 186, 206.

94. Ibid., 271, 278.

95. Jean Thompson, *Wide Blue Yonder* (New York: Simon and Schuster, 2002), 7.

96. Ibid., 52.

97. Ibid., 106.

98. Ibid., 245.

99. Ibid., 266–267.

100. Ibid., 362–363.

101. Clint McCown, *The Weatherman* (St. Paul, MN: Graywolf Press, 2004), 23.

102. Ibid., 76–77.

103. Ibid., 174.

104. Ibid., 216–218.

105. Ibid., 218.

106. Howard Nemerov, "Because You Ask about the Line between Poetry and Prose," in *The Vintage Book of Contemporary Poetry,* ed. J. D. McClatchy (New York: Vintage, 1990), 130.

Chapter Five. Suffering Weather

1. My thanks to storm chaser Jay Antle, assistant professor of history, Johnson County Community College, who recommended Tempest Tours. There are several reputable storm-chasing

outfits, but Martin Lisius and his organization proved exceptional. See http://www.tempest tours.com. For an excellent introduction to storm chasing, see David Robertson, "Beyond *Twister: A Geography of Recreational Storm Chasing on the Southern Plains," Geographical Review* 89, no. 4 (October 1999): 533–554.

2. On Colonel William Reid of the Royal Engineers, see Bob Sheets and Jack Williams, *Hurricane Watch: Forecasting the Deadliest Storms on Earth* (New York: Vintage, 2001), 24–30.

3. Statistics from "Tornado Forecasting and Warning," *Bulletin of the American Meteorological Society* 87 (2006): 972–976. A "weak" tornado is defined as one that lasts less than ten minutes and has a path less than 100 yards wide and a mile in length, with wind speeds below 150 mph. "While weak tornadoes account for less than 5% of all tornado fatalities, they are still potentially dangerous and are often more difficult to predict than larger storms because of their shorter lifetimes and sometimes innocuous precursor signals" (973).

4. Clifford Geertz, "Deep Play: Notes on the Balinese Cockfight," *Daedalus* 101 (1972): 1–37.

5. Joe Duggan, "Twisted Perspective," *Lincoln Journal Star,* June 24, 2003, 1A, 8A.

6. Thomas P. Grazulis, *The Tornado: Nature's Ultimate Windstorm* (Norman: University of Oklahoma Press, 2001), 58–60.

7. Doswell's thoughts on the problems associated with averaging unique phenomena are further spelled out on his Web site, http://www.cimms.ou.edu/~doswell.

8. Grazulis, *The Tornado,* 35, 185. *Stormtrack,* with more than 1,000 subscribers, is now edited by Tim Marshall in Flower Mound, Texas, and has been available only electronically since 2002. See http://www.stormtrack.org/ (accessed March 24, 2005). *Stormtrack* provides an extensive list of useful links to newsgroups and chat rooms, documents, educational resources, road and hotel guides, and chaser Web sites.

9. Sandra Coleman and Sam McCloud, "A Brief History of Storm Chasing," National Association of Storm Chasers and Spotters, http://www.chasingstorms.com/history.html (accessed June 22, 2005). Faidley's career can be followed on http://www.cyclonecowboy.com (accessed May 17, 2005) and in his book *Storm Chaser: In Pursuit of Untamed Skies* (Atlanta: Weather Channel/Bve Products, 1996). Mark Svenvold gives a slightly different history of storm chasing in *Big Weather: Chasing Tornadoes in the Heart of America* (New York: Henry Holt, 2005), and provides some colorful comments on Sean Casey, Warren Faidley, and other high-profile storm chasers.

10. Robertson, "Beyond *Twister*"; Robert S. Bristow and Heather Cantillon, "Tornado Chasing: The Ultimate Risk Tourism," *Parks and Recreation* 35, no. 9 (September 2000): 98–105; Steve Wilson, "Twisted?" *Travel Holiday,* May 2000, 68–71; J. Wise, "Playing Twister," *Women's Sports and Fitness News,* June 2000, 92–95; Ralph Blumenthal, "Hot on the Trail of Tornadoes, Where Too Close Is Just Right," *New York Times,* June 7, 2004, A1, A23; and Richard Conn, "The Aesthetics of Storm-Chasing," *Weatherwise* 42, no. 3 (June 1989): 143–147.

11. See, for example, "West Virginia Lightning," http://wvlightning.com/chasing/faq.shtml (accessed March 28, 2007).

12. All the weather observations from the diaries of William Clark and Meriweather Lewis have been excerpted and published in two volumes: Vernon Preston, ed., *Lewis and Clark: Weather and Climate Data from the Expedition Journals* (Boston: American Meteorological Society, 2007).

13. Charles A. Doswell III, "What Is a Tornado?" http://www.cimms.ou.edu/~doswell/a_tornado/atornado.html (accessed July 10, 2003); Charles A. Doswell III and Donald W. Burgess, "Tornadoes and Tornadic Storms: A Review of Conceptual Models," in *The Tornado: Its Structure, Dynamics, Prediction, and Hazards,* ed. C. Church et al. (Washington, DC: American Geophysical Union, 1993), 169.

14. David Kranz, Corrine Olson, Melanie Brandert, Randy Nascall, Ben Shouse, and Cory Beck, "Twister Levels Town; Wind Whips S. Dakota," *Sioux Falls Argus Leader,* June 25, 2003, 1A, 9A.

15. "Tornado Destroys Manchester, S.D.," *Sioux City Journal,* June 26, 2003, A1, A3.

16. Todd S. Glickman, *Glossary of Meteorology,* 2nd ed. (Boston: American Meteorological Society, 2000), 233; "What Is a Tornado?" http://www.cimms.ou.edu/~doswell/a_tornado/ atornado.html.

17. Priit J. Vesilind, "The Hard Science, Dumb Luck, and Cowboy Nerve of Chasing Tornadoes," *National Geographic* 205, no. 4 (April 2004): 2–37; John T. Snow, "Early Tornado Photographs," *Bulletin of the American Meteorological Society* 65 (1984): 360–364.

18. Henry J. Cordes, "An Auroran Hoards Still Bigger Hailstone," *Omaha World-Herald,* June 27, 2003, 1–2.

19. Grazulis, *The Tornado,* 77–116; Marlene Bradford, *Scanning the Skies: A History of Tornado Forecasting* (Norman: University of Oklahoma Press, 2001); Charles A. Doswell III, Steven J. Weiss, and Robert H. Johns, "Tornado Forecasting: A Review," in *The Tornado: Its Structure, Dynamics, Prediction, and Hazards,* 557–571. See also Steve Corfidi, "A Brief History of the Storm Prediction Center," with a timeline by Roger Edwards and Fred Ostby, http://www.spc.noaa .gov/history/early.html and http://www.spc.noaa.gov/history/timeline.html.

20. Larry Danielson, "Tornado Stories in the Breadbasket: Weather and Regional Identity," in *Sense of Place: American Regional Cultures,* ed. Barbara Allen and Thomas J. Schlereth (Lexington: University Press of Kentucky, 1990), 28–39. Danielson points out that weather lore is second only to health and medical lore in the number of entries collected in William E. Koch, *Folklore from Kansas: Customs, Beliefs, and Superstitions* (Lawrence: Regents Press of Kansas, 1980).

21. Joseph G. Galway, "J. P. Finley: The First Severe Storms Forecaster," *Bulletin of the American Meteorological Society* 66, nos. 11 and 12 (November and December 1985): 1389–1395, 1506–1510; John Park Finley, *The Subanu: Studies of a Sub-Visayan Mountain Folk of Mindanao* (Washington, DC: Carnegie Institution, 1913). This fieldwork included collecting Filipino meteorological terms.

22. John Park Finley, *Tornadoes: What They Are and How to Observe Them, with Practical Suggestions for the Protection of Life and Property* (New York: Insurance Monitor, 1887), 25.

23. Ibid., 30.

24. Ibid., 33, 114.

25. Ibid., 66.

26. Michael Roark, "Storm Cellars: Imprint of Fear on the Landscape," *Material Culture* 24, no. 2 (Summer 1992): 51; Sarah Ann Brown, "Storm Cellars of Cleburne County, Arkansas: A Study of a Vernacular Form" (M.A. thesis, George Washington University, 1982, 9–16). Part of this thesis was published in *Mid-America Folklore* 12, no. 2 (Fall 1984): 1–16.

27. Roark, "Storm Cellars," 45.

28. William Least Heat-Moon, *PrairyErth (a Deep Map)* (Boston: Houghton Mifflin, 1991), 40–45.

29. Finley, *Tornadoes,* 67–68.

30. Cleveland Abbe, "The Prediction of Tornadoes and Thunderstorms," *Monthly Weather Review* 27 (April 1899): 159; Abbe, "Unnecessary Tornado Alarms," *Monthly Weather Review* 27 (June 1899): 255. "As late as 1952, Ivan R. Tanhill, chief of the bureau's forecast division, defended the bureau's 'no tell' policy by stating that 'because such storms affect such a small area, more people were likely to lose their lives as a result of forecast-generated hysteria than from tornadoes themselves'"; see Stephen F. Corfidi, "The Birth and Early Years of the Storm Prediction Center," *Weather and Forecasting* 14, no. 4 (August 1999): 520.

31. Alfred J. Henry, "Spurious Tornado Photographs," *Monthly Weather Review* 27 (May 1899): 204; "A Newspaper Tornado Fake," *Monthly Weather Review* 27 (August 1899): 360–361.

32. Galway, "J. P. Finley," 1392; Donald Whitnah, *A History of the United States Weather Bureau* (Urbana: University of Illinois Press, 1961); W. E. Donaldson, "Tornado of June 6, 1904, at Binghamton, N.Y.," *Monthly Weather Review* 33 (May 1905): 239–240; C. M. Strong, "Tornado of May

10, 1905, at Snyder, Oklahoma," *Monthly Weather Review* 33 (August 1905): 355–356; S. D. Flora, "Tornadoes in Kansas," *Monthly Weather Review* 43 (December 1915): 615.

33. Bradford, *Scanning the Skies*, 52.

34. "Tornado Alley: The area of the United States in which tornadoes are most frequent. It encompasses the great lowland areas of the Mississippi, the Ohio, and lower Missouri River valleys." Ira W. Geer, ed., *Glossary of Weather and Climate* (Boston: American Meteorological Society, 1996), 233.

35. Herbert C. Hunter, "Tornadoes of the United States, 1916–1923," *Monthly Weather Review* 53 (May 1925): 199.

36. Ralph E. Huschke, ed., *Glossary of Meteorology* (Boston: American Meteorological Society, 1959).

37. Peter S. Felknor, *The Tri-State Tornado: The Story of America's Greatest Tornado Disaster* (Ames: Iowa State University Press, 1992); Wallace Akin, *The Forgotten Storm: The Great Tri-State Tornado of 1925* (Guilford, CT: Lyons Press, 2002).

38. Felknor, *The Tri-State Tornado*, 12–13, quotes the *St. Louis Post-Dispatch,* March 20, 1925, 1–2, on the destruction of De Soto, Illinois: "The scene resembles that of a World War battlefield, except that on a battlefield the victims are men. Here they are women and children. Many of the men escaped through the fact that they were away from home, mostly at work in the coal mines, and were out of the tornado's path." Akin quotes a more florid reporter in the *Illinois State Register,* March 21, 1925, 1, 7, "A World War I veteran stood gazing over newly made crosses, out into the garbled horizon, tangled by the fiendish assault of nature gone amuck. 'Four of those boys fought with me "over there." One of them is my brother. We saw Flanders together and he said to me: Bob, I'm glad we aren't buried in a mess like this.' Bob's brother sleeps here amid surrounds that rival Flanders in all its battle strewn glory" (70).

39. Felknor, *The Tri-State Tornado*, 80–97; Akin, *The Forgotten Storm*, 123–133.

40. Felknor, *The Tri-State Tornado*, 103.

41. Bradford, *Scanning the Skies*, 85–101; Grazulis, *The Tornado*, 89–90.

42. Charles A. Doswell III, "Societal Impacts of Severe Thunderstorms and Tornadoes: Lessons Learned and Implications for Europe," *Atmospheric Research* 67–68 (2003): 143–144.

43. Ibid., 142; Jane A. Bullock and George D. Haddow, "Emergency Management in the 21st Century," manuscript draft 1, March 12, author's possession; 2002; C. V. Anderson, *The Federal Emergency Management Agency (FEMA)* (Hauppauge, NY: Nova Science Publishers, 2002). For a more analytical history of disaster management, see Kevin Rozario, *The Culture of Calamity: Disaster and the Making of Modern America* (Chicago: University of Chicago Press, 2007).

44. Laura Kirkwood Plumb, "My Personal Experience with a Texas Twister," *Scribner's,* June 1925, 645–646.

45. Ibid., 648–649.

46. Joseph G. Galway, "Early Severe Thunderstorm Forecasting and Research by the United States Weather Bureau," *Weather and Forecasting* 7 (December 1992): 569; Doswell, Weiss and Johns, "Tornado Forecasting," 546.

47. The story of Fawbush and Miller's breakthrough in tornado forecasting has been told in several publications. See Grazulis, *The Tornado*, 86, who suggests that their approach was a refined version of Finley's; and C. C. Bates and J. F. Fuller, *America's Weather Warriors, 1814–1984* (College Station: Texas A&M University Press, 1986). For Miller's own account, see "The Unfriendly Sky," on the NOAA Web site, www.nssl.noaa.gov/GoldenAnniversary/Historic.html (accessed August 20, 2003).

48. Dwight W. Chapman, "A Brief Introduction to Contemporary Disaster Research," in *Man and Society in Disaster,* ed. George W. Baker and Dwight W. Chapman (New York: Basic Books, 1962), 5.

49. Galway, "Early Severe Thunderstorm Forecasting," 571–581; Corfidi, "The Birth and Early Years of the Storm Prediction Center," 507–525. On weather modification, see Horace R. Byers, "History of Weather Modification," in *Weather and Climate Modification,* ed. W. N. Hess (New York: Wiley, 1974), 3–44. Orville's article appears in *Collier's,* May 28, 1954, 25–26. I thank James R. Fleming for this reference.

50. Charles A. Doswell III, Alan R. Moller, and Harold E. Brooks, "Storm Spotting and Public Awareness since the First Tornado Forecasts of 1948," *Weather and Forecasting* 14 (1999): 552–554. An interesting assessment of the problems facing the Weather Bureau in establishing an integrated warning system is provided by the director of the NSSFC a few years after the Palm Sunday disaster. See Allen D. Pearson, "Tornado Warnings over Radio and Television," *Bulletin of the American Meteorological Association* 49 (April 1968): 361–363. For information from NOAA on becoming a SKYWARN spotter, see http://www.stormready.noaa.gov/contact.htm.

51. John H. Sims and Duane D. Baumann, "The Tornado Threat: Coping Styles of the North and South," *Science* 176 (June 30, 1972): 1386–1392.

52. Anthony F. C. Wallace, *Tornado in Worcester: An Exploratory Study of Individual and Community Behavior in an Extreme Situation,* National Research Council Publication 392 (Washington, DC: National Academy of Sciences, 1956).

53. For example, Anthony Oliver-Smith, "'What Is a Disaster?': Anthropological Perspectives on a Persistent Question," in *The Angry Earth: Disaster in Anthropological Perspective,* ed. Anthony Oliver-Smith and Susanna M. Hoffman (New York: Routledge, 1999), 18–34.

54. Wallace, *Tornado in Worcester,* 7.

55. Ibid., 8; Bob Pifer and H. Michael Mogil, "NWS Hazardous Weather Terminology," *Bulletin of the American Meteorological Society* 59 (December 1978): 1583–1588.

56. William H. Hooke and Roger A. Pielke Jr., "Short-Term Weather Prediction: An Orchestra in Need of a Conductor," in *Prediction: Science, Decision Making, and the Future of Nature,* ed. Daniel Sarewitz, Roger Pielke Jr., and Radford Byerly Jr. (Washington, DC: Island Press, 2000), 61–83.

57. Wallace, *Tornado in Worcester,* 10.

58. Dudley Lynch, *Tornado: Texas Demon in the Wind* (Waco, TX: Texian Press, 1976), 119–132.

59. Mhyra S. Minnis and A. Perry McWilliams, *Tornado: The Voice of the People in Disaster and After* (Lubbock: Texas Tech University Press, 1971). A similar study was conducted after the Topeka, Kansas, tornado of June 6, 1966; see James B. Taylor, Louis A. Zurcher, and William H. Key, *Tornado: A Community Response to Disaster* (Seattle: University of Washington Press, 1970).

60. Doswell, Moller, and Brooks, "Storm Spotting and Public Awareness," 548; Grazulis, *The Tornado,* 131–145.

61. Grazulis, *The Tornado,* 130–145.

62. Ibid., 131.

63. "Tetsuya Theodore Fujita (1920–1998)," http://www.tornadoproject.com/scale/tedfujita.htm.

64. John T. Snow and Theresa A. Leyton, "Reflections on Ted Fujita: The Relevance of His Many Contributions to Today's Wind Science" (draft of a paper delivered to the Eleventh International Conference of Wind Engineering, Lubbock, Texas, June 3, 2003), http://geosciences.ou.edu/~jsnow/present/Reflections%20on%20Ted%20Fujita.pdf (accessed January 30, 2004).

65. Snow and Leyton, "Reflections on Ted Fujita"; James R. McDonald, "T. Theodore Fujita: His Contribution to Tornado Knowledge through Damage Documentation and the Fujita Scale," *Bulletin of the American Meteorological Society* 82 (2001): 63–72; and Howard B. Bluestein, *Tornado Alley: Monster Storms of the Great Plains* (New York: Oxford University Press, 1999), 5. Various wind and wind damage scales have been developed in the century since Beaufort, for example, the Saffir-Simpson Damage Potential Scale used by the NWS to access hurricane damage since 1973. It is a five-point scale for winds 73 to 155+ mph.

66. Grazulis, *The Tornado,* 144.

67. NOAA's National Weather Service, Storm Prediction Center, "The Enhanced Fujita Scale (EF Scale)," http://www.spc.noaa.gov/efscale/ and http://www.spc.noaa.gov/faq/tornado/ef-scale.html (accessed February 8, 2007); Sean Potter, "Fine-Tuning Fujita," *Weatherwise* 60, no. 2 (March/April 2007): 64–71.

68. Grazulis, *The Tornado,* 91–92; David Atlas, "Severe Local Storms," *Bulletin of the American Meteorological Society* 57 (1976): 398–401; Doswell, Moller, and Brooks, "Storm Spotting and Public Awareness," 544–557; Doswell, Weiss, and Johns, "Tornado Forecasting," 557–571.

69. Valuable information on the history of disaster studies and the growth of public and private hazard management may be found in George W. Baker and Dwight W. Chapman, eds., *Man and Society in Disaster* (New York: Basic Books, 1962); Joanne M. Nigg and Dennis Mileti, "Natural Hazards and Disasters," in *Handbook of Environmental Sociology,* ed. Riley E. Dunlap and William Michelson (Westport, CT: Greenwood Press, 2002), 272–294; and Rutherford H. Platt, *Disasters and Democracy: The Politics of Extreme Natural Events* (Washington, DC: Island Press, 1999).

70. Howard Inglish, ed., *Tornado: Terror and Survival, The Andover Tornado—April 26, 1991* (Andover, KS: Counseling Center of Butler County, 1991), 11, 13, 57, 59; "Tornado Forecasting and Warning," 974.

71. Inglish, *Tornado,* 52; "Tornado Shelters," Sedgwick County Emergency Management, http://www.sedgwickcounty.org/emermgmt/tornado_shelters,html; Thomas Schmidlin et al., "Unsafe at Any (Wind) Speed? Testing the Stability of Motor Vehicles in Severe Winds," *Bulletin of the American Meteorological Society* 83 (2002): 1821–1830; and comments by Joseph Golden, Daniel McCarthy, and Robert G Goldhammer, *Bulletin of the American Meteorological Society* 83 (2002): 1831–1837; "Mobile Homes and Severe Windstorms," *Bulletin of the American Meteorological Society* 87 (2006): 976–977; Megan E. Perry, "Weatherfront: New Indiana Law Looks to Protect Mobile Home Residents," *Weatherwise* 60, no. 4 (July/August 2007): 10.

72. Inglish, *Tornado,* 11, 12, 16, 21, 26, 28, 37, 51, 53; Thomas Schmidlin and J. Knabe, "Tornado Shelters in Mobile Home Parks in the United States," *Journal of the American Society of Professional Emergency Planners,* 2001, 1–15.

73. Inglish, *Tornado,* 56–68.

74. Ibid., 77–85. For the early emphasis on children, see, for example, Stewart E. Perry, Earle Silber, and Donald A. Bloch, *The Child and His Family in Disaster: A Study of the 1953 Vicksburg Tornado,* Disaster Study Number Five (Washington, DC: Committee on Disaster Studies, National Research Council, 1956).

75. Doswell, "Social Impacts of Severe Thunderstorms, 147.

76. Stanley A. Changnon et al., "Effects of Recent Weather Extremes on the Insurance Industry: Major Implications for the Atmospheric Sciences," *Bulletin of the American Meteorological Society* 78 (1997): 425–435; Howard Kunreuther and Richard J. Roth, eds., *Paying the Price: The Status and Role of Insurance against Natural Disasters in the United States* (Washington, DC: Joseph Henry Press, 1998); Shelia Jasanoff, "The Songlines of Risk," *Environmental Values* 8, no. 2 (1999): 135–152.

77. Douglas Brinkley, *The Great Deluge: Hurricane Katrina, New Orleans, and the Mississippi Gulf Coast* (New York: William Morrow, 2006); Ronald J. Daniels, Donald F. Kettl, and Howard Kunreuther, eds., *On Risk and Disaster: Lessons from Hurricane Katrina* (Philadelphia: University of Pennsylvania Press, 2006); Michael Eric Dyson, *Come Hell or High Water: Hurricane Katrina and the Color of Disaster* (New York: Basic Civitas, 2006); Jed Horne, *Breach of Faith: Hurricane Katrina and the Near Death of a Great American City* (New York: Random House, 2006); Robert Polidori, *After the Flood* (Göttingen: Steidl, 2006); Chris Rose, *1 Dead in Attic* (New Orleans: Chris Rose Books, 2006); David Dante Troutt, *Black Intellectuals Explore the Meaning of Hurricane Katrina*

(New York: New Press, 2006); Ivor van Heerden and Mike Bryan, *The Storm: What Went Wrong and Why during Hurricane Katrina* (New York: Viking, 2006); Lola Vollen and Chris Young, *Voices from the Storm: The People of New Orleans on Hurricane Katrina and Its Aftermath* (New York: McSweeney's, 2006); Susan Zakin and Bill McKibben, with photographs by Chris Jordan, *In Katrina's Wake: Portraits of Loss from an Unnatural Disaster* (New York: Princeton Architectural Press, 2006); Eugene F. Provenzo and Asterie Baker Provenzo, *In the Eye of Hurricane Andrew* (Gainesville: University Press of Florida, 2002); Kristine Harper, *Hurricane Andrew* (New York: Facts-on-File, 2005); Roger A. Pielke Jr., *Hurricane Andrew in South Florida: Mesoscale Weather and Societal Responses* (Boulder, CO: ESIG, National Center for Atmospheric Research, 1995); Christopher Dyer and James R. McGoodwin, "'Tell Them We're Hurting': Hurricane Andrew, the Cultural Response, and the Fishing Peoples of South Florida and Louisiana," and Christopher Dyer, "The Phoenix Effect in Post-disaster Recovery: An Analysis of the Economic Development Administration's Culture of Response after Hurricane Andrew," both in *The Angry Earth,* 213–231 and 278–300. Hugo received a damage assessment in *Hurricane Hugo One Year Later,* ed. Benjamin L. Sill and Peter R. Sparks (New York: American Society of Civil Engineers, 1991).

78. On the 1900 hurricane, see John Edward Weems, *A Weekend in September* (College Station: Texas A&M University Press, 1987), and Erik Larson, *Isaac's Storm* (New York: Crown, 1999). For the 1928 storm, see L. E. Will, *Okeechobee Hurricane and the Hoover Dike,* 3rd ed. (Saint Petersburg, FL: Great Outdoors Publishing, 1971); Robert Mykle, *Killer 'Cane: The Deadly Hurricane of 1928* (New York: Cooper Square Press, 2002); Eliot Kleinberg, *Black Cloud: The Great Florida Storm of 1928* (New York: Carroll and Graff), 2003; and Russell L. Pfost, "Reassessing the Impact of Two Historical Florida Hurricanes," *Bulletin of the American Meteorological Society* 84 (2003): 1367–1372. The 1935 Labor Day hurricane is reported in detail in Willie Drye, *Storm of the Century: The Labor Day Hurricane of 1935* (Washington, DC: National Geographic, 2002); and Phil Scott, *Hemingway's Hurricane: The Great Florida Keys Storm of 1935* (New York: McGraw Hill, 2006). For the 1938 event, see Roger K. Brickner, *The Long Island Express: Tracking the Hurricane of 1938* (Batavia, NY: Hodgins Printing, 1988); Cherie Burns, *The Great Hurricane of 1938* (Boston: Atlantic Monthly Press, 2005); and R. A. Scotti, *Sudden Sea: The Great Hurricane of 1938* (Boston: Little, Brown, 2006). On "Camille," see Phillip D. Hearn, *Hurricane Camille: Monster Storm of the Gulf Coast* (Jackson: University Press of Mississippi, 2004); Ernest Zebrowski and Judith A. Howard, *Category 5: The Story of Camille—Lessons Unlearned from America's Most Violent Hurricane* (Ann Arbor: University of Michigan Press, 2005); and Stefan Bechtel, *Roar of the Heavens* (New York: Citadel Press, 2006). On Floyd, see Pete Davis, *Inside the Hurricane: Face to Face with Nature's Deadliest Storms* (New York: Henry Holt, 2000). For good state hurricane histories, see Jay Barnes, *Florida's Hurricane History* (Chapel Hill: University of North Carolina Press, 1998); Jay Barnes, *North Carolina's Hurricane History* (Durham: University of North Carolina Press, 1995); Walter J. Fraser Jr., *Lowcountry Hurricanes: Three Centuries of Storms at Sea and Ashore* (Athens: University of Georgia Press, 2006); John M. Williams and Iver W. Duedall, *Florida Hurricanes and Tropical Storms, 1871–2001* (Gainesville: University Press of Florida, 2002); and the sweeping history of hurricanes in North America and West Indies since Columbus by Marjorie Stoneman Douglas, *Hurricane* (New York: Rinehart, 1958).

79. The estimates of loss from extreme weather events come from *Workshop on the Social and Economic Impacts of Weather,* ed. Roger A. Pielke Jr. (Boulder, CO: UCAR, 1997), 4, 98. Whether more people die from extreme cold than extreme heat is a controversial topic and will be discussed later in this chapter. On Katrina, see NOAA, National Hurricane Center, "2005 Atlantic Hurricane Season," http://nhc.noaa.gov/2005atlan.shtml.

80. Roger A. Pielke Jr., "Disasters, Death, and Destruction: Making Sense of Recent Calamities," *Oceanography* 19, no. 2 (June 2006): 138; Pielke, "Reframing the U.S. Hurricane Problem," *Society and Natural Resources* 10 (1997): 485–499; Pielke, various blogs on "normalizing," http://

sciencepolicy.colorado.edu/prometheus; poverty and disability estimates from the U.S. Census Bureau, American Community Survey Profile, 2003, http://www.census.gove/acs/www/ Products/ Profiles/Single/2003/ACS/LA.htm. See also James B. Elsner and A. Birol Kara, *Hurricanes of the North Atlantic: Climate and Society* (New York: Oxford University Press, 1999), chap. 17, "People at Risk," 379–406; Lise Olsen, "City Had Evacuation Plan but Strayed from Strategy," *Houston Chronicle,* September 8, 2005, http://www.chron.com/disp/story.mpl/nation/3344347 .html (accessed January 3, 2007). On hurricane probabilities, see Sheets and Williams, *Hurricane Watch,* 292–294. On the complexities of determining property losses from hurricanes, see Charles C. Watson Jr. and Mark E. Johnson, "Hurricane Loss Estimation Models: Opportunities for Improving the State of the Art," *Bulletin of the American Meteorological Society* 85 (2004): 1713–1726.

81. Marita Sturken, "Weather Media and Homeland Security: Selling Preparedness in a Volatile World"; Russell R. Dynes and Havidán Rodriguez, "Finding and Framing Katrina: The Social Construction of Disaster"; and David Alexander, "Symbolic and Practical Interpretations of the Hurricane Katrina Disaster in New Orleans," all available at http://understandingkatrina .ssrc.org (accessed October 16, 2006). Disaster scholars are paying increasing attention to disasters in popular culture on the assumption that the "public" learns how to behave in catastrophic situations from movies, television, and other popular media. See Gary R. Webb, Tricia Wachtendorf, and Anne Eyre, "Bringing Culture Back In: Exploring the Cultural Dimensions of Disaster," *International Journal of Mass Emergency and Disasters* 18, no. 1 (March 2000): 5–19; and Jerry T. Mitchell, Deborah S. K. Thomas, Arleen A. Hill, and Susan L. Cutter, "Catastrophe in Reel Life versus Real Life: Perpetuating Disaster Myths through Hollywood Films," *International Journal of Mass Emergency and Disasters* 18, no. 3 (November 2000): 383–402.

82. Jeff Rosenfeld, "The Mourning after Katrina," *Bulletin of the American Meteorological Society* 86 (2005): 1555–1566.

83. John McPhee, *The Control of Nature* (New York: Farrar, Straus and Giroux, 1989); Scott Shalaway, *Charleston (WV) Gazette,* September 11, 2005. I am grateful to Dr. Shalaway for sending me a copy of his essay.

84. New Orleans 2006 Emergency Preparedness Plan, http://www.cityofno.com/Portals/ Porta146/portal.aspx?portal=46&tabid=38 and FEMA, http://www.ready.gov (both accessed January 3, 2007). There is a growing literature on pets in disasters. Brinkley begins his book with an account of efforts to rescue animals during Katrina, possibly an indication that compassion for pets now symbolizes the restoration of the social order. See also Sebastian E. Heath, Susan K. Voeks, and Larry T. Glickman, "A Study of Pet Rescue in Two Disasters," *International Journal of Mass Emergencies and Disasters* 18, no. 3 (November 2000): 361–381; and Katherine C. Grier, *Pets in America* (Chapel Hill: University of North Carolina Press, 2006). See also the Web site of United Animal Nation, an organization dedicated to pet rescue in disasters, www.una.org (accessed August 22, 2007).

85. Gina M. Eosco and William H. Hooke, "Coping with Hurricanes: It's Not Just about Emergency Response . . . ," *Bulletin of the American Meteorological Society* 87 (2006): 751–753; and Thomas W. Schmidlin, "On Evacuation and Deaths from Hurricane Katrina," *Bulletin of the American Meteorological Society* 87 (2006): 754–756.

86. The Nagin estimate comes from Brinkley, *The Great Deluge,* 108; the 1 million evacuees from Tim G. Frazier, "The Floyd Fiasco: Lessons Learned in Hurricane Evacuation Preparedness," *Pennsylvania Geographer* 43, no. 2 (Fall/Winter 2005): 23, quoting M. Pueschel, "Government Starts Medical Web Site for Katrina Evacuees," http://www.usmedicine.com/dailyNews .cfm?dailyID=255.

87. Kristin Dow and Susan L. Cutter, "Crying Wolf: Repeat Responses to Hurricane Evacuation Orders," *Coastal Management* 26, no. 4 (October–December 1998): 237–252.

88. Frazier, "The Floyd Fiasco," 22–44.

89. Roger A. Pielke Jr. and Roger A. Pielke Sr., *Hurricanes: Their Nature and Impacts on Society* (New York: Wiley, 1997), 57; "SLOSH Model," National Hurricane Center, http://www.nhc.noaa .gov/HAW2/english/surge/slosh.shtml (accessed January 11, 2007); Jim Reed, "Fleeing Floyd," *Scientific American Presents Weather* 11, no. 1 (Spring 2000): 47.

90. "What Is a Hurricane Party?" essortment, http://inin.essortment.com/hurricaneparty_ ttou.htm (accessed January 8, 2007). See also "Hurricane Party," in David Longshore, *Hurricanes, Typhoons, and Cyclones* (New York: Checkmark Books, 2000), 179–180.

91. Joy Butler, "Why Do We Have Hurricane Parties?" Suite 101, http://www.suite101.com/ article.cfm/_texas/117057/1 (accessed January 8, 2007).

92. Brinkley, *The Great Deluge,* 59–60. Victor Turner, *The Ritual Process: Structure and Anti-Structure* (Ithaca, NY: Cornell University Press, 1969). "Rituals of status reversal, either placed at strategic points in the annual circle or generated by disasters conceived of as being the result of grave social sins, are thought of as bringing social structure and communitas into right mutual relation once again" (178). Communitas refers to unstructured and fluid relationships arising during liminal periods, in contrast to the fixed hierarchies of stable social systems. Reversal of roles in communitas encourages participants to promote the ideals of equality and spontaneity. In the American context, citizen desire for a responsive and efficient local government after the Galveston hurricane of 1900 caused the Texas state legislature to replace the city's elected officials with a city commission. Elected city commissions were a popular part of the Progressive reform movement until the city manager model replaced them in the 1920s. See Charles N. Glaab and A. Theodore Brown, *A History of Urban America* (New York: Macmillan, 1967), 194–195; Kelly Frailing, "The Myth of a Disaster Myth: Potential Looting Should Be Part of Disaster Planning," *Natural Hazards Observer* 31, no. 4 (March 2007): 3–4.

93. Katherine Fry, examining the cultural aspects of the midwestern floods of 1993 and 1997, notes the plethora of souvenirs following these disasters, including "Floodweiser" beer. See *Constructing the Heartland: Television News and National Disaster* (Cresskill, NY: Hampton Press, 2003). Brinkley describes Chris Owens as "the most flamboyant diva ever to sashay down Bourbon Street." She handed out "Chris Owens" bottled water and chili in the French Quarter during the storm (*The Great Deluge,* 95, 135). "Katrina Creates Tattoo Demand," *Bulletin of the American Meteorological Society* 87 (2006): 22.

94. Sam Eifling, "Like a Hurricane," Slate, http://www.slate.com/id/2107244.

95. Carl Warden, "Hurricane House Turns with Wind," *Popular Science* 135 (October 1939): 68–69; Joseph B. Treaster, "Let a Hurricane Huff and Puff," *New York Times,* June 22, 2006, Business, 1, 4.

96. Stephen P. Leatherman, "Taking Hurricane Research to the Next Level," *Natural Hazards Observer* 32, no. 1 (September 2007): 2. See also the Florida International University Web site, http://www.ihc.fiu.edu/media/docs/WOWPhasesPoster.pdf (accessed October 3, 2007).

97. *The Second Hurricane,* libretto by Edwin Denby, music by Aaron Copland (Boston: C. C. Richard & Co., 1938), 106–107; Jessica Burr, "Copland, the West and American Identity," in *Copland Connotations: Studies and Interviews,* ed. Peter Dickinson (Suffolk, UK: Boydell Press, 2002), 22–28; Howard Pollack, *Aaron Copland: The Life and Work of an Uncommon Man* (New York: Henry Holt, 1999), 304–305. I thank William Pollak for sending me a recording of Copland's opera to listen to as I wrote this chapter.

98. Aaron Copland, *Copland: 1900 through 1942* (New York: St. Martin's/Marek, 1984), 252–265; Pollack, *Aaron Copland,* 306, 310. On Popular Front musicals of the 1930s, see Michael Denning, *The Cultural Front: The Laboring of American Culture in the Twentieth Century* (New York: Verso, 1997), 283–322. For Copland's later compositions in the New Deal and World War II collectivist ethos and the reaction against them in the 1950s, see Alex Ross, "Appalachian Autumn," *New Yorker,* August 27, 2007, 34–40.

99. William H. Chafe, "Hope for New Orleans," *Washington Post,* December 27, 2006, A19.

100. Dorman T. Shindler, "Weather the Storm," *Better Homes and Gardens,* August 2005, 164–167.

101. Chuck Doswell, "Thoughts after Hurricane Katrina," http://www.flame.org/~doswell/Katrina/Katrinas_message.html; Doswell, "Is There a *Good* Side to Severe Storms?" http://www.flame.org/~doswell/goodwx.html (both accessed December 28, 2005).

102. On the 1901 heat wave, see William B. Meyer, *Americans and Their Weather* (New York: Oxford University Press, 2000), 125. On Carrier, see Gail Cooper, *Air-Conditioning America: Engineers and the Controlled Environment, 1900–1960* (Baltimore: Johns Hopkins University Press, 1998), 23–28; and Marsha E. Ackermann, *Cool Comfort: America's Romance with Air-Conditioning* (Washington, DC: Smithsonian Institution Press, 2002), 17. On the American Lung Association, see its Web site, www.lungusa.org.

103. Edwin Grant Dexter, *Weather Influences: An Empirical Study of the Mental and Physiological Effects of Definite Meteorological Conditions* (New York: Macmillan, 1904), xix.

104. James Rotton and Ellen G. Cohn, "Climate, Weather, and Crime," in *Handbook of Environmental Psychology,* ed. Robert B. Bechtel and Arza Churchman (New York: Wiley, 2002), 485; Frederick Sargent II, "Changes in Ideas on the Climatic Origin of Disease," *Bulletin of the American Meteorological Society* 41 (1960): 238–244.

105. C.-E. A. Winslow, *Fresh Air and Ventilation* (New York: Dutton, 1926), 19. On Winslow's career, see Ackermann, *Cool Comfort,* 27–41. For a recent assessment of the dust problem, see Dale W. Griffin, Christina A. Kellogg, Virginia H. Garrison, and Eugene A. Shinn, "The Global Transport of Dust: An Intercontinental River of Dust, Microorganisms and Toxic Chemicals Flows through the Earth's Atmosphere," *American Scientist* 90, no. 3 (2002): 228, http://physics.ius.edu/~kyle/P310/articles/dust.html (accessed January 26, 2007). On Prudden, see Joseph A. Amato, *Dust: A History of the Small and the Invisible* (Berkeley: University of California Press, 2000), 107. On Charles Reed, see David Stradling, *Smokestacks and Progressives: Environmentalists, Engineers, and Air Quality in America, 1881–1951* (Baltimore: Johns Hopkins University Press, 1999), 59. William Scheppegrell, "Hay Fever and Hay Fever Pollens, *Archives of Internal Medicine* 19 (1917): 959–980. On hay fever as an American cultural and environmental phenomenon, see Gregg Mitman, "Hay Fever Holiday: Health, Leisure, and Place in Gilded Age America," *Bulletin of the History of Medicine* 77, no. 3 (Fall 2003): 600–635; and Mitman, *Breathing Space: How Allergies Shape Our Lives and Landscapes* (New Haven, CT: Yale University Press, 2007). Sixty-one years after Dr. Reed's declaration of the inalienable right to breathe pure air, President Lyndon Johnson reiterated that right. A century after Reed's declaration, President George W. Bush settled for "clear skies."

106. William F. Petersen, *The Patient and the Weather,* vol. 1, pt. 1 (Ann Arbor, MI: Edwards Brothers, 1935); Frederick Sargent II, *Hippocratic Heritage: A History of Ideas about Weather and Human Health,* appendix VIII, William Ferdinand Petersen (1887–1950) (New York: Pergamon Press, 1982), 51. Geer, *Glossary of Weather and Climate,* defines "kona" as "a stormy, rain-bringing wind from the southwest or south-southwest in Hawaii. It blows about five times a year on the southwest mountain slopes, which are in the lee of the prevailing northeast trade winds (*kona* is a Polynesian word meaning 'leeward')" (132). April Chan, "Scientists Still Mulling Causes of Weather-Related Pain," *USA Today,* February 21, 2005; and http://www.headache-migraine-release.com/barometric-pressure-headaches.html (accessed December 1, 2006). For the popularization of biometeorology, see Sallie Tisdale, "The Human Weather Vane," *Elle,* November 2002, 180, 182, 185, 262.

107. Ackermann, *Cool Comfort,* 77–102, 166; C. Theodore Larson, "Dialing for the Weather," *Survey Graphic* 23, no. 8 (August 1934): 362. On the Chicago and New York fairs, see Robert W. Rydell, *World of Fairs: The Century-of-Progress Expositions* (Chicago: University of Chicago Press,

1993). PitZips to ventilate underarms on clothing for strenuous outdoor activities are considered essential in many winter sports.

108. C. F. Talman, "Degree-Days," *Bulletin of the American Meteorological Society* 12 (1931): 69–70; "Climafy," *Bulletin of the American Meteorological Society* 17 (1936): 386; Anson D. Marston, "Meteorology in Air-Conditioning," *Bulletin of the American Meteorological Society* 20 (1939): 229.

109. Paul A. Siple, "Adaptation of the Explorer to the Climate of Antarctica" (Ph.D. diss., Clark University, 1939); Arnold Court, "Wind Chill," *Bulletin of the American Meteorological Society* 29 (1948): 487–493; "NWS Windchill Chart," http://www.weather.gov/os/windchill/index.shtml (accessed January 23, 2007).

110. Ackermann, *Cool Comfort,* 74, cites two articles in the *Washington Post* as sources for the Misery Index: Mike Causey, "No More Misery Index," June 16, 1994, D2, and Martha M. Hamilton, "D.C. without A.C.? Life Here Would Be Positively B.C.," June 16, 1994, A1+. "DI," *New Yorker,* June 6, 1959, 33–35; E. C. Thom, "The Discomfort Index," *Weatherwise* 12 (1959): 57–60.

111. "A New Service," *Commonweal,* May 29, 1959, 222; "Heat, Dampness, Discomfort," *America,* June 6, 1959, 403.

112. "Rename Discomfort Index," *Science News Letter,* August 1, 1959, 71; "Humidex," Glossary Weather Words, Keith Heidorn, http://www.islandnet.com/~see/weather/doctor.htm (accessed September 6, 2005); Robert L. Hendrick, "An Outdoor Weather-Comfort Index for the Summer Season in Hartford, Connecticut," *Bulletin of the American Meteorological Society* 40 (1959): 620–623. See also "Measuring the Misery," *Newsweek,* June 15, 1959, 29, which used pictures of four babies whose facial expressions reflected DI readings of 70, 75, 80, and 85.

113. For a good overview of current thinking among architects about thermal comfort, see Vivian Loftness, "Architecture and Climate," in *The Encyclopedia of Climatology,* ed. John E. Oliver and Rhodes W. Fairbridge (New York: Van Nostrand Reinhold, 1987), 68–82; Frederick H. Rohles Jr., "Temperature and Temperament: A Psychologist Looks at Comfort," *ASHREA Journal* 49 (January/February 2007): 14–19; Beppe Severgnini, *Ciao, America! An Italian Discovers the U.S.* (New York: Broadway Books, 1995), 65. My thanks to Shelly McKenzie for this and other references.

114. Allen Salkin, "Shivering for Luxury," *New York Times,* June 26, 2005, D1, D10.

115. "Smog," *The Compact Edition of the Oxford English Dictionary,* vol. 2 (New York: Oxford University Press, 1980), 4066. "Smog" began as a simple portmanteau of "smoke" and "fog," but on reaching California it took on a new identity as the haze created by the photochemical reaction of hydrocarbons and other chemicals to sunlight. The state's sunny sky became the serpent in the California Eden. On the smoke abatement crusade of the first half of the twentieth century, see Stradling, *Smokestacks and Progressives.* On El Niño, see Michael H. Glantz's excellent book *Climates of Change: Impacts of El Niño and La Niña on Climate and Society,* 2nd ed. (New York: Cambridge University Press, 2001).

116. Lynne Page Snyder, "'The Death-Dealing Smog over Donora, Pennsylvania': Industrial Air Pollution, Public Health, and the Politics of Expertise," *Environmental History Review* 18 (1994): 117–139; "Our Polluted Inheritance," *Science News Letter,* August 1, 1959, 70–71; H. E. Landsberg, "The Climate of Towns," in *Man's Role in Changing the Face of the Earth,* vol. 2, ed. William L. Thomas Jr. (Chicago: University of Chicago Press, 1955), 584–606; Kenneth Chang, "Scientists Watch Cities Make Their Own Weather," *New York Times,* August 15, 2000, Science, 1–2; Helmut Landsberg, "The Meteorologically Utopian City," *Bulletin of the American Meteorological Society* 54 (1973): 86–89.

117. Anne Whiston Spirn, *The Granite Garden: Urban Nature and Human Design* (New York: Basic Books, 1984), 69–81.

118. American Lung Association, "State of the Air 2007," http://www.lungaction.org/reports/

stateoftheart2007.html (accessed January 13, 2007); Matt Weiser, "Why Our Air Isn't Fit to Breathe," *Bakersfield Californian,* June 14, 2003, A1, A4–A6, and Weiser, "Who's In Charge Here?" *Bakersfield Californian,* June 15, 2003, A1, A4–A8. The articles engendered new discussions and the implementation of some of the plan's recommendations.

119. Morris Neiburger, "Can Air Pollution Be Stopped?" *Christian Century,* February 2, 1966, 139.

120. Leticia Kent, "Aerial Garbage," *New Republic,* February 12, 1966, 8.

121. Geer, *Glossary of Weather and Climate,* 18, 53, 108, 137, 203.

122. Richard J. Lazarus, *The Making of Environmental Law* (Chicago: University of Chicago Press, 2004), 71.

123. Douglas G. Fox, "Uncertainty in Air Quality Modeling," *Bulletin of the American Meteorological Society* 65 (1984): 27–36; Walter F. Dabberdt et al., "Meteorological Research Needs for Improved Air Quality Forecasting," *Bulletin of the American Meteorological Society* 85 (2004): 565.

124. "Information about the Air Quality Index (AQI)," http://www.air.dnr.state.ga.us/information/aqi.html.

125. Harold Fromm, "On Being Polluted," *Yale Review* 55, no. 4 (Summer 1976): 615, 629. On the difficulty of visually determining air quality, see Thomas R. Stewart, Paulette Middleton, and Daniel Ely, "Urban Visual Air Quality Judgments: Reliability and Validity," *Journal of Environmental Psychology* 3 (1983): 129–145.

126. Matt Weiser, "Smoky Air Remains a Hazard as Scientists Analyze Danger," *Sacramento Bee,* March 17, 2007, A1, A14.

127. Thomas Warner et al., "The Pentagon Shield Field Program," *Bulletin of the American Meteorological Society* 88 (2007): 167–176. Preliminary findings of a study of wind vectors in Manhattan, New York City, are presented in Steven R. Hanna et al., "Detailed Simulations of Atmospheric Flow and Dispersion in Downtown Manhattan," *Bulletin of the American Meteorological Society* 87 (2006): 1713–1726.

128. Kirk Johnson, "A Changing Climate in Ideas about Pollution," *New York Times,* May 20, 2001, 29–30; Joe Sherman, *Gasp! The Swift and Terrible Beauty of Air* (Washington, DC: Shoemaker and Hoard, 2004), 364.

129. Richard J. Wurtman and Judith J. Wurtman, "Carbohydrates and Depression," *Scientific American,* January 1989, 68; Jane Brody, "Getting a Grip on the Winter Blues," *New York Times,* December 5, 2006, 9. A lux is a unit equivalent to the illumination cast on a surface by one candle 1 meter away.

130. Michael Castleman, "Sunscam," *Mother Jones,* May/June 1998, http://www.motherjones.com (accessed February 10, 2007; U.S. Environmental Protection Agency, "What Is the UV Index?" http://www.epa.gov/sunwise/uvwhat.htm.

131. Jane E. Brody, "A Second Opinion on Sunshine: It Can Be Good Medicine," *New York Times,* June 17, 2003, D7.

132. John S. Westefeld, Aaron Less, Tim Ansley, and Hyun Sook-Yi, "Severe-Weather Phobia," *Bulletin of the American Meteorological Society* 87 (2006): 747–749; Stormphobia: www.stormphobia.org (accessed February 10, 2007).

133. Guy Hinsdale, "Climate and Disease (III)," *Bulletin of the American Meteorological Society* 18 (1937): 58; Clarence J. Root, "Deaths during the Heat Wave of July, 1936 at Detroit," *Bulletin of the American Meteorological Society* 18 (1937): 232–236; Frederick Sanders, "Towards Defining Human Needs: How Does the Atmosphere Hurt Us?" *Bulletin of the American Meteorological Society* 32 (1971): 446–449.

134. Stanley A. Changnon Jr., "How a Severe Winter Impacts on Individuals," *Bulletin of the American Meteorological Society* 60 (1979): 110–114.

135. Stanley A. Changnon, Kenneth E. Kunkel, and Beth C. Reinke, "Impacts and Responses to

the 1995 Heat Wave: A Call to Action," *Bulletin of the American Meteorological Society* 77 (1996): 1497–1506; Eric Klinenberg, *Heat Wave: A Social Autopsy of Disaster in Chicago* (Chicago: University of Chicago Press, 2002); P. G. Dixon, D. M. Brommer, B. C. Hedquist, A. J. Kalkstein, G. B. Goodrich, J. C. Walter, C. C. Dickerson IV, S. J. Penny, and R. S. Cerveny, "Heat Mortality versus Cold Mortality," *Bulletin of the American Meteorological Society* 86 (2005): 921–936.

136. Changnon, Kunkel, and Reinke, "Impact and Responses to the 1995 Heat Wave," 1497–1506; Klinenberg, *Heat Wave,* 79–128; Rodger Doyle, "Deaths from Excessive Cold and Excessive Heat," *Scientific American* 278, no. 2 (February 1998): 26.

137. Laurence S. Kalkstein and Kathleen M. Valimont, "An Evaluation of Summer Discomfort in the United States Using a Relative Climatological Index," *Bulletin of the American Meteorological Society* 67 (1986): 842–848; Kalkstein and Valimont, "An Evaluation of Winter Weather Severity in the United States Using the Weather Stress Index," *Bulletin of the American Meteorological Society* 68 (1987): 1535–1540.

138. University of Pittsburgh, Department of Neurological Surgery, "Research Finds Thunderstorm Deaths Occur Mainly in Men," http://www.neurosurgery.pitt.edu/s003/thunderstorm .html; E. Brian Curran, Ronald L. Holle, and Raúl E. López, "Lightning Casualties and Damages in the United States from 1959 to 1994," *Journal of Climate* 13, no. 19 (October 2000): 3448–3464; Ronald L. Lavoie, "Storm Research in the United States: Organizations and Major Facilities," in *The Thunderstorm in Human Affairs,* 2nd ed., revised and enlarged, ed. Edwin Kessler (Norman: University of Oklahoma Press, 1983), 161–168.

139. Sandra Blakeslee, "Lightning's Shocking Secrets," *New York Times,* July 18, 2000, D1, D6.

140. Joel Gratz and Erik Noble, "Lightning Safety and Large Stadiums," *Bulletin of the American Meteorological Society* 87 (2006): 1194–1187; Anne Eisenberg, "Watching the Fireworks (but Keeping Them Outside)," *New York Times,* August 20, 2006, Business, 5; Spectrum ad, inside cover of *Weatherwise,* January/February 2007.

Conclusion

1. William James, *The Varieties of Religious Experience* (New York: Modern Library, 2002), 535; emphasis added.

2. David E. Nye, ed., *Technologies of Landscape: From Reaping to Recycling* (Amherst: University of Massachusetts Press, 1999), 3.

3. Allen Guttmann, *From Ritual to Record: The Nature of Modern Sports* (New York: Columbia University Press, 1978).

4. For a full account of luck in American culture, see Jackson Lears, *Something for Nothing: Luck in America* (New York: Viking Penguin, 2003).

5. Stanley Kunitz with Genine Lentine, *The Wild Braid: A Poet Reflects on a Century in the Garden* (New York: Norton, 2005), 62–63.

6. Edward S. Casey, *Getting Back into Place: Toward a Renewed Understanding of the Place-World* (Bloomington: Indiana University Press, 1993), 219; emphasis in original.

Selected Bibliography

Printed Sources

Ackermann, Marsha E. *Cool Comfort: America's Romance with Air-Conditioning.* Washington, DC: Smithsonian Institution Press, 2002.

Advisory Committee on Weather Services for the U.S. Department of Commerce. *Weather Is the Nation's Business.* Washington, DC: Government Printing Office, 1953.

Bachelard, Gaston. *Air and Dreams: An Essay on the Imagination of Movement.* Dallas: Dallas Institute of Humanities and Culture, 1988 [French edition, 1943].

Barnes, Peter. *Who Owns the Sky? Our Common Assets and the Future of Capitalism.* Washington, DC: Island Press, 2001.

Batten, Frank, with Jeffrey Cruikshank. *The Weather Channel: The Improbable Rise of a Media Phenomenon.* Boston: Harvard Business School Press, 2002.

Baum, Marsha L. *When Nature Strikes: Weather Disasters and the Law.* Westport, CT: Praeger, 2007.

Boia, Lucian. *The Weather in the Imagination.* London: Reaktion Books, 2005.

Bulletin of the American Meteorological Society (monthly). Boston, 1920–.

Casey, Edward S. *Getting Back into Place: Toward a Renewed Understanding of the Place-World.* Bloomington: Indiana University Press, 1993.

Caskey, James E., ed. *A Century of Weather Progress: A Collection of Addresses Presented at a Joint Symposium Commemorating the Centennial of the United States Weather Services and the Golden Anniversary of the American Meteorological Society.* Boston: American Meteorological Society, 1970.

Changnon, Stanley D. "Measurement of Economic Impacts of Weather Extremes." *Bulletin of the American Meteorological Society* 84, no. 9 (September 2003): 1231–1235.

Changnon, Stanley A., J. M. Changnon, and Geoffrey J. D. Hewings. "Losses Caused by Weather and Climate Extremes: A National Index for the United States." *Physical Geography* 22 (2001): 1–27.

Changnon, Stanley A., and Geoffrey J. D. Hewings. "Losses from Weather Extremes in the United States." *Natural Hazards Review* 2, no. 3 (August 2001): 113–123.

Changnon, Stanley A., et al. "Effects of Recent Weather Extremes on the Insurance Industry: Major Implications for the Atmospheric Sciences." *Bulletin of the American Meteorological Society* 78, no. 3 (March 1997): 425–435.

Changnon, Stanley, et al. "Human Factors Explain the Increased Losses from Weather and Climate Extremes." *Bulletin of the American Meteorological Society* 81, no. 3 (March 2000): 437–442.

Churchill, William. *Weather Words of Polynesia.* Lancaster, PA: New Era Printing, 1907.

Committee on Partnerships in Weather and Climate Services, National Research Council. *Fair Weather: Effective Partnerships in Weather and Climate Services.* Washington, DC: National Academies Press, 2003.

Cooper, Gail A. *Air-Conditioning America: Engineers and the Controlled Environment, 1900–1960.* Baltimore: Johns Hopkins University Press, 1998.

Cox, John D. *Storm Watchers: The Turbulent History of Weather Prediction from Franklin to El Niño.* New York: Wiley, 2002.

———. *Weather for Dummies.* New York: Hungry Minds, 2000.

Cressman, George. American Meteorological Society Tape Recorded Interview Project. UCAR Archives, Nos. 84–86. Boulder, CO, 1992.

Davidson, Keay. "Our National Passion." *Scientific American Presents Weather* 11, no. 1 (Spring 2000): 6–11.

Day, John A. *The Book of Clouds.* New York: Barnes and Noble, 2003.

DeBlieu, Jan. *Wind: How the Flow of Air Has Shaped Life, Myth, and Land.* Boston: Houghton Mifflin, 1998.

Doswell, Charles A., III, ed., *Severe Convective Storms.* Boston: American Meteorological Society, 2001.

———. "Societal Impacts of Severe Thunderstorms and Tornadoes: Lessons Learned and Implications for Europe." *Atmospheric Research* 67–68 (2003): 135–152.

Doswell, Charles A., III, Alan R. Moller, and Harold E. Brooks. "Storm Spotting and Public Awareness since the First Tornado Forecasts of 1948." *Weather and Forecasting* 14 (1999): 544–557.

Edinger, James G. *Watching for the Wind: The Seen and Unseen Influences on Local Weather.* Garden City, NY: Anchor Books, 1967.

Emanuel, Kerry. *Divine Wind: The History and Science of Hurricanes.* New York: Oxford University Press, 2005.

Fine, Gary Alan. *Authors of the Storm: Meteorologists and the Culture of Prediction.* Chicago: University of Chicago Press, 2007.

Finley, John Park. *Tornadoes: What They Are and How to Observe Them, with Practical Suggestions for the Protection of Life and Property.* New York: Insurance Monitor, 1887.

Fleming, James Rodger. "Fixing the Weather and Climate: Military and Civilian Schemes for Cloud Seeding and Climate Engineering." In *The Technological Fix: How People Use Technology to Create and Solve Problems,* ed. Lisa Rosner, 175–200. New York: Routledge, 2004.

———. *Historical Essays on Meteorology, 1919–1995.* Boston: American Meteorological Society, 1996.

———. *Historical Perspectives on Climate Change.* New York: Oxford University Press, 1998.

Fleming, James Rodger, Vladimir Jankovic, and Deborah R. Coen, eds. *Intimate Universality: Local and Global Themes in the History of Weather and Climate.* Sagamore Beach, MA: Watson, 2006.

Franklin, William S. "A Much Needed Change of Emphasis in Meteorological Research." *Monthly Weather Review* 46 (October 1918): 449–453.

Gedzelman, Stanley David. *The Soul of All Scenery: A History of the Sky in Art.* http://www.sci .ccny.cuny.edu/~stan.

Geer, Ira W. ed. *Glossary of Weather and Climate.* Boston: American Meteorological Society, 1996.

Glantz, Michael H. *Climate Affairs: A Primer.* Washington, DC: Island Press, 2003.

Grazulis, Thomas P. *The Tornado: Nature's Ultimate Windstorm.* Norman: University of Oklahoma Press, 2001.

Grice, Gary K. "History of Weather Observing in Washington, DC, 1821–1950." http://mcc.sws .uiuc.edu/FORTS/histories/DC_Washington_DC_Grice.pdf.

Hamblyn, Richard. *The Invention of Clouds: How an Unknown Meteorologist Forged the Language of the Skies.* New York: Farrar, Straus and Giroux, 2001.

Harper, Kristine C. "Boundaries of Research: Civilian Leadership, Military Funding, and the International Network Surrounding the Development of Numerical Weather Prediction in the United States." Ph.D. diss., Oregon State University, 2003.

Henson, Robert. *Television Weathercasting: A History.* Jefferson, NC: McFarland, 1990.

Huler, Scott. *Defining the Wind: The Beaufort Scale, and How a Nineteenth-Century Admiral Turned Science into Poetry.* New York: Crown, 2004.

Janković, Vladimir. *Reading the Skies: A Cultural History of English Weather, 1650–1820.* Chicago: University of Chicago Press, 2000.

Kalkstein, Laurence S., and Robert E. Davis. "Weather and Human Mortality: An Evaluation of Demographic and Interregional Responses in the United States." *Annals of the Association of American Geographers* 79, no. 1 (1989): 44–64.

LaFollette, Marcel C. *Making Science Our Own: Public Images of Science 1910–1955.* Chicago: University of Chicago Press, 1990.

Luckiesh, Matthew. *The Book of the Sky.* New York: Dutton, 1933.

McAdie, Alexander. *A Cloud Atlas.* Chicago: Rand, McNally, 1923.

McCown, Clint. *The Weatherman.* St. Paul, MN: Graywolf, 2004.

Meinzer, Wyman. *Texas Sky.* Austin: University of Texas Press, 1998.

Mergen, Bernard. *Snow in America.* Washington, DC: Smithsonian Institution Press, 1997.

Meyer, William B. *Americans and Their Weather.* New York: Oxford University Press, 2000.

Misrach, Richard. *The Sky Book: Photographs of Richard Misrach.* Text by Rebecca Solnit. Santa Fe, NM: Arena Editions, 2000.

Mitman, Gregg. *Breathing Space: How Allergies Shape Our Lives and Landscapes.* New Haven, CT: Yale University Press, 2007.

Monmonier, Mark. *Air Apparent: How Meteorologists Learned to Map, Predict, and Dramatize Weather.* Chicago: University of Chicago Press, 1999.

Monthly Weather Review. Washington, DC, 1873–.

Moody, Rick. *The Ice Storm.* New York: Warner Books, 1994.

Morss, Rebecca E., and William H. Hooke. "The Outlook for U.S. Meteorological Research in a Commercializing World." *Bulletin of the American Meteorological Society* 86, no. 7 (July 2005): 921–936.

Moss, David A. *When All Else Fails: Government as the Ultimate Risk Manager.* Cambridge, MA: Harvard University Press, 2002.

National Research Council. *A Vision for the National Weather Service: Road Map for the Future.* Washington, DC: National Research Council, 1999.

Nelkin, Dorothy. *Selling Science: How the Press Covers Science and Technology.* Rev. ed. New York: Freeman, 1995.

Pielke, Roger A., Jr. "Asking the Right Questions: Atmospheric Sciences Research and Social Needs." *Bulletin of the American Meteorological Society* 78, no. 2 (February 1997): 255–264.

———. "Nine Fallacies of Floods." *Climatic Change* 42 (1999): 413–438.

Platt, Rutherford H. *Disasters and Democracy: The Politics of Extreme Natural Events.* Washington, DC: Island Press, 1999.

Quarrington, Paul. *Storm Chasers.* New York: St. Martin's, 2005.

Roberts, Elizabeth Madox. *Jingling in the Wind.* New York: Viking, 1928.

Rolt-Wheeler, Francis W. *The Boy with the U.S. Weather Men.* Boston, MA: Lothrop, Lee and Shepard, 1917.

Rotch, A. Lawrence. *Sounding the Ocean of Air; Being Six Lectures Delivered before the Lowell Institute of Boston in December 1898.* New York: E. & J. B. Young, 1900.

Rotton, James, and Ellen G. Cohn. "Climate, Weather, and Crime." In *Handbook of Environmental Psychology,* ed. Robert B. Bechtel and Arza Churchman, 481–498. New York: Wiley, 2002.

Schaefer, Vincent, and John Day. *A Field Guide to Clouds and Weather.* Boston: Houghton Mifflin, 1981.

Schneider, Stephen H., ed. *Encyclopedia of Climate and Weather.* 2 vols. New York: Oxford University Press, 1996.

Schwartz, Peter, and Doug Randall. "An Abrupt Climate Change Scenario and Its Implications for United States National Security." October 2003: http://www.environmentaldefense.org/documents/3566_AbruptClimateChange.pdf.

Seabrook, John. "Selling the Weather." *New Yorker,* April 3, 2000, 44–53.

Sheets, Bob, and Jack Williams. *Hurricane Watch: Forecasting the Deadliest Storms on Earth.* New York: Vintage, 2001.

Sherman, Joe. *Gasp! The Swift and Terrible Beauty of Air.* Washington, DC: Shoemaker and Hoard, 2004.

Sloane, Eric. *Clouds, Air, and Wind.* New York: Devlin-Adair, 1941.

———. *For Spacious Skies: A Meteorological Sketchbook of American Weather.* New York: Crowell, 1978.

Steinberg, Ted. *Acts of God: The Unnatural History of Natural Disaster in America.* 2nd ed. New York: Oxford University Press, 2006.

Steiner, Ralph. *In Pursuit of Clouds: Images and Metaphors.* Albuquerque: University of New Mexico Press, 1985.

Stewart, George R. *Storm.* New York: Random House, 1941.

Strauss, Sarah, and Ben S. Orlove, eds. *Weather, Climate, Culture.* New York: Berg, 2003.

Sturken, Marita. "Desiring the Weather: El Niño, the Media, and California Identity." *Public Culture* 13, no. 2 (Spring 2001): 161–189.

Svenvold, Mark. *Big Weather: Chasing Tornadoes in the Heart of America.* New York: Henry Holt, 2005.

Thompson, Jean. *Wide Blue Yonder.* New York: Simon and Schuster, 2002.

Vidal, Gore. *Williwaw.* New York: Dutton, 1946.

Watson, Lyall. *Heaven's Breath: A Natural History of the Wind.* New York: William Morrow, 1984.

Weatherwise (bimonthly). Washington, DC, 1948–.

White, Robert M. "The Making of NOAA, 1963–2005." *History of Meteorology* 3 (2006): 55–63. www.meteohistory.org.

Whitnah, Donald. *A History of the United States Weather Bureau.* Urbana: University of Illinois Press, 1961.

Internet Sites

American Lung Association, *State of the Air.* www.lungusa.org.

American Meteorological Society. AMS Journals Online. http://ams.allenpress.com/perlserv/?request=index-html.

Cloud Appreciation Society (England). www.cloudappreciationsociety.org.

Cloudman, John Day. www.cloudman.com.

Cooperative Weather Observers. http://www.weather.gov/om/coop.

Charles A. Doswell III. www.cimms.ou.edu/~doswell.

Environment Canada—Weather. http://www.weatheroffice.gc.ca/canada_e.html.

For Spacious Skies, Jack Borden. www.forspaciousskies.com.

Glossary of Meteorology, AMS. http://amsglossary.allenpress.com/glossary.

History of the National Weather Service. http://www.weather.gov/pa/history/index.php.

How the Weatherworks, H. Michael Mogil. www.weatherworks.com.

International Commission on History of Meteorology. http://www.meteohistory.org/.

National Association of Storm Chasers and Spotters. http://www.chasingstorms.com.

National Center for Atmospheric Research. www.ncar.ucar.edu/ncar/.

National Severe Storms Laboratory. http://www.nssl.noaa.gov.

NOAA Central Library, WINDandSEA: The Oceanic and Atmospheric Sciences Internet Guide. www.lib.noaa.gov/docs/windandsea.html.

NOAA Photo Library. www.photolib.noaa.gov.

NOAA History. www.history.noaa.gov.

Servicio Meteorológico Nacional (Mexico). http://smn.cna.gob.mx.

Societal Aspects of Weather (NCAR). www.sip.ucar.edu/socasp/.

Stormtrack. www.stormtrack.org.

Storm Prediction Center. http://www.spc.noaa.gov.

Tornado Project, Thomas Grazulis. www.tornadoproject.com.

Weather and Art, Stan Gedzelman. http://www.sci.ccny.cuny.edu/~stan/.

The Weather Channel. www.weather.com.

The Weather Doctor, Keith Heidorn. http://www.islandnet.com/~see/weather/doctor.htm.

Weather Matrix, Jesse Ferrell. www.weathermatrix.com.

Weather Prediction, Jeff Haby. http://theweatherprediction.com.

Index

National Council of Industrial
Meteorologists, 102, 105
National Digital Forecast Database (NDFD), 94
National Geographic, 67, 174, 260, 291
National Hurricane Center, 102, 289
National Hurricane Research Project, 85
National Institute for Atmospheric Research, 87
National Lightning Detection Network, 105
National Museum of American History
(Smithsonian), 189
National Oceanic and Atmospheric
Administration (NOAA), 30, 67, 87,
111, 123, 133, 188, 190, 223, 322
National Research Council (NRC), 63, 92,
105–106, 123, 275
National Science Foundation (NSF), 67, 86–
88, 121–122, 135, 190
National Severe Storms Laboratory, 166–167,
262, 318
National Severe Storms Project, 85, 255, 280
National Weather Analysis Center, 172
National Weather Association, 34, 105, 111,
116
National Weather Radio (NOAA), 62, 91, 258
National Weather Service, 8, 31–39, 59, 92–
97, 101–107, 115, 253, 265, 281, 305,
308. *See also* Weather Bureau, U.S.
National Weather Service Employees
Organization, 105
National Weather Service Modernization
Committee, 92–94
Nebeker, Frederick, 3
Neiburger, Morris, 171, 309
Nemerov, Howard, 5, 227, 232, 251
Nese, Jon, 59
Neubauer, John, 194
Neuberger, Hans, 29–30, 58, 175
Nevins, Allan, 235
New England hurricane, 288
New Hampshire, 1–2
Newman, Alfred, 164
New Mexico, 117–118
New Republic, The, 310
"News and the Weather, The," 221–222
NEXRAD (Next Generation Radar), 94, 282
New Yorker, The, 51, 306
"Night Clouds," 204–205
Nisbet, Peter Allen, 177, 179, 184
Nixon, Richard, 67, 242
Novak, Barbara, 176

numerical forecasting, 31, 89, 92–94, 121, 260
Nye, David, 166, 323

O'Brien, Robert, 58
Ohring, George, 88
O'Keeffe, Georgia, 143, 182–183
On Longing (Stewart), 188
Operation POPEYE, 121
Opper, F., 52
Orville, Howard T., 121, 276

Palmer Hydrological Drought Index, 48
Palm Sunday Tornado, 1965, 33–34, 262, 276
"Pam," Hurricane, 291–292
Parini, Jay, 203
Partridge, S., 82
Pasqualetti, Martin, 71
Patient and the Weather, The (Petersen), 304
Paxton, Bill, 165, 264
Peacock, Doug, 150
Pearson, Allen, 262, 280
Pearson, S. K., 82
Pei, I. M., 38–39, 83
Pennsylvania State University, 103, 105
Pepper, Lorena, 225
Peter, Carsten, 260
Peterson, William Ferdinand, 303–304
Petterssen, Sverre, 87
phenology, 14
Pielke, Roger A., Jr., 38–39, 96, 106, 288–
290, 294
"Pig Pen Effect," 313
Pitot tube, 149
Plumb, Laura Kirkwood, 274
Poincaré, Henri, 42, 107, 129
Polidori, Robert, 288
politics and weather, 48–49, 90–91
Pollution Standards Index, 311
Popular Science Monthly, 144
Porter, Eliot, 160
"President's Resignation, The," 228–229
Pretor-Pinnet, Gavin, 150
Probability of Precipitation (PoP), 35
"Problems with Hurricanes," 220–221
Project Cirrus, 117–118, 184
Project Climax, 121
Project Hailswrath, 121
Project Skywater, 121
Project Stormfury, 121–122
Project Whitetop, 121
Prudden, Mitchell, 303

Permissions